Phospholipids and Signal Transmission

NATO ASI Series

Advanced Science Institutes Series

A series presenting the results of activities sponsored by the NATO Science Committee, which aims at the dissemination of advanced scientific and technological knowledge, with a view to strengthening links between scientific communities.

The Series is published by an international board of publishers in conjunction with the NATO Scientific Affairs Division

A Life Sciences	Plenum Publishing Corporation
B Physics	London and New York
C Mathematical and Physical Sciences	Kluwer Academic Publishers
D Behavioural and Social Sciences	Dordrecht, Boston and London
E Applied Sciences	
F Computer and Systems Sciences	Springer-Verlag
G Ecological Sciences	Berlin Heidelberg New York
H Cell Biology	London Paris Tokyo Hong Kong
I Global Environmental Change	Barcelona Budapest

NATO-PCO DATABASE

The electronic index to the NATO ASI Series provides full bibliographical references (with keywords and/or abstracts) to more than 30 000 contributions from international scientists published in all sections of the NATO ASI Series. Access to the NATO-PCO DATABASE compiled by the NATO Publication Coordination Office is possible in two ways:

- via online FILE 128 (NATO-PCO DATABASE) hosted by ESRIN, Via Galileo Galilei, I-00044 Frascati, Italy.

- via CD-ROM "NATO Science & Technology Disk" with user-friendly retrieval software in English, French and German (© WTV GmbH and DATAWARE Technologies Inc. 1992).

The CD-ROM can be ordered through any member of the Board of Publishers or through NATO-PCO, Overijse, Belgium.

Series H: Cell Biology, Vol. 70

Phospholipids and Signal Transmission

Edited by

Raphaël Massarelli

Director of Research of CNRS
Centre de Neurochimie
Strasbourg, France

Lloyd A. Horrocks

Professor Emeritus
Department of Medical Biochemistry
College of Medicine
The Ohio State University
Columbus
Ohio, USA

Julian N. Kanfer

Professor at
Department of Biochemistry and Molecular Biology
The University of Manitoba
Winnipeg
Manitoba, Canada

Konrad Löffelholz

Professor at
Department of Pharmacology
University of Mainz
Mainz, Germany

 Springer-Verlag Berlin Heidelberg GmbH

Proceedings of the NATO Advanced Research Workshop on Phospholipids
and Signal Transmission, held at Wiesbaden, Germany, May 29–June 2, 1991

ISBN 978-3-662-02924-4   ISBN 978-3-662-02922-0 (eBook)
DOI 10.1007/978-3-662-02922-0

© Springer-Verlag Berlin Heidelberg 1993

Originally published by Springer-Verlag Berlin Heidelberg New York in 1993.
Softcover reprint of the hardcover 1st edition 1993

Typesetting: Camera ready by author
31/3145 - 5 4 3 2 1 0 - Printed on acid-free paper

PREFACE

Once Nietzsche said that human beings may be divided into two categories: Apollonians and Dionysians*. By this the philosopher meant that there are human beings a) who know what they are going to do in the long-term future (what we now call the grant application for the next 5 years), i.e., Apollonians, and b) who barely know what they are going to do tomorrow morning before breakfast, i.e., Dionysians.**

To organize a symposium, this symposium in particular, a committee had to be formed either of individuals sharing both Nietzschean characteristics or of individuals possessing either characteristic. Considering the rarity of the former type of subject, this organizing committee was spontaneously formed by a typical sample of both types of individuals.

We first met in Perugia in 1988. Those of us who were Apollonians had thus a chance to organize a programme. The Dionysians knew what was going to happen to them, but, of course, did not know yet how to cope with it. They duly did so every day of the meeting, after breakfast.

The organizers decided that it would be a useful exercise to assemble experts having different perspectives but all pursuing a very rapidly developing aspect of cell biology. They also hoped that these selected Apollonians and Dionysians would not merely recount their results but try to project the future through active interchanges of ideas and opinions with other attendees. The general purpose of the symposium was to increase our understanding of the subject and to attempt to get a bit ahead of the applications in an atmosphere different from the usual ritualistic encounters among scientists.

---

*Nowadays the alternatives Athenians and Demeterians must be used. This is what Nietzsche himself might have done, perhaps, had he met someone like Germaine Greer.

**A. Szent-Gyorgy sent a letter to <u>Science</u>, years ago, asking that Dionysians be exempted from writing grant applications. His initiative was not followed, to our knowledge, by roaring success. This indicates perhaps (but it is pure speculation) that Apollonians hold the majority on grant committees.

The specific aim of the symposium was to reinforce the concept that membrane components such as the phospholipids, as well as the sphingolipids, have the potential to serve as second messenger precursors. The formation of the various second messengers is provoked by the arrival at the external cell surface of a ligand.

The generation of an intracellular second messenger evokes a specific response of the system. Those of an immediate nature frequently are coupled with ion fluxes or enzyme activities. Those of a more delayed response may involve activation of either immediate/early or late genes.

These observations gradually evolved into the now classical intrinsic role of the polyphosphoinositides as precursors of two separate second messengers. Their liberation is catalyzed by the activation of a phosphoinositide-specific phospholipase C generating diacylglycerol, a protein kinase C activator, and InsP3, a Ca^{2+} mobilizer from intracellular stores. These aspects of phosphatidylinositol function were considered through formal presentations and posters. The contributions of two other major membrane phospholipids, phosphatidylcholine and phosphatidylserine, to the process of information transfer from the cellular environment into the intracellular compartment were also considered through formal presentations and numerous posters.

Symposium sessions were organized according to present knowledge of the most interesting phospholipids (and some of the sphingolipids) in signalling phenomena.

The choice of the site - Wiesbaden, Germany - and the ability to obtain a most essential ingredient - financial support - were an excellent demonstration that, provided with some goodwill, Apollonians and Dionysians can efficiently interact. It was, for the organizers, an interesting phenomenon that it was possible to gather 225 Apollonian and Dionysian participants of undetermined behavioural persuasion and 45 similar invitees through the cooperation of organizers situated in Mainz (Germany), Padua (Italy), Munich (Germany), Winnipeg (Canada), Columbus (USA) and Strasbourg (France) and still remain excellent friends. It was particularly complimentary to the organizers to have departing participants encouraging us to commence organizing a sequel to be staged in the next few years.

Whether the aims set by the organizers have been met with success is left to the judgment of the reader. We only wish to gratefully acknowledge the essential financial support of NATO, Fidia Research Laboratories,

Deutsche Forschungsgemeinschaft, Johannes Gutenberg University of Mainz and Louis Pasteur University of Strasbourg.

Particular thanks go to the staff of the Meeting Bureaus of Fidia in Abano Terme and Fidia Pharmaforschung in Munich. The local organizing committee was conducted efficiently by Dr. R. Lindmar, whom we warmly thank. The secretarial help of Ms. F. Brisach was essential for the survival of one of the Dionysians. The effort is warmly acknowledged.

We have mentioned above that the first meeting of the organizers took place in Perugia in 1988. This was during a symposium held in honor of the late G. Porcellati. We had the honor to be his friends, and it is without rhetoric that we wish to say how we had wished to have him as one of the Apollonian/Dionysian members of the committee if Nature had not decided differently. We regret once more that decision.

This book is dedicated to his memory.

The Editors

CONTENTS

GLYCOSYLPHOSPHATIDYLINOSITOL ANCHORED RECOGNITION MOLECULES THAT MEDIATE INTERCELLULAR ADHESION AND PROMOTE NEURITE OUTGROWTH

Patrick Doherty and Frank S Walsh
Department of Experimental Pathology
UMDS, Guy's Hospital
London SE1 9RT
UK

Introduction

Neuronal development can be divided into several well defined phases which include expansion of progenitor cells, cessation of mitosis, expression of genes that define the initial neuronal phenotype and, finally, elaboration of complex axonal and dendritic arbors. Much of this, and in particular the latter phase, is not pre-determined within the genome of individual cells, but is controlled by interactions of the cell with its microenvironment. A variety of strategies have been adopted to elucidate the molecular basis of the recognition events that underly development of the nervous system. Immunological techniques, often coupled with *in vitro* assays that measure perturbation of cell-cell interactions (for example inhibition of specific cell-cell adhesion or defasciculation of neurite bundles), have resulted in the identification of a large number of neuronal cell surface glycoproteins that may play key roles in the above. Two important concepts have recently emerged: firstly, most of the neuronal glycoproteins identified to date can be accommodated within one of three gene families, namely the integrin family of receptors that primarily interact with extracellular matrix components (Reichardt and Tomaselli, 1991), the family of calcium dependent cell adhesion molecules called cadherins (Takeichi, 1991) or the immunoglobulin gene family (Williams, 1987). The number of molecules in each family continues to increase and it is highly likely that many aspects of neural development, and in particular axonal growth and guidance, will involve the orchestration of function of several members of each family. The second important concept to emerge is that a large number of these glycoproteins are attached to the cell membrane via covalent linkage to membrane lipids and in particular glycosylphosphatidylinositol (GPI). In the present review we will discuss the evidence for and possible significance of the presence of GPI anchors in some of these glycoproteins.

NATO ASI Series, Vol. H 70
Phospholipids and Signal Transmission
Edited by R. Massarelli, L. A. Horrocks,
J. N. Kanfer, and K. Löffelholz
© Springer-Verlag Berlin Heidelberg 1993

Diversity of GPI-anchored recognition molecules

The most prominent GPI-anchored molecules implicated in axonal fasciculation, growth and guidance are listed in Table 1.

Name/Source	Apparent Molecular Weight (kDa)	Reference
Thy-1	25	Morris, 1985
Tectal membranes	33	Stahl et al., 1990
Fasciclin 1	72	Hortsch and Goodman, 1990
T-cadherin	90	Ranscht & Bronner-Fraser, 1991
NCAM	120-125	He et al., 1986 Walsh and Doherty, 1991
F3/F11	130-135	Gennarini et al., 1989 Brummendorf et al., 1989
TAG1\Axonin-1	135	Furley et al., 1990 Kuhn et al 1991

Table 1.   GPI-anchored recognition molecules implicated in axonal fasciculation, growth and guidance.

The Thy-1 glycoprotein is one of the simplest members of the immunoglobulin superfamily, and perhaps the prototype molecule for characterisation of GPI-anchors. The molecule was originally identified as a differentiation antigen on T-lymphocytes and more recently has been found in neurons (reviewed in Morris, 1985). The evidence for GPI-linkage to the membrane includes susceptibility to release by phosphatidylinositol specific phospholipase C (PIPLC). More recently, the GPI anchor from Thy-1 has also been isolated and its structure established. Modelling studies have shown that the GPI anchor is buried within the core of the Thy-1 protein and that the latter may sit directly on the surface of the membrane (Rademacher et al., 1991). Complete removal of Thy-1 from cells with PIPLC does not trigger the activation of intracellular secnd messenger pathways (Barboni et al., 1991), suggesting that the lipid portion is not involved in direct signal transduction. Antibodies to Thy-1 can induce increases in intracellular calcium and stimulate mitosis in T-cells. Considerable evidence suggests that this transmembrane signalling is not directly mediated by Thy-1, with some evidence indicating that accessory transmembrane spanning components may be required for the response (for review, see Robinson, 1991). The possibility that general membrane perturbation leads to changes in intracellular calcium must be considered in light of the observation that the clustering of other "lipid-linked" molecules and in particular the simple ganglioside GM1 can lead to similar responses (Gabellini et al., 1991). The function of Thy-1 in brain remains to be determined; however, it is of interest that its pattern of expression can initially be restricted to dendrites with perhaps the increased lateral mobility afforded by lipid anchoring in the membrane facilitating this process (Xue et al., 1991).

Many other molecules whose functions are perhaps better understood can also be released from membranes by PIPLC. For example the optic tectum expresses, in a graded manner, a component that can induce the collapse of neuronal growth cones; this component may play an important role in the establishment of the correct retinal-tectal projection. Although the responsible molecule has yet to be cloned and sequenced, considerable evidence suggests that it is a glycoprotein of ~ 33 kDa that can be released from the tectum by PIPLC (Stahl et al., 1990). It is not known if this molecule is related to the immunoglobulins; however, with the exception of the recently identified T-cadherin, all of the molecules listed in Table 1 are members of the immunoglobulin gene superfamily.

An important role for GPI anchors is perhaps suggested by the highly conserved nature of this form of membrane association. For example a glycoprotein originally identified by a monoclonal antibody that showed restricted binding to axon fascicles in the nervous system of drosophila and grasshoppers has recently been shown to be a lipid anchored member of the immunoglobulin superfamily (Bastiani et al., 1987; Zinn et al., 1988; Elkins et al., 1990) Fasciclin 1 has four immunoglobulin like domains and a very short hydrophobic region at the carboxy-terminal suggestive of possible attachment to the cell membrane by a GPI anchor (for example, see Low, 1985). Direct evidence for this was obtained with both a drosophila cell line that expressed Fasciclin 1 and one that only expressed Fasciclin 1 following transfection with the appropriate full length cDNA. In both instances PIPLC could release a fraction of Fasciclin 1 from the membrane and the latter could be labelled with [^{14}C] ethanolamine which is indicative of a lipid anchor.

Two additional molecules that were originally identified by antibodies are the F3/F11 glycoprotein (Brummendorf et al., 1988) Gennarini et al., 1989) and TAG1/Axonin-1 (Dodd et al., 1988; Furley et al., 1990; Kuhn et al 1991). Expression cloning and protein sequence deduced from full length cDNAs has established both to be members of the immunoglobulin superfamily. F3/F11 contains six immunoglobulin like domains and a short hydrophobic region at the carboxy terminus; it can be released from brain membranes and transfected cells by PIPLC. A restricted pattern of expression and in particular localisation to axon bundles is compatible with a role of the molecules in selective axonal fasciculation. TAG1/Axonin1 is structurally very similar to the F3/F11 glycoprotein in that it also has six immunoglobulin like domains and a short hydrophobic region at the carboxy terminus; it too can be released from membranes and transfected cells by PIPLC. This glycoprotein was shown to be transiently expressed on the axons of commissural neurons as they projected towards the ventral midline of the spinal cord, with a loss of antigenic sites after the axons crossed the floor-plate region.

A very interesting recent observation is that a novel member of the cadherin family, called T-cadherin, appears to be truncated at its carboxy terminal and is likely to be attached to the cell membrane by a GPI anchor (Ranscht 1991, Ranscht and Bronner-Fraser, 1991). All other cadherins are generally believed to be transmembrane proteins that constitutively associate with the cytoskeleton; indeed in the case of E-cadherin this association is required for its adhesive function (Takeichi, 1991). T-cadherin is expressed in a restricted manner in the developing chicken suggestive of a role in guiding initial neural crest cell migration and motor axon segregation (Ranscht and Bronner-Fraser, 1991).

The remaining molecule in Table 1 is the neural cell adhesion molecule NCAM. NCAM was originally identified as a molecule that promoted neuronal adhesion in the retina. NCAM is of particular interest for defining the role of GPI anchors as alternative splicing of a single gene encodes for three classes of this molecule that differ in their mode of association with the cell membrane (reviewed in Walsh and Doherty, 1991). All NCAM related mRNAs are derived from a single locus on human chromosome 11q23 (Nguyen et al., 1986; Walsh et al., 1986) and a syntenic region on mouse chromosome 9 (D'Eustachio et al., 1985). Alternative splicing of a total of 26 exons generates the three classes. Exons 1-12 are constitutively used by the three NCAM classes: however, differential use of a 239 base pair exon named SEC situated between exons 12 and 13 results in the synthesis of soluble NCAM. The remaining isoforms use exons 13 and 14. Attachment of NCAM to the cell membrane by a GPI anchor is determined by the use of exon 15 and is associated with the generation of NCAM proteins of 120-125 Kda. This exon encodes a 27 amino acid carboxy terminal region, which in common with other GPI-anchored proteins is probably cleaved prior to its attachment to the lipid anchor although the exact site of cleavage has not yet been identified. Use of exon 16 (rather than 15) is associated with the synthesis of a 70 amino acid sequence with α helical conformation over about 20 amino acids which is compatible with a transmembrane spanning domain. In addition to these three classes, alternative splicing of the VASE exon between exons 7 and 8, the MSD1a-c exons between exon 12 and the SEC exon, or exon 18 between exons 17 and 19 results in well over 20 distinct isoforms of NCAM being expressed in the brain.

The expression of GPI-anchored NCAM is highly regulated during development. For example skeletal muscle myoblasts express 140 kDa transmembrane NCAM while the myotubes that result from fusion of the myoblasts express a GPI-anchored NCAM isoform (Moore et al., 1987). As such, PIPLC treatment can essentially remove all NCAM from myotubes, but has little or no effect on NCAM immunoreactivity on myoblasts. Glial cells also express a GPI linked NCAM that can be released by PIPLC (He et al., 1986), but neurons appear to synthesise only transmembrane NCAM isoforms. The fact that NCAM exists both as transmembrane and GPI-linked isoforms suggests that it may be an ideal model system to elucidate the function of GPI-anchors.

GPI-anchored molecules can mediate intercellular adhesion.

A molecular genetic approached has been utilised to study the adhesive function of several of the recognition molecules listed in Table 1. When CAM deficient cells are transfected with cDNA for Fasciclin 1 (Elkins et al., 1990), F3/F11 (Gennarini et al., 1991) or both GPI and transmembrane NCAM (Pizzey et al., 1989), isoforms expression of the transgene can be associated with an increase in adhesion between transfected cells. In the case of Fasciclin 1 and NCAM, cells expressing the transfected molecules show a greater tendency to adhere to each other rather than to non-transfected cells suggesting, but not proving, that they operate via homophilic binding mechanisms (e.g. see Edelman, 1986). The above data clearly demonstrate that lipid-linked molecules can mediate specific cell-cell interactions; however in the case of both NCAM and F3/F11 increased adhesion was only found in "still media". The application of even very small shear forces disrupted cell-cell interactions mediated by these molecules, suggesting that their ability to act as recognition molecules may be more important than their ability to support adhesion.

At least three GPI-linked molecules stimulate neurite outgrowth.

When the TAG-1 glycoprotein is purified and immobilised on a nitrocellulose-coated tissue culture dish it can promote neurite extension from TAG-1 positive embryonic dorsal root ganglion neurons, but not superior cervical ganglion sympathetic neurons which do not express TAG-1 (Furley et al., 1990). Removal of TAG-1 from neurons with PIPLC does not inhibit their ability to grow on a TAG-

1 substratum, suggesting that a heterophilic interaction may mediate neurite outgrowth. Recent studies on the chick homolog of TAG-1, called Axonin-1, suggest that the L1 glycoprotein which is a transmembrane spanning adhesion molecule can act as a neuronal receptor for this CAM (Kuhn et al., 1991). When the F3/F11 neuronal cell surface protein is expressed in CHO cells via gene transfer, transfected cells show a markedly enhanced ability to promote neurite outgrowth from sensory neurons compared with non-transfected cells. Both studies suggest that TAG-1 and the F3/F11 glycoproteins may play a role in specifying positional information to neurons required for their correct growth and guidance.

The first GPI-linked molecule to be shown to promote neurite outgrowth was NCAM (Doherty et al., 1990a). The three NCAM isoforms described above have been transfected into 3T3 cells and stable clones selected and characterised. Providing NCAM is linked to the membrane in transfected cells it can enhance neurite outgrowth from a variety of neurons including human dorsal root ganglia, rat cerebellar granule cells and chicken retinal ganglion cells (Doherty et al., 1989; 1990a,b; 1991). When PIPLC is included in the culture media, NCAM-dependent neurite outgrowth can be fully inhibited on monolayers expressing GPI-linked, but not transmembrane isoforms, of NCAM. The cell clone expressing the highest level of NCAM had the greatest effect, increasing neurite length by approximately 2.5 times relative to (what was not insubstantial) growth on parental (untransfected) 3T3 cells. An interesting observation was the highly co-operative nature of this response. Clones that expressed 50% less NCAM had no effect on any measured morphological parameter. Thus a discrete threshold value of NCAM is required for the expression of this function; above this value relatively small changes in NCAM promote substantial increases in neurite outgrowth. GPI-linkage confers no advantage over transmembrane spanning domains. However, both NCAM isoforms are highly mobile in the plasma membrane. In contrast, N-cadherin which constitutively links to the cytoskeleton and would be expected to show restricted mobility, can also promote neurite outgrowth when expressed in 3T3 cells. However, N-cadherin does so in a highly linear rather than co-operative manner (Doherty et al., 1991).

CAMs may directly promote axonal growth via G-protein dependent activation of L- and N-type calcium channels.

The above postulate implies that adhesion *per se* is not sufficient for complex responses such as neurite growth. In order to determine the molecular basis of transduction of a recognition signal into a complex morphological response we have cultured the PC12 neuronal cell line on monolayers of control 3T3 cells or 3T3 cells expressing either transfected NCAM or N-cadherin. For comparison, PC12 cells were also grown in the presence of NGF or a collagen-coated substratum.

NCAM and N-cadherin in the monolayer directly induced a change in the morphology of PC12 cells from an adrenal to neuronal phenotype. For example after 40-48 hrs of co-culture, 68.5 ± 4.4 (7) % of PC12 cells extended a neurite greater than 20 μm on N-cadherin expressing monolayers as compared with 49.6 ± 2.6 (7) % and 18.3 ± 2.3 (8) % on NCAM and control monolayers, respectively. These responses could be inhibited by antibodies that block the function of NCAM and N-cadherin, but not only by antibodies that block the function of NGF or FGF. NGF promotes a similar morphological differentiation of PC12 cells grown on a collagen coated substratum (Greene, 1984); however, NCAM and N-cadherin dependent morphological differentiation differed from that of NGF/integrin receptor dependent differentiation in that the latter could be specifically inhibited by blocking gene transcription with cordycepin.

Morphological differentiation triggered by NCAM or N-cadherin also differed from that induced by NGF in that it could be fully inhibited by pre-treating PC12 cells with pertussis toxin. This observation suggests that a pertussis toxin-sensitive G protein that is not involved in transducing neuronotrophic factor signals may play a major role in transducing information arising from cell-cell interactions mediated by NCAM or N-cadherin. Pertussis toxin has previously been shown to inhibit muscarinic receptor induced (Inoue and Kenimer, 1988) and CAM antibody triggered (Schuch et al., 1989) calcium influx into PC12 cells. In PC12 cells whole cell calcium currents are carried by both L- and N-type calcium channels and each can be modulated by G proteins (e.g see Plummer et al., 1989). In PC12 cells verapamil and diltiazem specifically inhibit L-type calcium channels, whereas ω-conotoxin inhibits N-type calcium channels. On their own these agents inhibit NCAM/N-cadherin dependent morphological differentiation by ~50%. However, when N- and L-type channel antagonists are added together a more substantial inhibition (~ 80 - 100%) is obtained. It therefore seems likely that NCAM/N-cadherin dependent morphological differentiation can be accounted for by a G protein dependent activation of L- and N-type neuronal calcium channels. This contrasts with NGF dependent neurite outgrowth over collagen which is not inhibited by pertussis toxin or calcium channel antagonists. These data supports our hypothesis that adhesion *per se* is not sufficient to promote neurite outgrowth, and also implicate the cytoplasmic domains of CAMs in signal transduction processes that arise from cell-cell interactions (Doherty et al., 1991).

Can GPI-anchored molecules transduce a recognition signal into a cellular response?

Neurons correctly navigate complex pathways to find and innervate their appropriate target regions largely because they are able to recognise and interpret positional information within their microenvironment. The experiments described above provide unequivocal evidence that GPI-linked recognition molecules can provide positional information but do not test whether they can directly transduce a recognition signal into a complex cellular response such as neurite outgrowth. In a series of antibody perturbation studies the molecules on neurons that mediate neurite growth over a variety of complex cellular substrata have been identified. The results from a variety of studies (e.g Bixby et al., 1988; Tomaselli et al., 1988; Neugebauer et al., 1988; Seilheimer and Schachner, 1988; Doherty et al., 1991), suggest that up to four classes of recognition molecules can account for a substantial amount of neurite outgrowth over a variety of cellular substrata, with fibroblasts stimulating this function solely via neuronal integrin receptors and Schwann cells clearly activating integrins, NCAM, N-cadherin and L1 receptors in growth cones. However all of these CAMs in neurons, including NCAM exist as transmembrane isoforms and all the molecules are capable of directly interacting with the cytoskeleton. To our knowledge there has been no demonstration of lipid-anchored molecules in growth cones directly acting as receptors that mediate neurite outgrowth. For these reasons we would suggest that lipid-anchored molecules provide positional information, that such information may be all that is required for simple adhesive functions such as selective axonal fasciculation, but that transmembrane receptors may be required for more complex responses such as cell migration and axonal growth.

References

Barboni E, Gormley AM, Pliego Rivero FB, Vidal M and Morris RJ (1991) Activation of T
    lymphocytes by cross-linking of glycophospholipid anchored Thy-1 mobilises separate pools
    of intracellular second messengers to those induced by the antigen-receptor/CD3 complex.
    Immunology, 72:457-463
Bastiani MJ, Harrelson AL, Snow PM and Goodman CS (1987) Expression of fasciclin I and II
    glycoproteins on subsets of axon pathways during neuronal development in the grasshopper.
    Cell, 48:745-755
Bixby JL, Pratt RS, Lilien J and Reichardt LF (1987) Neurite outgrowth on muscle cell surfaces
    involves extracellular matrix receptors as well as Ca^{2+}-dependent and independent cell
    adhesion molecules. Proc. Natl. Acad. Sci., USA, 84:2555-2559
Brummendorf T, Wolff JM, Frank R and Rathjen F (1989) Neural cell recognition molecule F11,
    homology with fibronectin type III and immunoglobulin type C domains. Neuron, 2:1351-1361
Dodd J, Morton SB, Karagogeous D, Yamamoto M and Jessell TM (1988) Spatial regulation of
    axonal glycoprotein expression on subsets of embryonic spinal neurons. Neuron, 1:105-115
Doherty P and Walsh FS (1989) Neurite guidance molecules. Current Opinion in Cell Biology,
    1:1102-1106
Doherty P, Barton CH, Dickson G, Seaton P, Rowett LH, Moore SE, Gower HJ and Walsh FS (1989)
    Neuronal process outgrowth of human sensory neurons on monolayers of cells transfected
    with cDNAs for five human NCAM isoforms. J Cell Biol, 109:789-798
Doherty P, Fruns M, Seaton P, Dickson G, Barton CH, Sears TA and Walsh FS (1990a)
    A threshold effect of the major isoforms of NCAM on neurite outgrowth. Nature, 343:464-
    466
Doherty P, Cohen J and Walsh FS (1990b) Neurite outgrowth in response to transfected N-CAM
    changes during development and is modulated by polysialic acid. Neuron, 5:209-219
Doherty P, Rowett LH, Moore SE, Mann DA and Walsh FS (1991) Neurite outgrowth in response
    to transfected N-CAM and N-cadherin reveals fundamental differences in neuronal
    responsiveness to CAMs. Neuron, 6:247-258
Doherty P, Ashton SV, Moore SE and Walsh FS (1991) Morphoregulatory activities of NCAM and
    N-cadherin can be accounted for by G protein-dependent activation of L- and N-type neuronal
    Ca^{2+} channels. Cell, 67:21-33
D'Eustachio P, Owens GC, Edelman GM, and Cunningham BA (1985) Chromosomal location of the
    gene encoding the neural cell adhesion molecule (N-CAM) in the mouse. Proc. Natl. Acad. Sci.
    USA, 82:7631-7635
Edelman GM, (1984) Modulation of cell adhesion during induction, histogenesis and perinatal
    development of the nervous system. Ann Rev, Neurosci, 7:339-77
Elkins T, Hortsch M, Bieber AJ, Snow PM and Goodman CS (1990) Drosophila fasciclin I is a novel
    homophilic adhesion molecule that along with fasciclin III can mediate cell sorting. J Cell Biol,
    110:1825-1832

Furley AJ, Morton SB, Manolo D, Karagogeous D, Dodd J and Jessell TM (1990) The axonal glycoprotein TAG1 is an immunoglobulin superfamily member with neurite outgrowth promoting activity. Cell, 61:157-170

Gennarini G, Cibelli G, Rougon G, Mattei MG, and Goridis C (1989) The mouse neuronal cell surface protein F3: a phosphatidylinositol-anchored member of the immunoglobulin superfamily related to chicken contactin. J Cell Biol, 109:775-788

Gennarini G, Durbec P, Boned A, Rougon G and Goridis C (1991) Transfected F3/F11 neuronal cell surface protein mediates intercellular adhesion and promotes neurite outgrowth. Neuron, 6:595-606

Greene LA (1984) The importance of both early and delayed responses in the biological action of nerve growth factor. Trends Neurosci, 7:91-94

He HT, Barbet J, Chaix JC and Goridis C (1986) Phosphatidylinositol is involved in the membrane attachment of NCAM-120, the smallest component of the neural cell adhesion molecule. EMBO J, 5:2489-2494

Hortsch M and Goodman CS (1990) Drosophila fasciclin 1, a neural cell adhesion molecule, has a phosphatidylinositol lipid anchor that is developmentally regulated. J Biol Chem, 265-15104-15109

Inoue K and Kenimer JG (1988) Muscarinic stimulation of calcium influx and norepinephrine release in PC12 cells. J Biol Chem, 263:8157-8161

Kater SB and Mills LR (1991) Regulation of growth cone behaviour by calcium. J Neurosci, 11:891-899

Kuhn TB, ET Stoeckli, FG Rathjen and P Sonderegger (1991) Neurite outgrowth on immobilised axonin-1 is mediated by a heterophilic interaction with L1(G4). J Cell Biol, 115:1113-1126

Low MG (1989) The glycosyphosphatidylinositol anchor of membrane proteins. Biochim. Biophys. Acta, 988:427-454

Moore SE, Thompson J, Kirkness V, Dickson JG and Walsh FS (1987) Skeletal muscle neural cell adhesion molecule (NCAM): changes in protein and mRNA species during myogenesis of muscle cell lines. J Cell Biol, 105:1377-1366

Morris R (1985) Thy-1 in developing nervous tissue. Dev. Neurosci, 7:133-160

Neugebauer KM, Tomaselli KJ, Lilien J and Reichardt LF (1988) N-cadherin, N-CAM and integrins promote retinal neurite outgrowth on astrocytes in vitro. J Cell Biol, 107:1177-1187

Nguyen C, Mattei MG, Mattei JF, Santoni MJ, Goridis C and Jordan BR (1986) Localisation of the human NCAM gene to band q23 of chromosome 11; the third gene for a cell interaction molecule mapped to the distal portion of the long arm of chromosome 11. J Cell Biol, 102:711-715

Pizzey JA, Rowett LH, Barton CH, Dickson G and Walsh FS (1989) Intercellular adhesion mediated by human muscle neural cell adhesion molecule: effects of alternative exon use. J Cell Biol, 109:3465-3476

Plummer MR, Logothetis DE and Hess P (1989) Elementary properties and pharmacological sensitivities of calcium channels in mammalian peripheral neurons. Neuron, 2:1453-1463

Rademacher TW, Edge CJ and Dwek R (1991) Dropping anchor with the lipophosphoglycans. Current Biology, 1:41-42

Ranscht B (1991) Cadherin cell adhesion molecules in vertebrate neural development. Seminars in Neurosci, 3:285-286

Ranscht B and Bronner-Fraser M (1991) T-cadherin expression alternates with migrating neural crest cells in the trunk of the avian embryo. Development, 3:15-22

Reichardt LF and Tomaselli KJ (1991) Extracellular matrix molecules and their receptors: functions in neural development. Ann. Rev. Neurosci, 14:531-570

Robinson PJ (1991) Phosphatidylinositol membrane anchors and T-cell activation. Immunology Today, 12:35-41

Schuch U, Lohse MJ and Shachner M (1989) Neural cell adhesion molecules influence second messenger systems. Neuron, 3:13-20

Seilheimer B and Schachner M (1988) Studies of adhesion molecules mediating interactions between cells of peripheral nervous system indicate a major role for L1 in mediating sensory neuron growth on Schwann cells in culture. J Cell Biol, 107:341-351

Stahl B, Muller B, Von Buxberg Y, Cox EC and Bonhoeffer F (1990) Biochemical characterisation of a putative axonal guidance molecule of the chick visual system. Neuron, 5:735-743

Takeichi M (1991) Cadherin cell adhesion receptors as a morphogenetic regulator. Science, 251:1451-1455

Walsh FS, Putt W, Dickson JG, Quinn CA, Cox RD, Webb M, Spurr N and Goodfellow PN (1986). Human NCAM gene: mapping to chromosome 11 by analysis of somatic cell hybrids with mouse and human cDNA probes. Mol. Brain Res., 387:197-200

Walsh FS and Doherty P (1991) Structure and function of the gene for neural cell adhesion molecule. Seminar in Neurosci, 3:271-284

Williams AF (1987) A year in the life of the immunoglobulin superfamily. Immunology Today, 8:298-303

Xue GP, Rivero BP and Morris RJ (1991) The surface glycoprotein Thy-1 is excluded from growing axons during development: a study of the expression of Thy-1 during axogenesis in hippocampus and hindbrain. Development, 112:161-176

Zinn K, McAllister L and Goodman CS (1988) Sequence analysis and neuronal expression of fasciclin I in grasshopper and drosophila. Cell, 53:577-587

LIPID SYNTHESIS AND TARGETING TO THE MAMMALIAN CELL SURFACE

Alex Sandra*. Wouter van't Hof, Ida van Genderen, and Gerrit van Meer

Department of Cell Biology, AZU H02.314

Medical School, University of Utrecht

Heidelberglaan 100, 3584 CX Utrecht

Introduction

Eukaryotic cells are surrounded by a plasma membrane that delimits their cytoplasm, which itself is filled with numerous intracellular membranes sometimes enclosing yet other membranous systems. It is well-organized that the various organelles subsume different functions and possess unique protein and lipid composition. A fundamental questions is how the dynamic interplay between local synthesis, modification and degradation on the one hand, and the various modes of lipid traffic with their inherent sorting potential on the other hand, result in the stable intracellular lipid heterogeneity so similar among the different cell types. This review summarizes our knowledge on lipid traffic, addresses the localization of the sorting events, and provides some working hypotheses on the mechanisms of lipid sorting with primary emphasis on the appearance of lipids at the cell surface. Cellular lipid traffic will undoubtedly be more complicated than the schemes provided here. However, the complications may turn out to be variations on a common theme.

Of the various cellular membranes, the plasma membrane proteins and lipids determine the interactions of the cell with its external environment, and clearly play a crucial role in

*Permanent address: Dept. of Anatomy, College of Medicine, University of Iowa, Iowa City, IA 52242

NATO ASI Series, Vol. H 70
Phospholipids and Signal Transmission
Edited by R. Massarelli, L. A. Horrocks,
J. N. Kanfer, and K. Löffelholz
© Springer-Verlag Berlin Heidelberg 1993

processes in such diverse areas as embryonic development, cell growth, and signal transduction. The central question that we hope to resolve is how cells control the lipid composition of their surface, that is, the outer or exoplasmic leaflet of the plasma membrane bilayer, and modulate it with time. Especially interesting in this regard are the epithelia. In these tissues, the plasma membrane on one side of the cell, the apical domain, contacts the external environment, while the opposite basolateral domain interacts with the underlying tissue. These two domains, although parts of one continuous membrane, possess different protein and lipid compositions, posing the question of how these differences are generated and maintained.

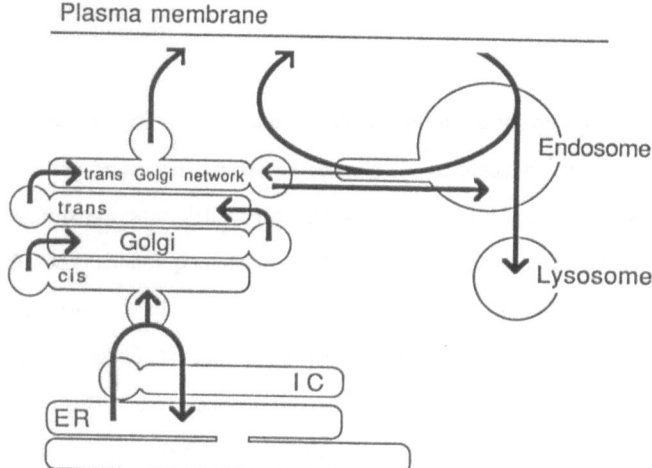

Figure 1: Pathways of vesicular traffic. IC: intermediate compartment; ER: endoplasmic reticulum

The available date suggest that the major process responsible for the delivery and retrieval of lipid molecules to and from the cell surface is vesicular trafficking via the routes of exocytosis and endocytosis (Fig. 1). In addition, in epithelial cells the two domains of the plasma membrane exchange membrane components by the vesicular pathway of transcytosis (Fig. 3). This review will stress the biosynthetic routing and sorting pathways and will focus primarily on glycosphingolipids (GSLs), sphingomyelin (SPH), and the major glycerophospholipids.

Synthesis and direct transport of lipids to the plasma membrane

Synthesis of sphingolipids on the luminal aspect of a Golgi membrane and direct transport to the plasma membrane

Any single cell expresses a specific spectrum of GSLs, usually one or two series derived from ceramide by the stepwise addition of monosaccharides. Apart from the addition of the first glucose (Coste et al., 1986), this assembly occurs on the luminal surface of subsequent cisternae of the Golgi complex (see Sasaki, 1981; van Echten et al., 1990; Young et al., 1990; Wattenberg, 1990). The implications of glycosltransferase activities at the cell surface (Pierce et al., 1980; Shur, 1989) for GSL biosynthesis are unclear.

SPH is synthesized by the energy independent transfer of phosphorylcholine from a phosphatidylcholine (PC) molecule onto ceramide. In the early studies (e.g. Marggraf et al., 1981; Voelker & Kennedy, 1982) the plasma membrane was proposed to be a major site of SPH biosynthesis. However, during biosynthetic labeling of SPH with radioactive choline the specific activity of plasma membrane SPH noticeably lagged behind that of SPH in a Golgi/ER fraction (Cook et al., 1988). The synthesis of SPH analogues has now been shown to occur on the luminal surface of the Golgi, specifically on the cis aspect of that organelle (Futerman et al., 1990; Wieland et al., 1990). Little synthesis was detected on the plasma membrane (Lipsky & Pagano, 1985; Kobayashi & Pagano 1989; van't Hof & van Meer, 1990). The enrichment of SPH in the plasma membrane therefore is not simply a result of local synthesis.

After synthesis in the Golgi, GSLs and SPH are transported to the plasma membrane with t1/2's of 20-30 min, a process which is inhibited at 15ÁC (Miller-Prodraza & Fishman, 1984;

Lipsky & Pagano, 1985; van Meer et al., 1987; Helms et al., 1990; van't Hof & van Meer, 1990). The appearance of SPH and GSLs on the cell surface is completely inhibited in mitotic cells, strongly supporting a vesicular transport process for both lipids, as vesicular transport is reversibly inhibited during mitosis (Kobayashi & Pagano, 1989). Transport of glycosylceramide (ClcCer) to the cell surface displayed a shorter lag period than that of SPH, which had led Karrenbauer et al. (1990) to suggest that GlcCer synthesis occurs in a trans Golgi cisterna (cf. Sasaki, 1990).

How GlcCer enters the vesicular transport pathway is presently unclear. Generally, GSLs and SPH display slow flip-flop rates (Devaux, 1991). Their luminal synthesis in the Golgi would, therefore, confine them to the exoplasmic leaflets of the various organelles. This would also explain why GSLs and SPH after assembly in the Golgi only reach organelles linked by vesicular transport (van Genderen et al., 1991).

Synthesis and direct transport of glycerophospholipids to the plasma membrane

Phosphatidylcholine distributions and transport

Cellular PC is synthesized at the ER and the Golgi by the CDP-choline pathway and to a smaller extent by the stepwise methylation of phosphatidylethanolamine (PE; Jelsema & Morré, 1978; Bell et al., 1981; Vance & Ridgway, 1988), whereas direct incorporation of choline by a base exchange reaction appears to be a minor route (see Yaffe & Kennedy, 1983). The creation of the ER bilayer is made possible by the presence of a PC-pore or "flippase" activity (see Bishop & Bell, 1988; Devaux, 1991), but somewhere in the Golgi stack this potential to translocate PC across the bilayer is lost. At least from the trans-Golgi onwards throughout the endocytic pathway transport has to account for two PC pools that do not rapidly intermix.

Kaplan & Simoni (1985a) have reported that newly synthesized PC rapidly equilibrated with that in the plasma membrane [$t_{1/2}$(37°C)<<2 min], suggesting monomeric transfer of PC (Fig. 2). However, when analyzing for two-pool kinetics, it is evident from the data that a

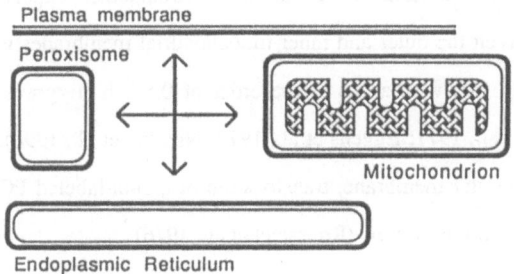

Plasma membrane

Peroxisome

Mitochondrion

Endoplasmic Reticulum

Figure 2: Transfer of lipid monomers through the cytoplasm. Bold lines: cytoplasmic bilayer leaflets

significant fraction of the newly synthesized PC reached the plasma membrane with a $t_{1/2}$ of 30-60 min. The process was influenced by energy poisons and inhibited at 15°C, characteristics of vesicular transport. The obvious prediction that this PC fraction represents the exoplasmic PC pool has not been specifically addressed. Transport of newly synthesized PC (and PE) by a special class of phospholipid-rich vesicles has been reported to be responsible for the rapid plasma membrane expansion after mitosis (Bluemink et al., 1983) and cell aggregation of *Dictyostelium* (De Silva & Siu, 1981).

Newly synthesized PC from the ER equilibrated with the outer mitochondrial membrane PC pool with a $t_{1/2}$ in the order of 30 min (McMurray & Dawson, 1969; Blok et al., 1971) or faster (Yaffe & Kennedy, 1983). Candidates for mediating this equilibration, which according to the present view is not vesicular, are two transfer proteins that possess the capability to transport PC monomers across an aqueous barrier, one strictly specific for PC, the other for phosphatidylinositol (PI) and PC (see Helmkamp, 1986). The need for such rapid equilibration is unclear at present. However, it is clear that a net transfer of PC into the mitochondria is required for their growth, since mitochondria do not synthesize PC themselves. Although the

PC-specific transfer protein can yield such net transfer in vitro, the experiment were performed under conditions where the PC concentrations of donor and acceptor membranes were far from equilibrium (discussed in Helmkamp, 1986). Therefore, the results cannot be extrapolated to intact cells, and it remains an enigma how mitochondria, or for that matter peroxisomes, expand their surface. Transfer of PC and PE between the outer and inner mitochondrial membranes is much slower than other intracellular PC traffic, with a $t_{1/2}$ in the order of 0.5-3 h (Bygrave, 1969; McMurray & Dawson, 1969; Blok et al., 1971; Eggens et al., 1979; Nicolay et al., 1990). In contrast with a rapid PC flip-flop in the outer membrane, translocation of a spin-labeled PC across the inner membrane was found to be extremely slow (Rousselet et al., 1976).

In a paradigm recently developed in our own laboratory, we have extended the observations of Pagano et al. (1981; 1983; 1985) in which fluorescently labeled NBD-phosphatidic acid (NBD-PA) was introduced into mammalian cells and converted into NBD-PC. The appearance of newly formed NBD-PC at the cell surface could then be directly monitored by depletion of the partially water soluble probe with BSA. The appearance of newly synthesized NBD-PC at the cell surface follows kinetics similar to that of newly synthesized NBD-GlcCer and is significantly faster than that of NBD-SPH. In addition, some cultured cell lines demonstrated two clearly distinct pools of newly synthesized NBD-PC. Besides the typical internal PC pool which can be detected at the cell surface upon warming in the presence of BSA, another large pool of fluorescent PC is present at the cell surface immediately upon incubation of NBD-PA with the cells at low temperature. The mechanisms of formation of these two spacially and temporally separate PC pools is currently under active investigation.

Phosphatidylethanolamine synthesis and transport

For PE synthesized by decarboxylation of phosphatidylserine (PS) in the inner mitochondrial membrane, transport out of the mitochondrion involves three steps: from the inner to the outer membrane (possibly involving "contact sites"), across the outer membrane, and to the ER or other organelles. The intramitochondrial exchange has been measured to have a rate similar to that of PC, with a $t_{1/2}$ of 0.5-3 h (Bygrave, 1969; McMurray & Dawson, 1969; Blok et al., 1971; Eggens et al, 1979), but the mechanism of this transport is still unclear. Transport from the outer membrane to the ER, according to the current views, occurs by monomeric exchange through the aqueous phase. This equilibration appears to be much slower than that of PI or PC, with a $t_{1/2}$ of about 70 min (McMurray & Dawson, 1969; Blok et al., 1971; Yaffe & Kennedy, 1983). A faster time course has been reported for the equilibration of newly synthesized PE with PE in the plasma membrane (Sleight & Pagano, 1983; Kobayashi & Pagano, 1989). The authors conclude that a rapid equilibration of PE over the two bilayer leaflets of the plasma membrane took place. A fast inside to outside translocation of PE was measured while in addition a fluorescent PE analogue underwent a rapid protein-mediated translocation in the opposite direction (Martin & Pagano, 1987). The measured effects of various inhibitors and temperatures on PE equilibration between membranes are best explained by a monomeric exchange process. This conclusion was recently corroborated by the observation that, in contrast to the vesicular transport of sphingolipids, PE transport was not inhibited in mitotic cells (Kobayashi & Pagano, 1989).

It has been suggested (Yorek et al., 1985; see Miller & Kent, 1986) that the CDP-ethanolamine pathway on the cytoplasmic surface of the ER and Golgi (Jelsema & Morré, 1978; Bell et al., 1981; Vance & Vance, 1988) is preferentially used for the synthesis of ethanolamine plasmalogens. Plasmalogens possess a long chain aldehyde in either linkage at the C1-position of the glycerol, instead of the esterified fatty acid of the regular phospholipids. In addition, the first two carbons of this chain contains a double bond. They constitute a significant fraction of the cellular phospholipids, especially of PE where they routinely make up 50% or more of that phospholipid species (see Lazarow, 1987; van den Bosch et al., 1988). The introduction of the ether bond has been localized to the luminal leaflet of the peroxisomal membrane, whereas acylation of the C2-position of the glycerol backbone and subsequent steps in plasmalogen biosynthesis take place only at the ER (Hardeman & van den Bosch, 1989). Peroxisomal alkylglycerophosphate (or alkyl-dihydroxyacetone phosphate) must thus be transported across the peroxisomal membrane, an as yet undefined process, and subsequently to the ER. The latter should present no mechanistic problem as this lysophosphatidic acid readily exchanges through the aqueous phase (see Bell et al., 1981; Lazarow, 1987).

Phosphatidylinositol and phosphatidylserine synthesis and transport

Essentially all PI is synthesized in the ER (Jelsema & Morré, 1978). It has been known for a long time that PI equilibrates between ER and mitochondrial membranes within minutes (McMurray & Dawson, 1969). Recent work using PI-specific phospholipase C digestion of the erythrocyte membrane has concluded that PI transbilayer movement is much slower than that of the aminophospholipids and comparable to that of PC (Bütikofer et al., 1990). Inositol lipids, as intermediates of the polyphosphoinositide signal transducing system, display a high turnover rate in the cytoplasmic leaflet of the plasma membrane. Rapid replenishment of plasma membrane PI is therefore required. A PI/PC transfer protein has been identified (Helmkamp, 1986), and its dual specificity may cause a net PI flux toward the plasmalemma. Since the PI concentration is lower in the plasma membrane than in the ER, the probability of the transfer protein picking up PI as compared to PC is lower in the plasmalemma than in the ER. However, the physiological function may be very different. Bankaitis et al. (1990) have determined that the *Saccharomyces cerevisiae* SEC14 gene product, which stimulates yeast Golgi secretory function, is the PI/PC transfer protein of yeast. Furthermore, mutations in the CDP-choline pathway of phospholipid synthesis bypass the requirement for the transfer protein (Cleves et al., 1991).

PS is synthesized by PS synthase primarily in the ER, but possibly also in the Golgi and plasma membrane (Jelsema & Morré, 1978; Vance & Vance, 1988; Voelker, 1989; 1990). PS is enriched more than three-fold in the plasma membrane and probably in all membranes in the endocytic pathway as compared to other intracellular membrane systems (Luzio & Stanley, 1983; Urade et al., 1988; cf. Evans & Hardison, 1985). In the plasma membrane, a specific amino-phospholipid translocator maintains PS and PE in the cytoplasmic leaflet (see Devaux, 1991). The relatively high concentration of PS in the cytoplasmic leaflet of the plasmalemma and membranes of the endocytic pathway may be due to an active intracellular sorting event, or alternatively, to the higher affinity of these membranes for PS. This difference in affinity may be based on the fact that such membranes have a potential that is positive on the exoplasmic surface (see Fuchs et al., 1989). The desorption of negatively charged PS into the cytoplasm would require more energy than diffusion from the other membrane systems. The exchange of PS between organelles displays some remarkable characteristics. It was found to be ATP-dependent (Voelker, 1989; 1990) and selective, in that an NBD-PS analogue partitioned into the Golgi apparatus preferentially (Kobayashi & Arakawa, 1991).

Synthesis and direct transport of cholesterol to the plasma membrane

Compared to the methodologies developed for monitoring the intracellular distribution and transport of phospholipids, those for cholesterol have been more indirect. The generally applied cholesterol oxidase method presumes that only the plasma membrane cholesterol is accessible to the enzyme but only circumstantial evidence has been presented that excludes the oxidation of cholesterol in at least some internal cellular compartments (Lange, 1991). Nevertheless, all evidence is consistent with a high concentration of cholesterol in the plasma membrane, which apparently is due to the presence of SPH (Pörn & Slotte, 1990). Transport of newly synthesized cholesterol from its site of synthesis in the ER to the plasma membrane appears vesicular. This process is inhibited at temperatures of 15°C and below, and is sensitive to energy inhibitors (Lange & Matthies, 1984; Kaplan & Simoni, 1985b; Slotte & Bierman, 1987). In some reports cholesterol transport appears to be vectorial with a short (10 min)

DeGrella & Simoni, 1982) or long (1-1.5 hr) half time (Lange & Matthies, 1984). However, in others (Kaplan & Simoni, 1985b) cholesterol transport is measured as an equilibration of plasma membrane and ER cholesterol with a $t_{1/2}$ of about 10 min. In contrast, equilibration of exogenously introduced plasma membrane cholesterol with that of the ER appear to have a $t_{1/2}$ of days (Lange & Matthies, 1984; Slotte & Bierman, 1987). Inhibitory effects of drugs on cholesterol transport have been reported (Liscum, 1990).

Indirect transport routes of lipids to the plasma membrane

The composition of the cell surface represents a state of dynamic equilibrium. New lipids and proteins are continuously inserted by the exocytic pathway and retrieved by endocytosis. In endocytosis, membrane vesicles pinch off from the cell surface and fuse with a subcompartment of a heterogeneous membrane system collectively called endosomes. From there, membrane components and contents can be transported onwards to the lysosomes where degradation occurs. Alternatively, they can bypass the lysosomal system to emerge on the same (recycling) or the opposite (transcytosis) membrane domain (Fig. 3). In addition, evidence from the last years has shown that the endocytic and exocytic routes are interconnected, most likely at the level of the TGN and endosomes (van Deurs et al., 1989; Fig. 1).

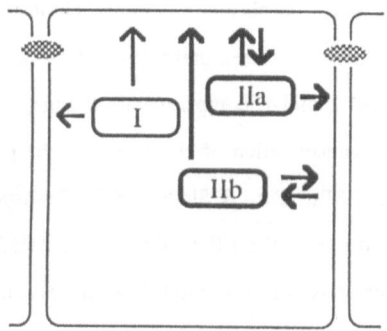

Figure 3: Scheme of endocytosis and transcytosis.

I: TGN, biosynthetic transport;

IIa: apical endosome;

IIb: basolateral   endosome, transcytotic transport

Lipid transport in the endocytic and recycling pathway has been demonstrated with a variety of fluorescent lipid analogues which were initially introduced into the exoplasmic leaflet of intact cultured cells. Fluorescent NBD-PC, NBD-SPH, NBD-GlcCer and N-Rh-PE were endocytosed into early, peripheral endosomes and then translocated into a perinuclear region. NBD-SPH and NBD-GlcCer recycled (Kok et al., 1989; Koval & Pagano, 1989; 1990), while the $t_{1/2}$ for a complete round of NBD-SPH recycling was estimated to be about 40 min (Koval & Pagano, 1989). It was concluded that both NBD-PC and NBD-GlcCer were able to reach the Golgi (Sleight & Pagano, 1984; Kok et al., 1991), whereas N-Rh-PE was channeled efficiently to the lysosomes (Kok et al., 1989; 1990). Also radiolabeled GlcCer and complex GSLs were found to recycle to Golgi (see Schwarzmann & Sandhoff, 1990; Trinchera et al., 1990).

In a study in which a variety of cultured cell lines were examined for the manner by which they translocated NBD-PC to internal cellular regions, Sleight & Abanto (1989) observed that the majority of cell lines follow the above described pattern of internalization, although others exhibited a diffuse internal fluorescence which appeared more consistent with monomer diffusion. The aminophospholipids NBD-PE and NBD-PS were also internalized after incubation with cultured cells, but the temperature at which this process was first evident was 7°C and above, as opposed to the higher temperatures required for PC and SPH internalization, implying transbilayer movement and monomeric exchange as the mechanism of internalization (Martin & Pagano, 1987; Kobayashi & Arakawa, 1991). Still other phospholipids are internalized by cells immediately upon addition, regardless of temperature. NBD-PA, for example, is hydrolyzed to NBD-diacyglycerol at the cell surface and then this lipid analogue is free to undergo spontaneous transbilayer movement. NBD-diacylglycerol is found throughout internal cell membranes whereupon it is metabolized to a variety of NBD-lipids, including PC, PE, fatty acid, and neutral glycerols (Pagano et al., 1981; 1983; 1985).

Epithelial cells display an endocytic system of greater complexity. In fact, they contain an essentially separate endocytic cycle or each plasma membrane domain (Bomsel et al., 1989;

1990; Parton et al., 1989; Cerneus & van der Ende, 1991). Furthermore, transport vesicles from one endocytic cycle may travel to the opposite cell surface, transcytosis (Fig. 3). In most epithelia studied to date, newly synthesized plasma membrane components are delivered to each of the two plasma membrane domains by a direct exocytic pathway from the Golgi. However, a special situation has been described by hepatocytes. In these cells, apical proteins are first directed to the basolateral sinusoidal surface. A direct route from the Golgi to the apical, bile canalicular, cell surface does not seem to exist. Subsequently, apical proteins are endocytosed and delivered to the bile canalicular surface by a vesicular pathway between both plasma membrane domains (Hubbard et al., 1989). Therefore, in these cells the transcytotic route is an integral part of the routing of newly synthesized membrane components to (one domain of) the plasma membrane. Intestinal cells apparently use both mechanisms (Matter et al., 1990; LeBivic et al., 1990), and it was thought that in these cells the size of the direct pathway from the TGN to the apical cell surface or the sorting machinery in the TGN has a limited capacity. However, we have found no differences in lipid sorting or pathway sizes in these cells as compared to dog kidney cells (van Meer et al., 1987; van't Hof and van Meer, 1990). As an alternative explanation, it has been suggested that independent of the presence of a direct apical pathway, some proteins follow the indirect pathway over the basolateral surface, due to specific information in their structure (Wessels et al., 1990).

Lipid sorting

Lipid sorting along the exocytic pathway of mammalian cells.

Sphingolipid sorting in cis-Golgi

The concentration of sphingolipids in the ER is low. Recent evidence suggests that vesicular transport between the cis-Golgi and the ER is bidirectional and that resident ER proteins that have entered the cis-Golgi are sorted into the return pathway (Pelham, 1989). Therefore, it GSLs and SPH are present in cis-Golgi, which is the present idea (see above), they must be excluded from that return traffic: thus these lipids would have to be segregated from other phospholipids into microdomains away from sites from where return vesicles would bud. In the case that sphingolipid biosynthesis were localized in a more distal cisterna, the question of whether lipid sorting occurs depends on the nature of the transport link between cis-Golgi and that cisterna. Two general alternatives for SPH (and GSL) transport and sorting are presented in Fig. 4: (A) is the sorting alternative. In (B) sphingolipids would be unable to return to the ER due to the lack of a return pathway. However, at present there is no direct evidence against vesicles returning through the Golgi stack. The sorting variant (A) has attractive implications for

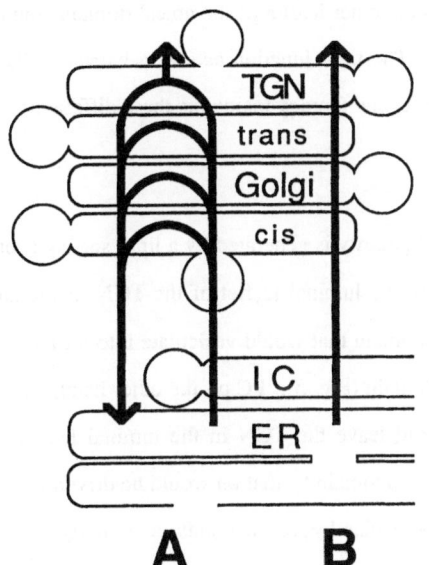

Figure 4: Models to explain concentration of sphingolipids into the exocytic route. (A): Selective recycling between subsequent cisternae (B) Unidirectional transport of membrane through the Golgi stack.

the mechanism by which cholesterol is concentrated on its way to the plasma membrane (van Meer, 1989a). An alternative mechanism to (A) that would also result in unidirectional transport of SPH has been proposed by Wieland et al. (1987; cf. Pagano, 1988). Whereas import of material into cis-Golgi would be vesicular, return traffic might occur by lipid monomers.

Although the data on protein recycling make this alternative unlikely, the model has interesting consequences: traffic of retained proteins, SPH, and possibly also cholesterol, would be vectorial and the components would be concurrently concentrated.

Glycosphingolipid sorting in the trans Golgi network of epithelial cells

A special type of lipid sorting has been observed in epithelial cells (Simons & van Meer, 1988; van Meer, 1989b). Epithelial cells display (protein and) lipid polarity. The apical plasma membrane domain on the side of the external milieu and the basolateral domain, facing the neighbor cells and the underlying tissue, possess unique lipid compositions. The general difference appears to be an enrichment of GSLs in the outer leaflet of the apical domain, and of PC in the outer leaflet of the basolateral domain. The tight junction acts as a barrier to lipid diffusion in this outer leaflet of the plasma membrane, and thereby maintains these differences in lipid composition.

Some years ago, we postulated that the lipid polarity is generated by a lipid sorting event in the trans Golgi reticulum or network (TGN). In the luminal leaflet of the TGN membrane GSLs would spontaneously aggregate into a microdomain that would vesiculate into a transport vesicle destined for the apical plasma membrane domain (Fig. 5). PC on the other hand, would be excluded from the GSL microdomain and would leave the TGN in the luminal leaflet of vesicles traveling to the basolateral surface. The microdomain formation would be driven by the physiochemical property of GSLs to form intermolecular hydrogen bonds, a property absent from PC (Pascher, 1976; Thompson & Tillack, 1985).

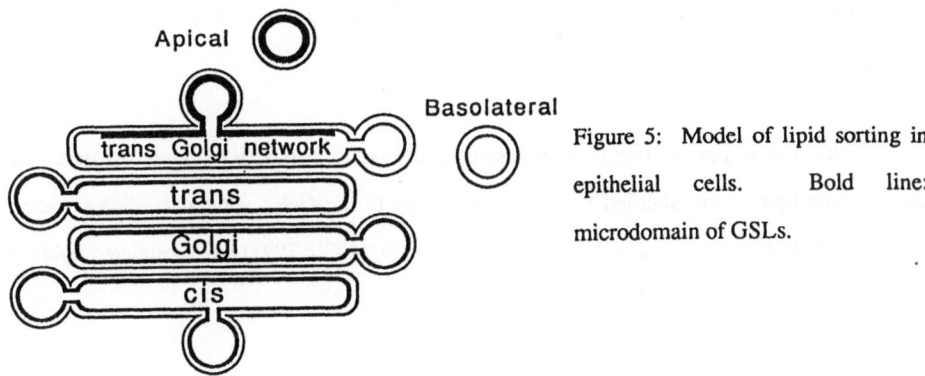

Figure 5: Model of lipid sorting in epithelial cells. Bold line: microdomain of GSLs.

The third lipid that is abundant in the exoplasmic leaflet of plasma membranes, SPH, would display a potential preference for either direction depending on the polarity of the cell type in questions. SPH is present in equal concentrations on both surfaces of some epithelia (MDCK cells; van Meer & Simons, 1982; 1986), more basolateral in others (intestinal cells: Kawai et al., 1974; Cacao-2 cells: van't Hof & van Meer, 1990), and more apical in yet others (Molitoris & Simon, 1985; discussed in van Meer, 1989b).

Lipid sorting in the endocytic and transcytotic routes

Many studies over the last 15 years have addressed the pathways of endocytosis and transcytosis and their selectivity (see reviews by Caplan & Matlin, 1989; Hubbard, 1989; Rodman et al., 1990), while the signals responsible for recognition by the sorting machinery and the sorting mechanism itself have lately received emphasis (see Breitfeld et al., 1989; Caplan & Matlin, 1989). Protein sorting has been found to occur at the level of the plasma membrane during the formation of the endocytic vesicles, and in the endosomes where the decision is taken whether a protein is preferentially shuttled to the lysosomes, transported to the TGN, recycled to the plasma membrane, or sent to the opposite plasma membrane domain. The TGN seems not to

28

be of major importance as a sorting compartment along the transcytotic pathway (Caplan & Matlin, 1989).

Evidence for lipid sorting in the endocytic pathway has also been reported. As described above, some lipids were shuttled from the endosomes to the Golgi apparatus whereas others were efficiently concentrated in the lysosomes. Kok et al. (1990; 1991) have compared lipids of the various classes within single studies.

Lipids also must be sorted during transcytosis. If lipid transport during transcytosis were random, this would result in the intermixing of apical and basolateral lipids, which clearly does not occur. As for proteins, the endosomes seem the best candidate for the site of transcytotic lipid sorting, since lipids sorting at the plasma membrane has never been observed. Because apical/basolateral sorting in endosomes concerns a number of the same proteins and lipids and the same proteins and lipids and the same vectorial processes as that in the TGN, sorting in the two compartments may be mechanistically related.

Concluding remarks

For a complete understanding of the steady state lipid composition of the plasma membrane, we need to know the kinetics of the transport of each lipid class into and out of that membrane. For this, we have to define the sites of synthesis and degradation for the various lipids, the transport pathways to and from the plasma membrane, and the kinetic properties of all these processes. Furthermore, we would need a full characterization of metabolic interconversions between lipid classes. We have come a long way, still much needs to be done.

Acknowledgements: The present work was made possible by (senior) fellowships from the Fogarty International Fellowship program (F06 TW 01534; A.S.), the Royal Netherlands

Academy of Arts and Sciences (G.v.M.), and the Netherlands Heart Foundation (grant 88.045, I.v.G.).

References

Bankaitis VA, Aitken, JF, Cleves AE, Dowhan W (1990) An essential role for a phospholipid transfer protein in yeast Golgi function. Nature 347:561-562

Bell RM, Ballas LM, Coleman RA (1981) Lipid topogenesis. J Lipid Res 22:391-403

Bishop WR, Bell RM (1988) Assembly of phospholipids into cellular membranes: Biosynthesis, transmembrane movement and intracellular translocation. Annu Rev Cell Biol. 4:579-610

Blok MC Wirtz KWA, Scherphof GL (1971) Exchange of phospholipids between mirosomes and inner and outer mitochondrial membranes of rat liver. Biochim Biophys Acta 233:61-75

Bluemink JG, van Maurik PAM, Tertoolen LGJ, van der Saag PT, de Laat SW (1983) Ultrastructural aspects of rapid plasma membrane growth in mitotic neuroblastoma cells. Eur J Cell Biol. 32:7-16

Bomsel M, Prydz K, Parton RG, Gruenberg J, Simons K (1989) Endocytosis in filter-grown Madin-Darby canine kidney cells. J Cell Biol. 109:3243-3258

Bomsel M, Parton R, Kuznetsov SA, Schroer TA, Gruenberg J. (1990) Microtubule- and motor-dependent fusion in vitro between apical and basolateral endocytic vesicles from MDCK cells. Cell 62:719-731

Breitfeld, PP, Casanova JE, Simister NE, Ross SA, McKinnon WC, Mostov KE (1989) Sorting signals. Curr Opin Cell Biol 1:617-623

Bütikofer P, Lin ZW, Chiu DT-Y, Lubin B, Kuypers FA (1990) Transbilayer distribution and mobility of phosphatidylinositol in human red blood cells. J Biol Chem 265:16035-16038

Bygrave FL (1969) Studies on the biosynthesis and turnover of phospholipid components of the inner and outer membranes of rat liver mitochondria. J Biol Chem 244:4768-4772

Caplan M, Matlin KS (1989) Sorting of membrane and secretory proteins in polarized epithelial cells. Modern Cell Biology, Vol. 8: Functional Epithelial Cells in Culture. (Satir BH, ed,; Vol: Matlin KS, Valentich DJ, eds.) Alan R. Liss, Inc., New York 71-127

Cerneus DP, van der Ende A (1991) Recycling of apical and basolateral transferrin receptors through separate endosomes in polarized Bewo cells. J Cell Biol, in press

Cleves AE, McGee TP, Whitters EA, Champion KM, Aitken JR, Dowhan W, Goebl M, Bankaitis VA (1991) Mutations in the CDP-choline pathway for phospholipid biosynthesis bypass the requirement for an essential phospholipid transfer protein. Cell 64:789-800

Cook HW, Palmer FBS, Byers DM, Spence MW (1988) Isolation of plasma membranes from cultured glioma cells and application to evaluation of membrane sphingomyelin turnover. Anal Biochem 174:552-560

Coste H, Martel MB, Got R (1986) Topology of glucosylceramide synthesis in Golgi membranes from procine submaxillary glands. Biochim Biophys Acta 858:6-12

DeGrella RF, Simoni RD (1982) Intracellular transport of cholesterol to the plasma membrane. J Biol Chem 257:14256-14262

De Silva NS, Siu C-H (1981) Vesicle-mediated transfer of phospholipids to plasma membrane during cell aggregation of Dictyostelium discoideum. J Biol Chem 256:5845-5850

Devaux PF (1991) Static and dynamic lipid asymmetry in cell membranes. Biochemistry 30:1163-1173

Eggens I, Valtersson C., Dallner G., Ernster L (1979) Transfer of phospholipids between the endoplasmic reticulum and mitochondria in rat hepatocytes in vivo. Biochem Biophys Res Commun 91:709-714

Evans WH, Hardison WGM (1985) Phospholipid, cholesterol, polypeptide and glycoprotein composition of hepatic endosome subfractions. Biochem J 232:33-36

Fuchs R, Mâle P, Mellman I (1989) Acidification and ion permeabilities of highly purified rat liver endosomes. J Biol Chem 264:2212-2220

Futerman AH, Stieger B, Hubbard AL, Pagano RE (1990) Sphingomyelin synthesis in rat liver occurs predominantly at the cis and medial cisternae of the Golgi apparatus. J Biol Chem 265:8650-8657

Hardeman D, van den Bosch H (1989) Topopgraphy of ether phospholipid biosynthesis. Biochim Biophys Acta 1006:1-8

Helmkamp Jr. GM (1986) Phospholipid transfer proteins: Mechanism of action. J Bioenerg Biomembr 18:71-91

Helms JB, Karrenbauer A, Wirtz KWA, Rothman JE, Wieland FT (1990) Reconstitution of steps in the constitutive secretory pathway in permeabilized cells. Secretion of glycosylated tripeptide and truncated sphingomyelin. J Biol Chem 265:20027-20032

Hubbard AL, Stieger B, Bartles JR (1989) Biogenesis of endogenous plasma membrane proteins in epithelial cells. Annu Rev Physiol 51:755-770

Jelsema CL, Morré DJ (1978) Distribution of phospholpid biosynthetic enzymes among cell components of rat liver. J Biol Chem 253:7960-7971

Kaplan MR, Simoni RD (1985a) Intracellular transport of phosphatidylcholine to the plasma membrane. J Cell Biol 101:441-445

Kaplan MR, Simoni RD (1985b) Transport of cholesterol from the endoplasmic reticulum to the plasma membrane. J Cell Biol 101:446-453

Karrenbauer A, Jeckel D, Just W, Birk R, Schmidt RR, Rothman JE, Wieland FT (1990) The rate of bulk flow from the Golgi to the plasma membrane. Cell 63:259-267

Kobayashi T, Arakawa Y (1991) Transport of exogenous fluorescent phosphatidylserine analogue to the Golgi apparatus in cultured fibroblasts. J Cell Biol 113:235-244

Kobayshi T, Pagano RE (1989) Lipid transport during mitosis. Alternative pathways for delivery of newly synthesized lipids to the cell surface. J Biol Chem 264:5966-5973

Kok JW, Eskelinen S, Hoekstra K, Koekstra D (1989) Salvage of glucosylceramide by recycling after internalization along the pathway of receptor-mediated endocytosis. Proc Natl Acad

Sci USA 86:9896-9900

Kok JW, ter Beest M, Scherphof G., Hoekstra D (1990) A non-exchangeable fluorescent phospholipid analog as a membrane traffic marker of the endocytic pathway. Eur J Cell Biol. 53:173-184

Kok JW, Babia T, Hoekstra D (1911) Sorting of sphingolipids in the endocytic pathway of HT29 cells. J Cell Biol, in press

Koval M, Pagano RE (1989) Lipid recycling between the plasma membrane and intracellular compartments: Transport and metabolism of fluorescent sphingomyelin analogues in cultured fibroblasts. J Cell Biol. 108:2169-2181

Koval M, Pagano RE (1990) Sorting of an internalized plasma membrane lipid between recycling and degradative pathways in normal and Niemann-Pick, Type A fibroblasts. J Cell Biol 111:429-442

Lange Y (1991) Disposition of intracellular cholesterol in human fibroblasts. J Lipid Res 32:329-339

Lange Y, Matthies HJG (1984) Transfer of cholesterol from its site of synthesis to the plasma membrane. J Biol Cehm 259:14624-14630

Lazarow PB (1987) The role of peroxisomes in mammalian cellular metabolism. J Inher Metab Dis 10 Suppl 1:11-22

Le Bivic A, Quaroni A., Nichols B, Rodriguez-Boulan E (1990) Biogenetic pathways of plasma membrane protein in Caco-2, a human intestinal epithelial cell line. J Cell Biol 111:1351-1361

Lipsky NG, Pagano RE (1985) Intracellular translocation of fluorescent sphingolipids in cultured fibroblasts: Endogenously synthesized sphingomyelin and glucocerebroside analogues pass through the Golgi apparatus en route to the plasma membrane. J Cell Biol 100:27-34

Liscum L (1990) Pharmacological inhibition of the intracellular transport of low-density lipoprotein-derived cholesterol in Chinese hamster ovary cells. Biochim Biophys Acta

1045:40-48

Luzio JP, Stanley KK (1983) The isolation of endosome-derived vesicles from rat hepatocytes. Biochem J 216:27-36

Marggraf WD, Anderer FA, Kanfer JN (1981) The formation of sphingomyelin from phosphatidylcholine in plasma membrane preparations from mouse fibroblasts. Biochim. Biophys Acta 664:61-73

Martin OC, Pagano RE (1987) Transbilayer movement of fluorescent analogs of phosphatidylserine and phosphatidylethanolamine at the plasma membrane of cultured cells. J Biol Chem 262:5890-5898

Matter K, Brauchbar M, Bucher K, Hauri H-P (1990) Sorting of endogenous plasma membrane proteins occurs from two sites in cultured human intestinal epithelial cells (Caco-2). Cell 60:429-437

McMurray WC, Dawson RMC (1969) Phospholipid exchange reactions within the liver cell. Biochem J 112:91-108

Miller-Prodraza H, Fishman PH (1984) Effect of drugs and temperature on biosynthesis and transport of glycosphingolipids in cultured neurotumor cells. Biochim Biophys Acta 805:44-51

Miller MA, Kent C (1986) Characterization of the pathways for phosphatidylethanolamine biosynthesis in Chinese hamster ovary mutant and parental cell lines. J Biol Chem 261:9753-9761

Molitoris BA, Simon FR (1985) Renal cortical brush-border and basolateral membranes: cholesterol and phospholipid composition and relative turnover. J Membr Biol 83:207-215

Nicolay K, Hovius R, Bron R,Wirtz K, de Kruijff B (1990) The phosphatidylcholine-transfer protein catalyzed import of phosphatidylcholine into isolated rat liver mitochondria. Biochim Biophys acta 1025:49-59

Pagano RE (1990) Lipid traffic in eukaryotic cells: mechanisms for intracellular transport and organelle-specific enrichment of lipids. Curr Opin Cell Biol 2:652-663

Pagano RE, Longmuir KJ, Martin OC (1983) Intracellular translocation and metabolism of a fluorescent phosphatidic acid analogue in cultured fibroblasts. J Biol Chem 258:2034-2040

Pagano RE, Longmuir KJ, Martin OC, Struck DK (1981) Metabolism and intracellular localization of a fluorescently labeled intermediate in lipid biosynthesis within cultured fibroblasts. J Cell Biol 91:872-877

Pagano RE, Longmuir KJ (1985) Phosphorylation, transbilayer movement, and facilitated intracellular transport of diacylglycerol are involved in the uptake of a fluorescent analog of phosphatidic acid by cultured fibroblasts. J Biol Chem 260:1909-1916

Parton RG, Prydz K, Bomsel M, Simons K, Griffiths G (1989) Meeting of the apical and basolateral endocytic pathways of the Madin-Darby canine kidney cell in late endosomes. J Cell Biol 109:3259-3272

Pascher I (1976) Molecular arrangements in sphingolipids. Conformation and hydrogen bonding of ceramide and their implication on membrane stability and permeability. Biochim Biophys Acta 455:433-451

Pelham HRB (1989) Control of protein exit from the endoplasmic reticulum. Annu Rev Cell Biol 5:1-23

Pierce M, Turley EA, Roth S (1980) Cell surface glycosyltransferase activities. Int Rev Cytol 65:1-47

Pörn MI, Slottte JP (1990) Reversible effects of sphingomyelin degradation on cholesterol distribution and metabolism in fibroblasts and transformed neuroblastoma cells. Biochem J 271:121-126

Rodman JS, Mercer RW, Stahl PD (1990) Endocytosis and transcytosis. Curr Opin Cell Biol. 2:664-472

Rousselet A, Coubeau A,Vignais PM, Devaux PF (1976) Study on the transverse diffusion of spin-labeled phospholipids in biological membranes. II. Inner mitochondrial membrane of rat liver: use of phosphatidylcholine exchange protein. Biochim Biophys Acta 426:372-384

Sasaki T (1981) Sequential glycosylations of endogenous glycosphingolpipids in hamster fibroblasts incubated with nucleotide sugars. Biochim. Biophys Acta 666:426-432

Sasaki T (1990) Glycolipid transfer protein and intracellular traffic of glucosylceramide. Experientia 46:611-616

Schwarzmann G, Sandhoff K (1990) Metabolism and intracellular transport of glycosphingolipids. Biochemistry 29:10865-10871

Shur B (1989) Expression and function of cell surface galactosyltransferase. Biochim Biophys Acta 988:389-409

Simons K, van Meer G (1988) Lipid sorting in epithelial cells. Biochemistry 27:6197-6202

Sleight RG, Abanto MN (1989) Differences in intracellular transport of a fluorescent phosphatidylcholine analog in established cell lines. J Cell Sci 93:363-374

Sleight RG, Pagano RE (1984) Rapid appearance of newly synthesized phosphatidylethanolamine at the plasma membrane. J Biol Chem 258:9050-9058

Sleight RG, Pagano RE (1984) Transport of a fluorescent phosphatidylcholine analog from the plasma membrane to the Golgi appartus. J Cell Biol 99:742-751

Stotte JP, Bierman EL (1987) Movement of plasma membrane sterols to the endoplasmic reticulum in cultured cells. Biochem J 248:237-242

Thompson TE, Tillack TW (1985) Organization of glycosphingolipids in bilayers and plasma membranes of mammalian cells. Ann Rev Biophys Biophys Chem 14:361-386

Trinchera M, Ghidoni R, Sonnino S, Tettamanti G (1990) Recycling of glucosylceramide and sphingosine for the biosynthesis of gangliosides and sphingomyelin in rat liver. Biochem J 270:815-820

Urade R, Hayashi Y, Kito M (1988) Endosomes differ from plasma membranes in the phospholipid molecular species composition. Biochim Biophys Acta 946:151-163

Vance DE, Ridgway ND (1988) The methylation of phosphatidylethanolamine. Prog Lipid Res

27:61-79

Vance JE, Vance DE (1988) Does rat liver Golgi have the capacity to synthesize phospholipids for lipoprotein secretion J Biol Chem 263:5898-5909

van den Bosch H, Schalkwijk CG, Schrakamp G, Wanders RJA, Schutgens RBH, et al (1988) Abberation in de novo ether lipid biosynthesis in peroxisomal disorders. In Biological Membranes: Aberrations in Membrane Structure and Function (Karnovsky ML, Leaf A., Bolis LC, eds) pp. 139-150, New York: Liss

van Deurs B, Petersen OW, Olsnes S, Sandvig K (1989) The ways of endocytosis. Int Rev Cytol 117:131-177

van Echten G, Iber H, Stotz H, Takatsuki A, Sandhoff K (1990) Uncoupling of ganglioside biosynthesis by Brefeldin A. Eur J Cell Biol 51:135-139

van Genderen I, van Meer G. Geuze HJ, Slot J-W, Voorhout W (1991) Subcellular localization of Forssman glycolipid in epithelial MDCK cells by immuno electronmicroscopy after freeze-substitution. Submitted.

van Meer G (1989a) Lipid traffic in animal cells. Annu Rev Cell Biol 5:247-275

van Meer G (1989b) Polarity and polarized transport of membrane lipids in a cultured epithelium. Modern Cell Biology (Satir BH, ed.), Functional Epithelial Cells in Culture. (Matlin KS, Valentich JD, eds.), Alan R. liss, Inc., New York 8:43-69

van Meer G, Simons K (1982) Viruses budding from either the apical or the basolateral plasma membrane domain of MDCK cells have unique phospholipid compositions. EMBO J 1:847-852

van Meer G, Stelzer EHK, Wijnaendst-van-Resandt RW, Simons K (1987) Sorting of sphingolipids in epithelial (Madin-Darby canine kidney) cells. J Cell Biol 105:1623-1635

van't Hof W, van Meer G (1990) Generation of lipid polarity in intestinal epithelial (Caco-2) cells: Sphingolipid synthesis in the Golgi complex and sorting before vesicular traffic to the plasma membrane. J Cell Biol 111:977-986

Voelker DR (1989) Phosphatidylserine translocation to the mitochondrion is an ATP-dependent process in permeabilized animal cells. Proc Natl Acad Sci USA 86:9921-9925

Voelker DR (1990) Characterization of phosphatidylserine synthesis and translocation in permeabilized animal cells. J Biol Chem 265:14340-14346

Voelker DR, Kennedy EP (1982) Cellular and enzymic synthesis of sphingomyelin. Biochemistry 21:2753-1759

Wattenberg BW (1990) Glycolipid and glycoprotein transport through the Golgi complex are similar biochemically and kinetically. Reconsitution of glycolipid transport in a cell free system. J Cell Biol 111:421-428

Wessels HP, Hansen GH, Fuhrer C, Look AT, Sjhöström H, Norén O, Spiess M (1990) Aminopeptidase N is directly sorted to the apical domain in MDCK cells. J Cell Biol 111:2923-2930

Wieland FT, Gleason ML, Serafini TA, Rothman JE (1987) The rate of bulk flow from the endoplasmic reticulum to the cell surface. Cell 50:289-300

Yaffe MP, Kennedy EP (1983) Intracellular phospholipid movement and the role of phospholipid transfer proteins in animal cells. Biochemistry 22:1497-1507

Yorek MA, Rosario RT, Dudley DT, Spector AA (1985) The utilization of ethanolamine and serine for ethanolamine phosphaglyceride synthesis by human Y79 retinoblastoma cells. J Biol Chem 260:2930-2936

Young JR WW, Lutz MS, Mills SE, Lechler-Osborn S (1990) Use of brefeldin A to define sites of glycosphingolipid synthesis; GA2/GM2/GD2 synthase is trans to the brefeldin A block. Proc Natl Acad Sci USA 87:6838-6842

THE AMINOPHOSPHOLIPID TRANSPORTER FROM HUMAN RED BLOOD CELLS

Peter E. Coderre and Alan J. Schroit
Department of Cell Biology
The University of Texas M.D. Anderson Cancer Center
1515 Holcombe Blvd.
Houston, Texas 77030

I. Introduction

Nearly twenty years ago, Bretscher (1972) proposed that the distribution of phospholipids in the erythrocyte membrane is asymmetric. It is now well established that certain phospholipids preferentially reside in either the inner or outer membrane leaflets of most cell types. For example, phosphatidylcholine and sphingomyelin are predominantly localized in the plasma membrane's outer leaflet, while phosphatidylinositol and the aminophospholipids, phosphatidylethanolamine (PE) and phosphatidylserine (PS), are primarily localized in the inner leaflet (Gordesky et al., 1975; Verkleij et al., 1973). Recent studies have provided convincing evidence that membrane lipid asymmetry is generated and probably maintained by specific transport proteins or 'flippases' (Backer and Dawidowicz, 1987; Bishop and Bell, 1985). One of these transport proteins, the aminophospholipid transporter, is responsible for the movement of PS and PE from the outer leaflet to the inner leaflet of the erythrocyte membrane (Seigneuret and Devaux, 1984; Daleke and Huestis, 1985; Tilley et al., 1986; Connor and Schroit, 1987).

In this article, we summarize recent developments on the characterization of the aminophospholipid specific transporter from human red blood cells. We also discuss related developments from studies on Rh polypeptides that support the hypothesis that the aminophospholipid transporter and Rh proteins are either identical or closely associated with one another.

II. Aminophospholipid Transport in Red Blood Cells

Protein-mediated movement of aminophospholipids was first described in erythrocytes by Seigneuret and Devaux (1984). When spin-labeled PS and PE were added to red blood cells, a temperature- and time-dependent decrease

NATO ASI Series, Vol. H 70
Phospholipids and Signal Transmission
Edited by R. Massarelli, L. A. Horrocks,
J. N. Kanfer, and K. Löffelholz
© Springer-Verlag Berlin Heidelberg 1993

in the fraction of spin-labeled lipid that could be reduced by ascorbate was observed. Since ascorbate cannot penetrate the cell's bilayer membrane, these findings indicated that the analogs were transported from their initial site of insertion in the outer leaflet to the inner leaflet. Similar observations have been obtained with unlabeled, spin-labeled, fluorescent or isotopically labeled lipids in red blood cells (Daleke and Huestis, 1985; Tilley et al., 1986; Connor and Schroit, 1987, 1988), nucleated cells (Zachowski et al., 1987; Martin and Pagano, 1987), intracellular membranes (Zachowski et al., 1989), and membrane vesicles (Zachowski and Morot, 1990). Because PS does not translocate between leaflets of artificially generated vesicles (Tanaka and Schroit, 1986), these findings indicated that lipid transport in cell membranes is mediated by a facilitated transport mechanism involving lipid-specific transporters. Indeed, the translocation of PS from the outer to the inner leaflet of red blood cells is ATP-dependent, since transport does not occur in ATP-depleted cells but can be reconstituted in erythrocyte ghosts resealed in the presence of Mg^{2+}-ATP (Seigneuret and Devaux, 1984). Additional evidence that the transport of anionic phospholipids depends on lipid-specific protein transporters was obtained from experiments showing that transport is inhibited by agents that oxidize protein sulfhydryls (Seigneuret and Devaux, 1984; Daleke and Huestis, 1985; Tilley et al., 1986; Connor and Schroit, 1987; Zachowski et al., 1987; Martin and Pagano, 1987). While these observations strongly suggested the involvement of a specific protein, conclusive evidence of the transporter's specificity was shown by the inability to transport D-isomers of PS (Martin and Pagano, 1987). Substrate specificity is also dictated, however, by the glyceride backbone and esterification of the *sn*-2 position. Substitution of glyceride with ceramide abolished translocation even when the polar head group contained the phosphoserine moiety (Morrot et al., 1989), and lysoPS translocated across the bilayer membrane at very slow rates (Bergmann et al., 1984). Polar head group specificity was further supported by the observation that progressive methylation of PE resulted in a concomitant decrease of specific transport (Morrot et al., 1989).

III. Characterization of the Aminophospholipid Translocase

Currently, two proteins are candidates for the aminophospholipid-specific flippase in red blood cells. These are based on (1) crosslinking experiments using appropriately labeled substrate and inhibitor analogs (Schroit et al., 1987; Connor and Schroit, 1988), and (2) theoretical arguments based on the biochemical requirements for lipid transport (Zachowski and Devaux, 1990).

Recent studies indicated that a component of erythrocyte membrane band 7 (polypeptides in the 30 kDa to 32 kDa region) might be involved in the transmembrane movement of PS. Analysis of membrane proteins from human red blood cells incubated with a photoactivatable ^{125}I-labeled PS analog, under conditions conducive to PS transport, revealed that a large fraction of this probe preferentially labeled a 32 kDa polypeptide after photolysis (Schroit et al., 1987). Additional evidence for the involvement of this protein in lipid transport was obtained by using a ^{125}I-labeled transport inhibitor. Because the intrabilayer movement of PS in red blood cells requires that membrane sulfhydryls be reduced (Daleke and Huestis, 1985, Zachowski et al., 1986), an iodinated derivative of pyridyldithioethylamine (PDA), a thiol reagent which specifically and reversibly inhibits PS transport, was used (Connor and Schroit, 1988). The results from this study showed that the inhibitor labeled the same 32 kDa protein that was labeled with the iodinated substrate analog, strongly suggesting that this protein is involved in the transport of PS. This polypeptide was recently identified as part of a set of closely related isoforms that are associated with the Rh blood group system, since both substrate- and inhibitor-labeled polypeptides were specifically immunoprecipitated with monoclonal Rh antibodies (Schroit et al., 1990). This suggests that the Rh blood group system is involved in the maintenance of membrane lipid asymmetry. Indeed, the Rh protein is a multispanning membrane polypeptide (Cherif-Zahar et al., 1990, Avent et al., 1990), a property common to other membrane proteins associated with transport and channel functions.

On the other hand, Devaux and colleagues proposed that the aminophospholipid translocase is a Mg^{2+}-ATPase with a molecular weight in the range of 115 to 150 kDa (Zachowski et al., 1989, Zachowski and Devaux, 1990, Morrot et al., 1990). This is based on the observation that transport in red blood cells requires hydrolyzable ATP (Seigneuret and Devaux, 1984), that partially purified Mg^{2+}-ATPase from erythrocytes is stimulated by PS and inhibited by fluoride, vanadate, and Ca^{2+} ions, and that the Mg^{2+}-ATPase has no known function (Forgac and Cantley, 1984), in contrast to the Ca^{2+}- and Na+/K+- ATPases. Although the Rh protein and Mg^{2+}-ATPase are very different from one another, recent findings suggest that, in fact, both proteins might be involved in the transport of aminophospholipids. Since the transmembrane movement of PS requires hydrolyzable ATP, and sequence analysis indicates that the Rh protein does not contain ATP binding sites (Avent et al., 1990), it is reasonable to assume that its transport function would depend on the activity of a distinct ATP-utilizing enzyme. Thus, observations suggesting the involvement of both Rh protein and Mg^{2+}-ATPases in PS transport are not mutually exclusive. Indeed, the transport of fluorescent labeled PS analogs was shown to require the coordinated and complimentary participation of a 32 kDa polypeptide and a presumably distinct protein located at the endofacial membrane surface (Connor and Schroit, 1990).

IV. Molecular Biology of the Transporter/Rh Protein

A gene encoding for the Rh polypeptide has recently been independently cloned in Paris (Cherif-Zahar et al., 1990) and Bristol (Avent et al., 1990). This was accomplished by generating primers from N-terminal amino acid sequence data for PCR amplification. The PCR products were then used to screen bone marrow cDNA libraries. Analysis of the nucleotide sequence for the Rh gene revealed an open reading frame capable of encoding a 417 amino acid, 45,500 dalton polypeptide (Avent et al., 1990; Cherif-Zahar et al., 1990). These results are surprising since the apparent Mr of Rh polypeptides obtained from analysis of SDS-polyacrylamide gels is approximately 32,000 daltons. One possible explanation for this discrepancy is that the extremely hydrophobic nature of the Rh protein results in the binding of an excessive amount of detergent thus altering the mobility of the protein, a phenomenon known to occur with very hydrophobic proteins (Helenius and Simons, 1975). Alternatively, the apparent smaller size of the Rh polypeptide on SDS gels could be the result of posttranslational proteolytic cleavage of the intact polypeptide, although Suyama et al. (1991) have recently provided evidence against this possibility.

The deduced amino acid sequence of the Rh protein includes six cysteine residues (Avent et al., 1990; Cherif-Zahar, et al., 1990) whose positions with regard to their neighboring amino acids may be of importance. One cysteine, residue 284, is predicted to lie at the exofacial surface in the sequence cys-his-leu-ile-pro, while the other five cysteine residues are also predicted to lie on the endofacial surface in the sequence motif cys-leu-pro (CLP). These CLP patterns are flanked by basic amino acids and precede or follow a region of predicted alpha-helix entering or emerging from the inner leaflet of the membrane. It is possible that these cysteine residues are palmitylation sites (Agre and Cartron, 1991).

V. Are the PS Transporter and the Rh Polypeptide the Same?

Several 32,000 dalton polypeptides have been identified as proteins bearing defined Rh epitopes (Gahmberg, 1982; Moore et al., 1982; Blanchard et al., 1988; Bloy et al., 1988; Hughes-Jones et al., 1988; Saboori et al., 1988). These Rh polypeptides possess many of the same properties characteristic of the aminophospholipid transporter. For example, polypeptides bearing the Rh D, c and E epitopes are non-glycosylated (Gahmberg, 1983), possess cysteine residues (Green, 1967, 1983) and are associated with the cytoskeleton (Gahmberg and Karhi, 1984; Ridgwell et al., 1984; Paradis et al., 1986; Bloy et al., 1987).

Evidence for the involvement of Rh polypeptides in the asymmetric distribution of PS in the red blood cell membrane has been presented by Schroit et al. (1990). The PS translocase was labeled with the transport inhibitor, ^{125}I-PDA, or with the transportable substrate, ^{125}I-azido-PS, and the ability of monoclonal Rh antibodies to immunoprecipitate the labeled translocase was determined. These immunoprecipitated 32 kDa Rh proteins were found to be labeled with the iodinated probes implicating the involvement of the Rh polypeptides in the movement and maintenance of PS asymmetry in the red blood cell membrane.

Although it is unclear whether Rh polypeptides and the aminophospholipid transporter are the same molecule, recent studies suggest that a region of the ^{125}I-labeled transporter shares homology among the various Rh phenotypes. Proteins labeled with PDA or azide-lipid subjected to limited proteolysis by a number of proteases and analyzed by SDS-PAGE have revealed peptides of similar size irrespective of Rh phenotype (unpublished observations). This observation is consistent with the existence of a highly conserved region in all Rh polypeptides. This has been shown by iodopeptide maps from surface labeled cells (Krahmer and Prohaska, 1987; Blanchard et al., 1988), and by the ability of antiserum to denatured RhD protein to cross react with Rh c and Rh E polypeptides on immunoblots (Suyama and Goldstein, 1988). These data indicate that Rh polypeptides contain a highly conserved domain which may be a region involved in aminophospholipid transport.

The notion that aminophospholipid transport is independent of phenotype indicates that the epitopes responsible for Rh antigenicity play no role in aminophospholipid transport. This hypothesis is supported by results which have shown that transport is not inhibited by Rh antibodies (Schroit et al., 1990, Smith and Daleke, 1990) and that Rh null cells, cells that do not contain any known Rh epitopes, transport PS normally (Schroit et al., 1990). Although the absence of Rh polypeptides in null cells has been interpreted to indicate that the aminophospholipid transporter cannot be Rh (Smith and Daleke, 1990), it should be noted that null cells are defined by lack of immunoreactivity with Rh antibody. Since these cells contain a 32-kDa aminophospholipid transporter that is indistinguishable from aminophospholipid transporter isolated from cells bearing Rh antigen (Schroit, et al., 1990), these data point to the possibility that null cells contain an "Rh-like" polypeptide devoid of known Rh epitopes. Thus, Rh null cells have a functional aminophospholipid transporter which is immunologically unrelated to known Rh phenotypes.

VI. The Rh Complex/Transporter Model

Based on our current understanding of the biochemical requirements for translocation of PS across the plasma membrane bilayer and from predictions of the protein structure from the nucleotide sequence of the Rh gene, the following model for aminophospholipid transport has been constructed.

Hydropathy analysis of the cDNA sequence (Avent et al., 1990; Cherif-Zahar et al., 1990) suggests that the Rh polypeptide is extremely hydrophobic and spans the bilayer 12 or 13 times with very limited exposure to the external or internal hydrophilic environments. The lack of solubility and the limited availability to proteolytic digestion of the Rh protein (Agre and Cartron, 1991) support the hydrophobic nature predicted by the nucleotide sequence of the Rh gene. Much like the glucose transporter, another red cell integral membrane protein, the aminophospholipid transporter is a polytopic protein having multiple transmembrane hydrophobic stretches. The glucose transporter is classified as a type IV protein, believed to involve a transmembrane water-filled channel whose function is to mediate specific transport of small ions and hydrophilic molecules through the membrane (Fischbarg et al., 1989). Considering the low molecular weight of PS (approximately 700 daltons) and the structural similarities of the Rh protein to the glucose transporter, an attractive hypothesis would be that the 32-kDa polypeptide responsible for the movement of aminophospholipids across the bilayer membrane is a pore that operates in a manner similar to that of the glucose transporter.

While the specific amino acids comprising the transporter's active site have not been identified, it is likely that one or more cysteine and histidine residues participate. The sensitivity of the transporter to sulfhydryl oxidants implicates the involvement of one or more cysteine residues (Connor and Schroit, 1988). These cysteine residues are likely candidates for palmitylation (de Vetten and Agre, 1988), suggesting their involvement in transport regulation. Histidine residues may also be involved in the active site since histidine-specific reagents abolish PS transport (unpublished observations) and alter Rh immunoreactivity (Victoria et al., 1986). The involvement of histidine residues may suggest the involvement of a charge relay system in PS movement.

Since lipid movement requires hydrolyzable ATP, and the Rh nucleotide sequence does not encode a consensus ATP binding site, it stands to reason that the energy needs for active transport must be fulfilled by another protein such as an ATPase or protein kinase. This indicates that enzymes distinct from Rh are involved in lipid transport (Connor and Schroit, 1990, Schroit and Zwaal, 1990) and suggests that Rh polypeptides are complexed with these enzymes thereby fulfilling the transporter's energy requirements. This possibility is supported by unrelated studies which have shown that the transporter/Rh protein does not exist as a single polypeptide in the red cell membrane (Gahmberg, 1983; Moore and Green, 1987). Indeed, it has been suggested that Rh proteins form a complex or "Rh cluster" (Bloy et al., 1989) with other membrane components one of which may be an ATP-utilizing enzyme.

References

Agre P, Cartron JP (1991) Molecular biology of the Rh antigens. Blood 78:551-563

Avent ND, Ridgwell K, Tanner MJA, Anstee DJ (1990) cDNA cloning of a 30 kDa erythrocyte membrane protein associated with Rh (Rhesus) blood group antigen expression. Biochem J 271:821-825

Backer JM, Dawidowicz EA (1987) Reconstitution of a phospholipid flippase from rat liver microsomes. Nature 327:341-343

Bergmann WL, Dresssler V, Haest CWM, Deuticke B (1984) Reorientation rates and asymmetry of distribution of lysophospholipids between the inner and outer leaflet of the erythrocyte membrane. Biochim Biophys Acta 772:328-336

Bishop WR, Bell RM (1985) Assembly of the endoplasmic reticulum phospholipid bilayer: The phosphatidylcholine transporter. Cell 42:51-60

Blanchard D, Bloy C, Hermand P, Cartron JP, Saboori A, Smith BL, Agre P (1988) Two-dimensional iodopeptide mapping demonstrates erythrocyte Rh D, c, and E polypeptides are structurally homologous but nonidentical. Blood 72:1424-1427

Bloy C, Blanchard D, Dahr W, Beyreuther K, Salmon C, Cartron JP (1988) Determination of the N-terminal sequence of human red cell Rh(D) polypeptide and demonstration that the Rh(D), (c) and (E) antigens are carried by distinct polypeptide chains. Blood 72:661-666

Bloy C, Blanchard D, Lambin P, Goossens D, Rouger P, Salmon C, Cartron JP (1987) Human monoclonal antibody against Rh(D) antigen: Partial characterization of Rh(D) polypeptide from human erythrocytes. Blood 69:1491-1497

Bloy C, Blanchard D, Hermand P, Kordowicz M, Sonneborn HH, Cartron, JP (1989) Properties of the blood group LW glycoprotein and preliminary comparison with Rh proteins. Mol Immunol 26:1013-1019

Bretscher MS (1972) Asymmetric lipid bilayer structure for biological membranes. Nature (London) New Biol 236:11-12

Cherif-Zahar B, Bloy C, Le Van Kim C, Blanchard D, Bailly P, Hermand P, Salmon C, Cartron JP, Colin Y (1990) Molecular cloning and protein structure of a human blood group Rh polypeptide. Proc Nat Acad Sci USA 87:6243-6247

Connor J, Schroit AJ (1987) Determination of lipid asymmetry in human red cells by resonance energy transfer. Biochemistry 26:5099-5105

Connor J, Schroit AJ (1988) Transbilayer movement of PS in erythrocytes: Inhibition of transport and preferential labeling of a 31,000 dalton protein by sulfhydryl reactive reagents. Biochemistry 27:848-851

Connor, J, Schroit AJ (1990) Aminophospholipid translocation in erythrocytes: Evidence for the involvement of a specific transporter and an endofacial protein. Biochemistry 29:37-43

Daleke DL, Huestis WH (1985) Incorporation and translocation of aminophospholipids in human erythrocytes. Biochemistry 23:5406-5416

de Vetten MP, Agre P (1988) The Rh polypeptide is a major fatty acid acylated erythrocyte membrane protein. J Biol Chem 263:18193-18196

Fischbarg J, Kuang K, Hirsch J, Lecuona S, Rogozinski L, Silverstein SC, Loike J (1989) Evidence that the glucose transporter serves as a water channel in J774 macrophages. Proc Natl Acad Sci USA 86:8397-8401

Forgac M, Cantley L (1984) The plasma membrane (Mg^{2+})-dependent adenosine triphosphatase from the human erythrocyte is not an ion pump. J Membr Biol 80:185-190

Gahmberg CG (1982) Molecular identification of the human Rho(D) antigen. FEBS Lett 140:93-97

Gahmberg CG (1983) Molecular characterization of the human red cell Rho(D) antigen. EMBO J 2:223-228

Gahmberg CG, Karhi KK (1984) Association of Rho(D) polypeptides with the membrane skeleton in Rho(D)-positive human red cells. J Immunol 133:334-337

Gordesky SE, Marinetti GV, Love R (1975) The reaction of chemical probes with the erythrocyte membrane. J Memb Biol 20:111-132

Green FA (1967) Erythrocyte membrane sulfhydryl groups and Rh antigen activity. Immunochemistry 4:247-257

Green FA (1983) The mode of attenuation of erythrocyte membrane Rho(D) antigen activity by 5,5'-dithiobis-(2-nitrobenzoic acid) and protection against loss of activity by bound anti-Rho(D) antibody. Mol Immunol 20:769-775

Helenius A, Simons K (1975) Solubilization of membranes by detergents. Biochim Biophys Acta 415:29-79

Hughes-Jones NC, Bloy C, Gorick B, Blanchard D, Doinel C, Rouger P, Cartron JP (1988) Evidence that the c, D and E epitopes of the human Rh blood group system are on separate polypeptide molecules. Mol Immunol 25:931-936

Krahmer M, Prohaska R (1987) Characterization of human red cell Rh (Rhesus-) specific polypeptides by limited proteolysis. FEBS Lett 226:105-108

Martin OC, Pagano RE (1987) Transbilayer movement of fluorescent analogs of phosphatidylserine and phosphatidylethanolamine at the plasma membrane of cultured cells. J Biol Chem 262:5890-5898

Moore S, Green C (1987) The identification of specific Rhesus polypeptide blood group ABH active glycoprotein complexes in the human red-cell membrane. Biochem J 244:735-741

Moore S, Woodrow CF, McClelland DBL (1982) Isolation of membrane components associated with human red cell antigens Rho(D), (c) (E), and Fya. Nature (London) 295:529-531

Morrot G, Herve P, Zachowski A, Felmann P, Devaux PF (1989) Aminophospholipid translocase of human erythrocytes: Phospholipid substrate specificity and effect of cholesterol. Biochemistry 28:3456-3462

Morrot G, Zachowski A, Devaux PF (1990) Partial purification and characterization of the human erythrocyte Mg^{2+}-ATPase. A candidate aminophospholipidtranslocase. FEBS Lett 266:29-32

Paradis G, Bazin R, Lemieux R (1986) Protective effect of the membrane skeleton on the immunologic reactivity of the human red cell Rho(D) antigen. J Immunol 137:240-244

Ridgwell K, Tanner MJA, Anstee DJ (1984) The Rhesus(D) polypeptide is linked to the human erythrocyte cytoskeleton. FEBS Lett 174:7-10

Saboori AM, Smith BL, Agre P (1988) Polymorphism in the Mr 32,000 Rh protein purified from Rh(D) positive and negative erythrocytes. Proc Natl Acad Sci USA 85:4042-4045

Schroit AJ, Bloy C, Connor J, Cartron JP (1990) Involvement of the Rh blood group polypeptides in the maintenance of aminophospholipid asymmetry. Biochemistry 29:10303-10306

Schroit AJ, Madsen J, Ruoho AE (1987) Radioiodinated, photoactivatable phosphatidylcholine and phosphatidylserine: Transfer properties and differential photoreactive interaction with human erythrocyte membrane proteins. Biochemistry 26:1812-1819

Schroit AJ, Zwaal RFA (1991) Transbilayer movement of phospholipids in red cell and platelet membranes. Biochim Biophys Acta (in press)

Seigneuret M, Devaux PF (1984) ATP-dependent asymmetric distribution of spin-labeled phospholipids in the erythrocyte membranes relation to shape changes. Proc Natl Acad Sci USA 81:3751-3755

Smith RE, Daleke DL (1990) PS transport in Rh null erythrocytes. Blood 76:1021-1027

Suyama K, Goldstein J (1988) Antibody produced against isolated Rh(D) polypeptide reacts with other Rh-related antigens. Blood 72:1622-1626

Suyama K, Goldstein J, Aebersold R, Kent S (1991) Regarding the size of Rh proteins. Blood 77:411 (letter)

Tanaka Y, Schroit AJ (1986) Calcium/phosphate induced immobilization of fluorescent PS in synthetic bilayer membranes: Inhibition of lipid transfer between vesicles. Biochemistry 25:2141-2148

Tilley L, Cribier S, Roelofsen B, Op den Kamp JAF, van Deenen LLM (1986) ATP-dependent translocation of aminophospholipids across the human erythrocyte membrane. FEBS Lett 194:21-27

Verkleij AJ, Zwaal RFA, Roelofsen B, Comfurius P, Kastelijn D, van Deenen LLM (1973) The asymmetric distribution of phospholipids in the human red blood cell membrane. Biochim Biophys Acta 323:178-193

Victoria EJ, Branks MJ, Masouredis SP (1986) Rh antigen immunoreactivity after histidine modification. Mol Immunol 23:1039-1044

Zachowski A, Devaux PF (1990) Transmembrane movement of lipids. Experientia 46:644-656

Zachowski A, Favre E, Cribier S, Herve P, Devaux PF (1986) Outside-inside translocation of aminophospholipids in the human erythrocyte membrane is mediated by a specific enzyme. Biochemistry 25:2585-2590

Zachowski A, Herrmann A, Paraf A, Devaux PF (1987) Phospholipid outside-inside translocation in lymphocyte plasma membranes is a protein-mediated phenomenon. Biochim Biophys Acta 897:197-200

Zachowski A, Henry JP, Devaux PF (1989) Control of transmembrane lipid asymmetry in chromaffin granules by an ATP-dependent protein. Nature 340:75-76

Zachowski A, Morot Gaudry-Talarmain YM (1990) Phospholipid transverse diffusion in synaptosomes: Evidence for the involvement of the aminophospholipid translocase. J Neurochem 55:1352-1356

LIPIDS (COMPOSITION AND TOPOGRAPHY) IN ADRENAL CHROMAFFIN GRANULES AND THEIR RELEVANCE FOR THE FUNCTION OF THESE ORGANELLES

H. Winkler
Department of Pharmacology
University of Innsbruck
Peter-Mayr-Strasse 1a
A-6020 Innsbruck
Austria

The catecholamine-storing organelles of adrenal medulla, the chromaffin granules, can be isolated in high purity and good yield (see Winkler & Carmichael, 1982). Therefore the biochemical characterization of these organelles has advanced significantly.

Composition and topography of lipids in the membranes of chromaffin granules.

As far as the lipid composition is concerned early studies already established the major facts (see Winkler, 1976). Most, if not all, of the lipid in chromaffin granules (for discussion see Winkler, 1976) are present in the membranes of these organelles. As shown by freeze-etching (Plattner et al., 1969) the basic structure of these membrane lipids is a bimolecular leaflet (see also Marsh et al., 1976). Major lipids of bovine chromaffin granules (Blaschko et al., 1967) are cholesterol and phospholipids (2 μmol/mg protein: see Winkler & Westhead, 1980), whereas only traces of triglycerides and cholesterol esters are found (Winkler et al., 1967). Fatty acids are found in a concentration of 100 nmol/mg protein (Husebey & Flatmark, 1984).

A relatively high cholesterol/phospholipid ratio of 0.6 is a typical feature of these organelles (see Winkler, 1976). In 1967 we reported that chromaffin granules were characterized by a high lysolecithin content (16.7% of the total lipid phosphorus: Blaschko et al., 1967). This is not only true for bovine granules but also for those of other species including rat (Winkler et al., 1967). This finding has been confirmed in several subsequent studies (see Winkler & Carmichael, 1982). A claim that lysolecithin in chromaffin granules was caused by a post mortem artifact (Arthur & Sheltawy, 1980) has been disproven (Frischenschlager et al.,

NATO ASI Series, Vol. H 70
Phospholipids and Signal Transmission
Edited by R. Massarelli, L. A. Horrocks,
J. N. Kanfer, and K. Löffelholz
© Springer-Verlag Berlin Heidelberg 1993

1983; De Oliveira-Filgueiras et al., 1981). The fatty acid composition of the major phospholipids has been determined, for lysolecithin it was concluded that it is present as the 1-acyl isomer (Winkler & Smith, 1968).

Two groups reported independently that chromaffin granules contained gangliosides (Geissler et al., 1977; Dreyfus et al., 1977) in a molar phospholipid/ganglioside ratio of 32 (Geissler et al., 1977). The gangliosides consist of two major GM_3 components containing either N-acetylneuraminic acid or N-glycolylneuraminic acid (Sekine et al., 1984).

The topography of the membrane phospholipids of chromaffin granules has been investigated by several groups (Buckland et al., 1978; Voyta et al., 1978; De Oliveira-Filgueiras et al., 1979; Phillips, 1973). Major conclusions are: Phosphatidylethanolamine and phosphatidylinositol are preferentially localized in the outer leaflet of the granule membrane. A similar localization seems likely for phosphatidylserine, since an enzyme translocating this lipid is present in these organelles (see below). On the other hand, lysolecithin appears concentrated in the inner lipid leaflet.

Most gangliosides are found in the inner leaflet (Izumi et al., 1979; Westhead & Winkler, 1982), as demonstrated by neuraminidase digestion of intact and lysed chromaffin granules. On the other hand (Lazarovici et al., 1989) tetanus toxin binds to the outside of intact chromaffin granules suggesting that a polysialoganglioside or a glycoprotein-proteoglycan is present there (see also Meyer & Burger, 1976).

Enzymes of lipid metabolism in chromaffin granules

Membranes of chromaffin granules contain several enzymes involved in the metabolism of lipids. The presence of phosphatidylinositol kinase was already established by Buckley et al. in 1971 (for further references see Winkler & Carmichael, 1982). This enzyme which phosphorylates phosphatidylinositol in the 4-position has been characterized and purified about 200-fold (Husebye et al., 1990).

For the metabolism of phospholipids early studies found enzyme activities in various cell organelles, but not in chromaffin granules. Thus the major activity of lysophospholipase was localized in microsomal membranes (Hörtnagl et al., 1969), whereas phospholipase A activities were found in lysosomes (Smith & Winkler, 1968). More recent studies discovered these enzyme activities also in chromaffin granules. The lysophospholipase activity in these organelles becomes rapidly inactivated by dialysis (Franson & van den Bosch, 1982) explaining the previous failure to detect it.

Phospholipase A_2 activities with an alkaline pH optimum were also found in chromaffin granules (Frischenschlager, 1985, and Husebye & Flatmark, 1987). However, activities were low and it was difficult to exclude that a contamination with plasma membranes accounted for the enzyme activity present in purified chromaffin granules (Frischenschlager, 1985). An unequivocal demonstration of phospholipase A_2 in chromaffin granules was made recently (Hildebrandt & Albanesi, 1991). These authors discovered that a glycerol stimulated phospholipase A_2 was specifically confined to the membranes of these organelles. It had its pH optimum at 7.8, retained its activity in the absence of calcium and was inactivated by p-bromophenyl bromide.

Finally chromaffin granules contain an ATP-dependent aminophospholipid translocase which selectively transports phospatidylserine from the inner to the cytoplasmic monolayer (Zachowski et al., 1989). This enzyme may be identical with the previously described (Apps & Percy, 1987; Moriyama & Nelson, 1988) vanadate sensitive ATPase II.

Involvement of lipids in the function of chromaffin granules

Membrane permeability: A characteristic feature of the membranes of chromaffin granules is that they are less permeable to various cations than those of any previously studied subcellular organelles (Beers et al., 1982). Thus the conductance to H^+ is at least one order of magnitude less than that measured in mitochondria. Obviously the properties of the lipid leaflet are essential for these findings. At present one can only speculate that the relatively high cholesterol content (see above) of the granule membranes might contribute to this low permeability.

Protein-lipid ·interactions: Obviously the properties of proteins embedded or bound to the membranes will be influenced by their interaction with lipids. Two major and functionally important intrinsic membrane proteins are the cytochrome b-561 and the proton-pumping ATPase. The amino acid sequence of the former has been determined. It is a hydrophobic protein spanning the membranes six times (Perin et al., 1988). The molecule is covalently fatty acylated through ester linkages of serine or threonine residues (Kent & Fleming, 1990). Interactions of this hydrophobic molecule with the membrane lipids will certainly contribute to its function as an electron transporting system (Njus et al., 1983; Srivastava et al., 1984).

The proton-pumping ATPase of chromaffin granules, which belongs to the V-type of this enzyme (Nelson, 1991), consists as shown by biochemical characterization of a peripheral part and a membrane spanning sector. Thus the overall structure resembles closely that of the F-ATPase of mitochondria (Nelson, 1991; Apps et al., 1989). In agreement on the surface of chromaffin granules "stalked particles" appearing like the F_1-ATPase particles in mitochondria could be visualized by negative staining (Schmidt et al., 1982). At least one of the membrane spanning proteins probably responsible for the proton translocation through the membranes has been cloned (Mandel et al., 1988). This protein of 15849 Da spans the membranes four times. Nothing is yet known about its interaction with membrane lipids. On the other hand, the vanadate sensitive ATPase (ATPase II) from chromaffin granules is activated by phosphatidylserine (Moriyama et al., 1991).

It is a characteristic feature of chromaffin granules that several proteins are bound to the membrane but are also found in the soluble content (see Winkler & Westhead, 1981). This applies to dopamine β-hydroxylase (see Winkler et al., 1986), glycoprotein III (Fischer-Colbrie et al., 1984) and carboxypeptidase H (Fricker et al., 1990). For dopamine β-hydroxylase several hypotheses explaining its binding to the membranes have been advanced (see Winkler et al., 1986). However several mechanisms are by now excluded: There is no phosphatidylinositol glycolipid anchor (see Taljanidicz et al., 1989). Furthermore in the primary amino acid sequence there is no hydrophobic part with the exception of the N-terminal signal peptide (Lamouroux et al., 1987). This uncleaved signal sequence may in fact provide the answer to a long discussion since the membrane bound form of this enzyme has obviously retained this sequence (Taljanidiscz et al., 1989). It is not clear whether the ability of purified soluble dopamine β-hydroxylase to bind to phosphatidylserine containing vesicles contributes somehow to the distribution of this enzyme in chromaffin granules (Taylor & Fleming, 1989). For glycoprotein III there is no evidence that the membrane bound form contains an uncleaved signal sequence, however also this molecule has no obvious hydrophobic stretches in its sequence (Palmer & Christie, 1990). For carboxypeptidase H it has been suggested that only the membrane bound form contains a 21-residue COOH-terminal peptide which can form an amphiphilic alpha-helix providing a pH dependent membrane binding (Fricker et al., 1990).

Involvement of lipids in the exo/endocytotic cycle

It would go beyond the scope of this short essay to discuss this question exhaustively. Only two points will be dealt with in more detail. Secretion from the adrenal medulla occurs by exocytosis (for a review see Winkler et al., 1988). During this process the membrane of the chromaffin granule becomes incorporated in the plasma membrane.

What is the fate of the granule lipids during this process? Data directly dealing with the lipids are not available. We have therefore to draw conclusions based on the behaviour of the membrane proteins of these organelles. During exocytosis these membrane proteins become exposed on the surface of chromaffin cells (Wildmann et al., 1981; Dowd et al., 1983; Phillips et al., 1983). They do, however, not diffuse laterally (Patzak et al., 1986; Lotshaw et al., 1986) and subsequently the membrane patches with the granule antigens are efficiently and specifically retrieved (Lingg et al., 1983; Patzak et al., 1984; Patzak & Winkler, 1986). Apparently this endocytotic mechanism ens··res a complete removal of the membrane antigens of chromaffin granules but does this also involve a specific retrieval of the granule lipids? This seems likely considering two points (i) After exocytosis the membrane antigens of granules stay together in patches most likely within their lipid domain. (ii) Chromaffin granule membranes have a high lysolecithin content whereas that of the plasma membrane is low (Nijjar & Hawthorne, 1974; Malviya et al., 1986). A constant mixing of the membrane lipids during the exocytosis/endocytosis cycle seems therefore unlikely. If we accept these arguments then we are left with an unresolved question. What prevents the lateral diffusion of granule lipids in the plasma membrane during the exo/endocytotic cycle?

During exocytosis the two involved membranes have first to attach and then to fuse (see Schweizer et al., 1990). In chromaffin cells (Schmidt et al., 1983) as in other secretory cells (Plattner, 1989), exocytosis probably starts as a focal event with a small pore finally leading to complete fusion and incorporation of the granule membrane into that of the plasma membrane. What role do the lipids play in this mechanism? Recent studies have in fact concentrated more on the role of proteins like calmodulin, calpactin, annexin or G_E proteins (Burgoyne, 1990) including those which have been termed docking proteins (Schweizer et al., 1990) or fusoproteins like synexin or in paramecium cells a specific phosphoprotein (see Plattner, 1989 for an exhaustive discussion). However, whatever role belongs to these proteins, the final stages of fusion must also involve the membrane lipids. Are these components only passive partners of proteins or do they play a more critical role? Or more specifically: Does in chromaffin granules the high concentration of lysolecithin, of this fusogenic lipid, indicate some function of this component in exocytosis? This question is now usually answered in the negative (see e.g. Plattner, 1989) since high concentrations of lysolecitihin are only found in chromaffin granules but not in several other secretory organelles. This is obviously a convincing argument but we cannot exclude that e.g. in chromaffin granules the removal of lysolecithin after exocytosis is less sufficient than in other organelles and therefore this lipid stays on and can therefore be found in the recycling (see Winkler, 1988) membranes. In any case during exocytosis in adrenal medulla arachidonic acid is released possibly by the action of a phospholipase A_2. Inhibition of this enzyme lowered both catecholamine secretion and arachidonic acid release (Frye & Holz, 1985). However, in contrast to these results Morgan & Burgoyne (1990) reported a dissociation of these two events thus arguing against a concept of a connection between phospholipase A_2 and exocytosis.

A participation of phospholipase A_2 activity has also been implicated for the fusion process between chromaffin granules and the plasma membrane in an in vitro system (Karli et al., 1990). The action of this enzyme was thought to provide fatty acids or a derivative of them as fusogenic agents. Obviously there are many loose ends but no coherent picture on phospholipase A participation in exocytosis has emerged. Thus, twenty-five years after the discovery of a high lysolecithin content in chromaffin granules we are still left with the old question: does its presence hint at some participation of phospholipid turnover in exocytosis or has it to be explained by some other mechanisms unrelated to exocytosis?

Conclusions

The membranes of catecholamine-storing vesicles in adrenal medulla have a specific lipid composition. A characteristic feature is the high lysolecithin content. These lipids are present in a membrane with a characteristic topography with phosphatidylserine in the outer and ganglioside and lysolecithin preferentially localized in the inner leaflet. These membranes contain enzymes for lipid metabolism, e.g. a lysophospholipase and a phospholipase A_2. Thus it is obvious that the lipids in these membranes are not present in a static state but are likely to have a significant turnover. The properties of the membrane lipids reemphasize an old, but still unresolved question: Which function of chromaffin granules depends on the characteristic composition, topography and turnover of the membrane phospholipids?

References

Apps DK, Percy JM (1987) The H^+-translocating ATPase of chromaffin granule membranes. Ann NY Acad Sci USA 493:178-179

Apps DK, Percy JM, Perez-Castineira JR (1989) Topography of a vacuolar-type H^+-translocating ATPase: chromaffin-granule membrane ATPase I. Biochem J 263:81-88

Arthur G, Sheltawy A (1980) The presence of lysophosphatidylcholine in chromaffin granules. Biochem J 191:523-532

Beers MF, Carty SE, Johnson RG, Scarpa A (1982) H^+-ATPase and catecholamine transport in chromaffin granules. Ann NY Acad Sci USA 402:116-133

Blaschko H, Firemark H, Smith AD, Winkler H (1967) Lipids of the adrenal medulla: lysolecithin, a characteristic constituent of chromaffin granules. Biochem J 104:545-549

Buckland RM, Radda GK, Shennan CD (1978) Accessibility of phospholipids in the chromaffin granule membrane. Biochim Biophys Acta 513:321-337

Buckley JT, Lefebvre YA, Hawthorne JN (1971) Identification of actively phosphorylated component of adrenal medulla chromaffin granules. Biochim Biophys Acta 239:517-519

Burgoyne RD (1990) Secretory vesicle-associated proteins and their role in exocytosis. Annu Rev Physiol 52:647-659

De Oliveira-Filgueiras OM, Van den Besselaar AMHP, Van den Bosch H (1979) Localization of lysophosphatidylcholine in bovine chromaffin granules. Biochim Biophys Acta 558:73-84

De Oliveira-Filgueiras OM, Van den Bosch H, Johnson RG, Carty SE, Scarpa A (1981) Phospholipid composition of some amine storage granules. FEBS Lett 129:309-313

Dowd DJ, Edwards C, Englert D, Mazurkiewicz JE (1983) Immunofluorescent evidence for exocytosis and internalization of secretory granule membrane in isolated chromaffin cells. Neuroscience 10:1025-1033

Dreyfus H, Aunis D, Harth S, Mandel P (1977) Gangliosides and phospholipids of the membranes from bovine adrenal medullary chromaffin granules. Biochim Biophys Acta 489:89-97

Fischer-Colbrie R, Zangerle R, Frischenschlager I, Weber A, Winkler H (1984) Isolation and immunological characterization of a glycoprotein from adrenal chromaffin granules. J Neurochem 42:1008-1016

Franson RC, Van den Bosch H (1982) Lysophospholipase activity of bovine adrenal medulla. Biochim Biophys Acta 711:75-82

Fricker LD, Das B, Hogue Angeletti R (1990) Identification of the pH-dependent membrane anchor of carboxypeptidase E. J Biol Chem 265:2476-2482

Frischenschlager I (1985) Lysolecithin und Phospholipase A_2 in den chromaffin Granula des Nebennierenmarks und Stimulus-Secretion-Coupling. Ph D Thesis, Innsbruck

Frischenschlager I, Schmidt W, Winkler H (1983) Is lysolecithin an in vivo constituent of chromaffin granules? J Neurochem 41:1480-1483

Frye RA, Holz RW (1985) Arachidonic acid release and catecholamine secretion from digitonin-treated chromaffin cells: effects of micromolar calcium, phorbol ester, and protein alkylating agents. J Neurochem 44:265-273

Geissler D, Martinek A, Margolis RU, Margolis RK, Skrivanek JA, Ledeen R, König P, Winkler H (1977) Composition and biogenesis of complex carboxyhydrates of ox adrenal chromaffin granules. Neuroscience 2:685-693

Hildebrandt E, Albanesi JP (1991) Identification of a membrane-bound, glycol-stimulated phospholipase A_2 located in the secretory granules of the adrenal medulla. Biochemistry 30:464-472

Hörtnagl H, Winkler H, Hörtnagl H (1969) The subcellular distribution of lysophospholipase in bovine adrenal medulla. Eur J Biochem 10:243-248

Husebye ES, Flatmark T (1984) The content of long-chain free fatty acids and their effect on energy transduction in chromaffin granule ghosts. J Biol Chem 259:15272-15276

Husebye ES, Flatmark T (1987) Characterization of phospholipase activities in chromaffin granule ghosts isolated from the bovine adrenal medulla. Biochim Biophys Acta 920:120-130

Husebye ES, Letcher AJ, Lander DJ, Flatmark T (1990) Purification and kinetic properties of a membrane-bound phosphatidylinositol kinase of the bovine adrenal medulla. Biochim Biophys Acta 1042:330-337

Izumi F, Miyashita T, Kashimoto T, Wada A (1979) Chromaffin granule as a model for the study of exocytosis. In (Usdin, Kopin & Barchas, eds.) Basic and Clinical Frontiers, pp 322-324, Pergamon Press, NY

Karli UO, Schäfer T, Burger MM (1990) Fusion of neurotransmitter vesicles with target membrane is calcium independent in a cell-free system. Proc Natl Acad Sci USA 87:5912-5915

Kent UM, Fleming PJ (1990) Cytochrome b_{561} is fatty acylated and oriented in the chromaffin granule membrane with its carboxyl terminus cytoplasmically exposed. J Biol Chem 265:16422-16427

Lamouroux A, Vigny A, Faucon Biguet N, Darmon MC, Franck R, Henry J-P, Mallet J (1987) The primary structure of human dopamine β-hydroxylase: insights into the relationship between the soluble and the membrane-bound forms of the enzyme. EMBO J 6:3931-3937

Lazarovici P, Fujita K, Contreras ML, DiOrio JP, Lelkes PI (1989) Affinity purified tetanus toxin binds to isolated chromaffin granules and inhibits catecholamine release in digitonin-permeabilized chromaffin cells. FEBS Lett 253:121-128

Lingg G, Fischer-Colbrie R, Schmidt W, Winkler H (1983) Exposure of an antigen of chromaffin granules on cell surface during exocytosis. Nature 301:610-611

Lotshaw DP, Ye HZ, Edwards C (1986) Endocytosis of surface bound dopamine β-hydroxylase and plasma membrane following catecholamine secretion by bovine adrenal chromaffin cells. Neurochem Int 9:391-399

Malviya AN, Gabellec MM, Rebel G (1986) Plasma membrane lipids of bovine adrenal chromaffin cells. Lipids 21:417-419

Mandel M, Moriyama Y, Hulmes JD, Pan Y-CE, Nelson H, Nelson N (1988) cDNA sequence encoding the 16-kDa proteolipid of chromaffin granules implies gene duplication in the evolution of H^+-ATPases. Proc Natl Acad Sci USA 85:5521-5524

Marsh D, Radda GK, Ritchie GA (1976) A spin-label study of the chromaffin granule membrane. Eur J Biochem 71:53-61

Meyer DI, Burger MM (1976) The chromaffin granule surface. Localization of carbohydrate on the cytoplasmic surface of an intracellular organelle. Biochim Biophys Acta 443:428-436

Morgan A, Burgoyne RD (1990) Relationship between arachidonic acid release and Ca^{2+}-dependent exocytosis in digitonin-permeabilized bovine adrenal chromaffin cells. Biochem J 271:571-574

Moriyama N, Nelson N (1988) Purification and properties of a vanadate- and N-ethylmaleimide-sensitive ATPase from chromaffin granule membranes. J Biol Chem 263:8521-8527

Moriyama N, Nelson N, Maeda M. Futai M (1991) Vanadate-sensitive ATPase from chromaffin granule membranes formed a phosphoenzyme intermediate and was activated by phosphatidylserine. Arch Biochem Biophys 286:252-256

Nelson N (1991) Structure and pharmacology of the proton-ATPases. Trends Pharmacol Sci 12:71-75

Nijjar MS, Hawthorne JN (1974) A plasma membrane fraction from bovine adrenal medulla: preparation, marker enzyme studies and phospholipid composition. Biochim Biophys Acta 367:190-201

Njus D, Knoth J, Cook C, Kelley PM (1983) Electron transfer across the chromaffin granule membrane. J Biol Chem 258:27-30

Palmer DJ, Christie DL (1990) The primary structure of glycoprotein III from bovine adrenal medullary chromaffin granules. J Biol Chem 265:6617-6623

Patzak A, Winkler H (1986) Exocytotic exposure and recycling of membrane antigens of chromaffin granules: ultrastructural evaluation after immunolabelling. J Cell Biol 102:510-515

Patzak A, Böck G, Fischer-Colbrie R, Schauenstein K, Schmidt W, Lingg G, Winkler H (1984) Exocytotic exposure and retrieval of membrane antigens of chromaffin granules: quantitative evaluation of immunofluorescence on the surface of chromaffin cells. J Cell Biol 98:1817-1824

Perin MS, Fried VA, Slaugther CA, Südhof TC (1988) The structure of cytochrome b561, a secretory vesicle-specific electron transport protein. EMBO J 7:2697-2703

Phillips JH (1973) Phosphatidylinositol kinase: a component of the chromaffin granule membrane. Biochem J 136:579-687

Phillips JH, Burridge K, Wilson SP, Kirshner N (1983) Visualization of the exocytosis/endocytosis secretory cycle in cultured adrenal chromaffin cells. J Cell Biol 97:1906-1917

Plattner H (1989) Regulation of membrane fusion during exocytosis. Int Rev Cyt 119:197-285

Plattner H, Winkler H, Hörtnagl H, Pfaller W (1969) A study of the adrenal medulla and its subcellular organelles by the freeze-etching method. J Ultrastruct Res 28:191-202

Schmidt W, Winkler H, Plattner H (1982) Adrenal chromaffin granules: evidence for an ultrastructural equivalent of the proton-pumping ATPase. Eur J Cell Biol 27:96-104

Schmidt W, Patzak A, Lingg G, Winkler H, Plattner H (1983) Membrane events in adrenal chromaffin cells during exocytosis: a freeze-etching analysis after rapid cryofixation. Eur J Cell Biol 32:31-37

Schweizer FE, Schäfer T, Burger MM (1990) Intracellular mechanisms in exocytotic secretion. Biochem Pharmacol 41:163-169

Sekine M, Ariga T, Miyatake T, Kuroda Y, Suzuki A, Yamakawa T (1984) Ganglioside composition of chromaffin granule membrane in bovine adrenal medulla. J Biochem 95:155-160

Smith AD, Winkler H (1968) Lysosomal phospholipases A_1 and A_2 of bovine adrenal medulla. Biochem J 108:867-874

Srivastava M, Duong LT, Fleming PJ (1984) Cytochrome b-561 catalyzes transmembrane electron transfer. J Biol Chem 259:8072-8075

Taljanidisz J, Stewart L, Smith AJ, Klinman JP (1989) Structure of bovine adrenal dopamine β-monooxygenase, as deduced from cDNA and protein sequencing: evidence that the membrane-bound form of the enzyme is anchored by an uncleaved signal peptide. Biochemistry 28:10054-10061

Taylor CS, Fleming PJ (1989) Conversion of soluble dopamine β-hydroxylase to a membrane binding form. J Biol Chem 264:15242-15246

Voyta JC, Slakey LL, Westhead EW (1978) Accessibility of lysolecithin in catecholamine secretory vesicles to acyl CoA: lysolecithin acyl transferase. Biochem Biophys Res Commun 80:413-417

Westhead E, Winkler H (1982) The topography of gangliosides in the membrane of the chromaffin granule of bovine adrenal medulla. Neuroscience 7:1611-1614

Wildmann J, Dewair M, Matthaei H (1981) Immunochemical evidence for exocytosis in isolated chromaffin cells after stimulation with depolarizing agents. J Neuroimmunol 1:353-364

Winkler H (1976) The composition of adrenal chromaffin granules: an assessment of controversial results. Neuroscience 1:65-80

Winkler H (1988) Occurrence and Mechanism of Exocytosis in Adrenal Medulla and Sympathetic Nerve. In: Handbook of Experimental Pharmacology, Vol.90/I, Trendelenburg U, Weiner N (eds) Springer Verlag, Berlin, Heidelberg, p 43

Winkler H, Carmichael SW (1982) The Chromaffin Granule. In: Poisner and Trifaro (eds) The Secretory Granule, Elsevier Biomedical Press, Amsterdam

Winkler H, Smith AD (1968) Lipids of chromaffin granules: fatty acid composition of phospholipids, in particular lysolecithin. Naunyn-Schmiedebergs Arch Pharmacol 261:379-388

Winkler H, Westhead EW (1980) The molecular organization of adrenal chromaffin granules. Neuroscience 5:1803-1823

Winkler H, Strieder N, Ziegler E (1967) Uber Lipide, insbesondere Lysolecithin, in den chromaffinen Granula verschiedener Species. Naunyn-Schmiedebergs Arch Pharmacol 256:407-415

Winkler H, Apps DK, Fischer-Colbrie R (1986) The molecular function of adrenal chromaffin granules: established facts and unresolved topics. Neuroscience 18:261-290

Zachowski A, Henry J-P, Devaux PF (1989) Control of transmembrane lipid asyymetry in chromaffin granules by an ATP-dependent protein. Nature 340:75-76

ACCELERATED CELL MEMBRANE DEGRADATION IN ALZHEIMER'S DISEASE BRAIN: RELATIONSHIP TO AMYLOID FORMATION?

Roger M. Nitsch[1,2], Barbara E. Slack[2], John H. Growdon[1] and Richard J. Wurtman[2].

Dept. of Neurology[1]
Massachusetts General Hospital and
Harvard Medical School
ACC 830 Fruit Street
Boston, MA 02114

CHARACTERISTICS OF ALZHEIMER'S DISEASE

Alzheimer's disease is a degenerative brain disease first described by the German psychiatrist Alois Alzheimer (Alzheimer, 1907). His initial report was based upon the histological examination of brain tissue from a 51 year old female patient who experienced a rapid progressive deterioration of memory and died four and a half years after the onset of the first symptoms with total loss of mental functions. The brain of this woman showed generalized atrophy without macroscopic lesions. Using a novel silver staining method developed by Bielschowsky, Alzheimer found that more than a quarter of the cortical neurons contained fibrillary tangles, and that this abnormality was accompanied by the extracellular deposition of "multiple miliary lesions", containing a "unique substance" which was later recognized to be amyloid. In his initial report, Alzheimer also described severe loss of nerve cells, which was most prominent in the upper layers of the brain cortex. More recently, Terry and colleagues used computerized image processing to confirm the loss of cortical neurons, and showed that large cortical neurons are particularly affected (Terry et al., 1981). Neuronal atrophy and the presence of degenerating neurites in the immediate vicinity of amyloid plaques has generated two controversial hypotheses: First, that amyloid is toxic to neurons, and, second, that amyloid is a byproduct of neuronal atrophy formed by neurons during their degeneration.

Using sequence analysis, it was shown that amyloid fibrils in AD brain consist of a 39-42 amino acid peptide (Glenner & Wong, 1984, Masters et al., 1985, Joachim et al., 1988), termed beta/A4. The finding that beta/A4 is synthesized as part of a membrane-spanning 695-771 amino acid

[2]Dept. of Brain and Cognitive Sciences, Massachusetts Institute of Technology, E25-604, Cambridge, MA 02139.

NATO ASI Series, Vol. H 70
Phospholipids and Signal Transmission
Edited by R. Massarelli, L. A. Horrocks,
J. N. Kanfer, and K. Löffelholz
© Springer-Verlag Berlin Heidelberg 1993

amyloid precursor protein (Kang et al., 1987, Tanzi et al., 1988, Ponte et al., 1988, Kitaguchi et al., 1988), which under normal conditions is processed by cleavage within the beta/A4 domain (Esch et al., 1990, Sisodia et al., 1990, Oltersdorf et al., 1990, Anderson et al., 1991), prompted the suggestion that proteolytic cleavage of this protein is abnormal in AD brain (for review, see Ishiura, 1991). It has been hypothesized that abnormal APP processing could result from an alteration in the protease activity or the substrate specificity of an APP-cleaving protease yet to be discovered (Selkoe, 1990), or by mutations of the normal beta/A4 transmembrane sequence, as described in some cases with familiar AD (Goate et al., 1991). A third possibility is that the normally protected transmembrane domain of APP is exposed to abnormal proteolytic cleavage as a consequence of defective membrane structure. Initial evidence that membrane phospholipid catabolites may contribute to the regulation of APP processing stem from the observations that protein kinase C is involved in the phosphorylation of APP (Gandy et al., 1988), and that activation of protein kinase C results in the stimulation of APP cleavage (Buxbaum et al., 1990).

PHOSPHOLIPID METABOLISM IN ALZHEIMER'S DISEASE

Membrane phospholipid metabolism in AD brain is abnormal. In a recent postmortem study of 10 patients with histopathologically verified AD and 10 normal control subjects matched according to age of death (75.7 ± 3.6 years for AD, and 74.4 ± 4.6 years for controls) and postmortem time interval (9.0 ± 1.3 hours for AD, and 10.3 ± 1.0 hours for controls), we examined levels of phospholipids and their metabolites in the frontal, parietal, and primary auditory cortex (Nitsch et al., 1992). We found that levels of the major phospholipid classes, phosphatidylcholine and phosphatidylethanolamine, are significantly decreased in AD cortex by 15-20%. These results are consistent with the previously reported decreases in phosphatidylinositol levels in AD temporal cortex (Stokes & Hawthorne, 1987). Given that cellular degeneration in AD brain is confined to the neuronal, rather than the glial cell population, the per cent decrease of neuronal phospholipids is certainly underestimated by the measurements of whole brain homogenates used in the study reported here. The phospholipid depletion we found in AD brain is thus compatible with the concept of accelerated neuronal membrane degeneration in AD brain cortex. In the same tissue samples, we found 60-80% increases in the phospholipid catabolite glycerophosphocholine (GPC), and, in some brain areas, glycerophosphoethanolamine (GPE) (Nitsch et al., 1991, Blusztajn et al., 1990). This observation confirmed the initial findings of Barany et al. (1986), Miatto et al. (1986), and Pettegrew et al. (1987), who also found increased levels of glycerophosphodiesters in postmortem AD brain samples using ^{31}P NMR spectroscopy. The levels of glycerophosphodiesters in AD brain were reported to correlate with the number of senile plaques and neurofibrillary tangles (Pettegrew et al., 1988). GPC and GPE are exclusively formed by deacylation of phosphatidylcholine and phosphatidylethanolamine, respectively (Blusztajn & Wurtman, 1983), and accumulation of these compounds thus indicates accelerated phospholipid breakdown. This evidence further strengthens the hypothesis that membrane degeneration is accelerated in AD brain. To rule out the possibility that the accumulation of GPC in AD brain was due to decreased activity of its degrading enzyme, we measured GPC cholinephosphodiesterase (EC 3.1.4.38) activity (Kanfer & McCartney, 1987) in the same brain samples with increased GPC levels. We found normal enzyme activity in AD brain, and concluded that GPC breakdown was not impaired in AD (Nitsch et al., 1992). The accumulation of GPC and GPE in AD was accompanied by a near-stoichiometric decrease in levels of choline and ethanolamine. Decreased phospholipid levels along with the concomitant accumulation of phosphodiester catabolites and with a stoichiometric depletion of the precursors choline and ethanolamine are consistent with the hypothesis that neuronal membrane degeneration in AD brain is accompanied by increased turnover of phospholipids. Increased membrane turnover may involve stimulation of lipolytic enzymes, in particular phospholipase A_2 (PLA2) and lysophospholipase, since these enzymes are involved in the deacylation of phospholipids to form glycerophosphodiesters. In fact, it has been reported that the activity of lysophospholipase in the cytosolic fraction of freshly assayed AD brain tissue is increased 2 to 4-fold in the nucleus basalis and the hippocampus (Farooqui et al., 1988), a finding that could well account for increased formation of GPC and GPE in AD brain. Furthermore, activities of monoacylglycero- and diacylglycerolipases have also been reported to be increased in AD brain, supporting the concept that membrane degradation is elevated in AD. This alteration in brain cell membrane metabolism may contribute to the pathophysiology of neuronal degeneration in AD brain. To rule out the possibility that abnormal phospholipid metabolism arises secondarily to neuronal degeneration, we have shown that a similar pattern of alterations does not occur in other neurodegenerative diseases (Nitsch et al., 1992).

PHOSPHOLIPID ABNORMALITIES HAVE A CHARACTERISTIC PATTERN IN AD

Phospholipid metabolism in other neurodegenerative diseases such as Down's syndrome, Huntington's disease, and Parkinson's disease differs markedly from AD: Brain tissue levels of GPC were not increased in any of these diseases. Choline levels were also reduced in Huntington's, but not in Parkinson's disease and Down's syndrome. Phosphocholine levels were normal in AD, but were significantly decreased in Huntington's disease brain, indicating that phosphatidycholine metabolism is differentially affected in these two diseases, but apparently is normal in Down's syndrome and in Parkinson's disease brains.

IS AMYLOID DEPOSITION IN AD RELATED TO MEMBRANE DEGENERATION?

The observation that phospholipid metabolism is normal in brains of patients with Down's syndrome (Nitsch et al., 1992), despite its AD-like histopathology (Olson & Shaw, 1969) indicates that deposition of amyloid alone does not account for abnormal membrane metabolism, and rules out the possibility that amyloid formation in Down's syndrome is promoted by the membrane defect observed in AD. The most likely explanation for amyloid generation in Down's syndrome brain is the 1.5-fold increase in dosage of the APP gene which maps to chromosome 21 (Goldgaber et al., 1987). The increased APP gene dosage in Down's syndrome is accompanied by a similar increase of APP serum levels (Rumble, et al., 1989). In AD, however, the APP gene dose is normal (Tanzi et al., 1987, Podlisny et al., 1987, St. George-Hyslop, 1987), necessitating different explanations for amyloid formation in Down's syndrome and AD.

The hypothesis that APP fragments are secreted subsequent to cleavage of the extracellular N-terminal close to the transmembrane domain is supported by two observations: First, water-soluble, carboxy terminal-truncated APP fragments appear in the CSF of healthy humans and AD patients (Palmert et al., 1989, Weidemann et al., 1989). Second, APP-transfected human and non-human cell-lines, as well as non-transfected human SY5Y cells and differentiated PC-12 cells in culture secrete water-soluble N-terminal fragments into the cell culture medium (Weidemann et al., 1989, Esch et al., 1990, Wang et al., 1991, Anderson et al., 1991). The exact cleavage site of the secreted fragments from the precursors is localized extracellularly within the beta/A4 domain at a site 12-14 amino-acids from the membrane. In APP-transfected monkey kidney COS-1 cells, membrane-associated cleavage of the beta/A4 peptide also occurred in the vicinity of the membrane (Sisodia et al., 1990). These findings strongly suggest that normal APP cleavage precludes the formation of amyloidogenic products containing intact beta/A4 fragments. Consequently, it was proposed that the generation of amyloidogenic APP fragments and their subsequent self-aggregation to amyloid fibrils could be due to the failure of normal cleavage within the beta/A4 domain (for review, see Ishiura et al., 1991).

Based upon these data, we hypothesized that the defects in membrane composition and metabolism discovered in AD brain contribute to the abnormal cleavage of APP, the production of amyloidogenic fragments, and the deposition of amyloid in the parenchyma. A possible functional role of membranes in preventing amyloidogenic fragments from aggregating has been presented by Dyrks et al. (1988) who showed that the presence of membranes inhibited self-aggregation of APP fragments in vitro. That cell membrane integrity is important in maintaining normal APP processing was suggested by the finding that APP-containing fragments are released following detergent-induced nonspecific membrane damage in PC-12 cells (Baskin et al., 1991). In order to determine whether a pathophysiological disturbance of membrane phospholipid metabolism affects APP processing, we have established an in vitro model system in which phospholipase activities can be specifically altered. In preliminary experiments, we observed that stimulation of PLA2 in cell culture results in a 10-fold increase in cellular GPC release, along with a concomitant reduction of normal APP cleavage and secretion (Nitsch et al., unpublished observation). These results suggest a central role of PLA2 in the pathophysiology of altered APP processing associated with abnormal membrane phospholipid metabolism. We are currently investigating the molecular mechanisms of this relationship.

Initial evidence for abnormal APP processing in humans suffering from AD was provided by the finding of reduced levels of immunoreactive APP fragments in the CSF of AD patients (Prior et al., 1991, Henriksson et al., 1991). Although the interpretation of this observation is difficult as long as the exact cleavage sites of APP fragments found in CSF are not known, these results may be indicative of reduced cleavage of beta/A4-containing fragments in AD, and the consequent shifting of these amyloidogenic products into the alternative intracellular processing pathways (Estus et al., 1992; Golde et al., 1992).

Phospholipid catabolites might be involved in APP processing as suggested by the observation that phosphorylation and cleavage of APP in differentiated PC-12 cells is stimulated by protein kinase C (Gandy et al., 1988, Buxbaum et al., 1990), which reportedly is decreased in AD brain (Cole et al., 1988). The physiological activator of protein kinase C is diacylglycerol, a phospholipid catabolite generated by phospholipase C (for review, see Nishizuka, 1986). Since the generation of diacylglycerol is a receptor-coupled event mediated by phospholipase C activation (for review, see Berridge & Irvine, 1989, Exton et al., 1990), regulation of APP processing may in fact be receptor-mediated. This contention is supported by preliminary results from our laboratory which show that activation of the bradykinin receptor in differentiated PC-12 cells leads to potent stimulation of phospholipase C, along with rapid cleavage and release of APP fragments, as early as 15 minutes after stimulation (Nitsch et al., unpublished observation). We suggest that a phospholipase C-mediated second messenger pathway may be important in the cleavage, and thus the removal, of beta/A4-containing amyloid precursors.

Changes in phospholipase activities in the brain could therefore be a pivotal event leading to abnormal phospholipid metabolism, paralleled by alterations in APP processing in AD.

ACKNOWLEDGMENTS

We thank Fionnualla Doyle and Anastassios Pittas for technical assistance. We greatfully acknowledge the McLean Hospital Brain Tissue Resource Center (Director: Dr. Edward D. Bird), supported by MH-NS 31862, for providing us with Huntington's disease and Down's syndrome brain tissue samples. R.M.N. is the Hoffman fellow in Alzheimer's disease at Massachusetts General Hospital. This work was supported by the National Institutes of Health grants AG-07906, AG-05134, MH-28783, and the Center for Brain Sciences and Metabolism Charitable Trust.

REFERENCES

Anderson JP, Esch FS, Keim PS, Sambamurti K, Lieberburg I and Robakis NK (1991) Exact cleavage site of Alzheimer amyloid precursor in neuronal PC-12 cells. Neurosci Lett 128:126-128

Alzheimer A (1907) Ueber eine eigenartige Erkrankung der Hirnrinde. Allg Z Psych & Psych-Gerichtl Med 64:146-148.

Berridge MJ and Irvine RF (1989) Inositol phosphates and cell signalling. Nature 341:197-204

Barany M, Chang Y-C and Arus C (1985) Increased glycerol-3-phosphoryl-choline in post-mortem Alzheimer's brain. Lancet i:517

Baskin F, Rosenberg RN and Greenberg BD (1991) Increased release of an amyloidogenic C-terminal Alzheimer amyloid precursor protein fragment from stressed PC-12 cells. J Neurosci Res 29:127-132

Blusztajn JK, Lopez Gonzalez-Coviella I, Logue M, Growdon JH and Wurtman RJ (1990) Levels of phospholipid catabolic intermediates, glycerophosphocholine and glycerophosphoethanolamine, are elevated in brains of Alzheimer's disease but not of Down's syndrome patients. Brain Res 536:240-244

Blusztajn JK and Wurtman RJ (1983) Choline and cholinergic neurons. Science 221:614-620

Buxbaum JD, Gandy SE, Cicchetti P, Ehrlich ME, Czernik AJ, Fracasso RP, Ramabhadran TV, Unterbeck AJ and Greengard P (1990) Processing of Alzheimer beta/A4 amyloid precursor protein: Modulation by agents that regulate protein phosphorylation. Proc Natl Acad Sci USA 87:6003-6006

Cole G, Dobkins KR, Hansen LA, Terry RD and Saitoh T (1988) Decreased levels of protein kinase C in Alzheimer brain. Brain Res 452:165-174

Dyrks T, Weidemann A, Multhaup G, Salbaum JM, Lemaire H-G, Kang L, Mueller-Hill B, Masters CL and Beyreuther K (1988) Identification, transmembrane orientation and biogenesis of the amyloid A4 precursor of Alzheimer's disease. EMBO J 7:949-957

Esch FS, Keim PS, Beattie EC, Blacher RW, Culwell AR, Oltersdorf T, McClure D and Ward P (1990) Cleavage of amyloid beta peptide during constitutive processing of its precursor. Science 248:1122-1124

Estus S, Golde TE, Kunishita T, Blades D, Lowery D, Eisen M, Usiak M, Qu X, Tabira T, Greenberg BD and Younkin SG (1992) Potentially amyloidogenic, carboxyl-terminal derivatives of the amyloid protein precursor. Science 255:726-728.

Exton JH (1990) Signaling through phosphatidylcholine breakdown. J Biol Chem 265:1-4

Farooqui AA, Liss L and Horrocks L (1988) Stimulation of lipolytic enzymes in Alzheimer's disease. Ann Neurol 23:306-308

Glenner GG and Wong CW (1984) Alzheimer's disease: Initial report of the purification and characterization of a novel cerebrovascular amyloid protein. Biochem Biophys Res Commun 120:885-890

Gandy S, Czernik AJ and Greengard P (1988) Phosphorylation of Alzheimer's disease amyloid precursor peptide by protein kinase C and Ca^{2+}/calmodulin-dependent protein kinase II. Proc Natl Acad Sci USA 85:6218-6221

Goate A, Chartier-Harlin M-C, Mullan M, Brown J, Crawford F, Fidani L, Giuffra L, Haynes A, Irving N, James L, Mant R, Newton P, Rooke K, Roques P, Talbot C, Pericak-Vance M, Roses A, Williamson R, Rossor M, Owen M and Hardy J (1991) Segregation of a missense mutation in the amyloid precursor protein gene with familial Alzheimer's disease. Nature 349:704-706

Golde TE, Estus S, Younkin LH, Selkoe DJ and Younkin SG (1992) Processing of the amyloid protein precursor to potentially amyloidogenic derivatives. Science 255:728-730.

Goldgaber D, Lerman M, McBride OW, Saffiotti U and Gajdusek DC (1987) Characterization and chromosomal localization of a cDNA encoding brain amyloid of Alzheimer's disease. Science 235:877-880

Henriksson T, Barbour RM, Braa S, Ward P, Fritz LC, Johnson-Wood K, Chung HD, Burke W, Reinikainen KJ, Riekkinen P and Schenk DB (1991) Analysis and quantitation of the beta-amyloid precursor protein in the cerebrospinal fluid of Alzheimer's disease patients with a monoclonal antibody-base immunoassay. J Neurochem 56:1037-1042

Ishiura S (1991) Proteolytic cleavage of the Alzheimer's disease amyloid A4 precursor protein. J Neurochem 56:363-369

Kanfer JN and McCartney DG (1989) Glycerophosphorylcholine phosphodiesterase activity of rat brain myelin. J Neurosci Res 24:231-240

Kang J, Lemaire H-G, Unterbeck A, Salbaum JM, Masters CL, Grzeschik K-H, Multhaup G, Beyreuther K and Mueller-Hill B (1987) The precursor of Alzheimer's disease amyloid A4 protein resembles a cell surface receptor. Nature 325:733-736

Kitaguchi N, Takahashi Y, Tokushima Y, Shiojiri S and Ito H (1988) Novel precursor of Alzheimer's disease amyloid protein shows protease inhibitory activity. Nature 331:530-532

Masters CL, Simms G, Weinman NA, Multhaup G, McDonald BL and Beyreuther K (1985) Amyloid plaque core protein in Alzheimer's disease and Down's syndrome. Proc Natl Acad Sci USA 82:4245-4249

Miatto O, Gonzalez G, Buonanno F and Growdon JH (1986) In vitro 31P NMR spectroscopy detects altered phospholipid metabolism in Alzheimer's disease. Can J Neurol Sci 13:535-539

Nishizuka Y (1986) Studies and perspectives of protein kinase C. Science 233:305-312

Nitsch RM, Blusztajn JK, Pittas AG, Slack BE, Growdon JH and Wurtman RJ (1992) Evidence for a membrane defect in Alzheimer's disease brain. Proc Natl Acad Sci USA 89:1671-1675

Olson MI and Shaw C-M (1969) Presenile dementia and Alzheimer's disease in mongolism. Brain 92:147-156

Oltersdorf T, Ward PJ, Henricksson T, Beattie EC, Neve R, Lieberburg I and Fritz LC (1990) The Alzheimer amyloid precursor protein. J Biol Chem 265:4492-4497

Palmert MA, Berman Podlisny M, Witker DS, Oltersdorf T, Younkin LH, Selkoe DJ and Younkin SG (1989) The beta-amyloid protein precursor of Alzheimer's disease has soluble derivatives found in human brain and cerebrospinal fluid. Proc Natl Acad Sci USA 86:6338-6342

Pettegrew JW, Panchalingam K, Moosy J, Martinez J, Rao G and Boller F (1988) Correlation of phosphorus-31 magnetic resonance spectroscopy and morphologic findings in Alzheimer's disease. Arch Neurol 45:1093-1096

Pettegrew JW, Kopp SJ, Minshew NJ, Glonek T, Feliksik JM, Tow JP and Cohen MM (1987) 31P nuclear magnetic resonance studies of phosphoglyceride metabolism in developing and degenerating brain: Preliminary observations. J Neuropathol Exp Neurol 46:419-430

Podlisny MB, Lee G and Selkoe DJ (1987) Gene dosage of the amyloid-beta precursor protein in Alzheimer's disease. Science 238:669-671

Prior R, Moenning U, Schreiter-Gasser U, Weidemann A, Blennow K, Gottfries CG, Masters CL and Beyreuther K (1991) Quantitative changes in the amyloid beta/A4 precursor protein in Alzheimer cerebrospinal fluid. Neurosci Lett 124:69-73

St. George-Hyslop PH, Tanzi RE, Polinsky RJ, Neve RL, Pollen D (1987) absence of duplication of chromosome 21 genes in familial and sporadic Alzheimer's disease. Science 235:885-889

Selkoe D (1990) Deciphering Alzheimer's disease: The amyloid precursor protein yields clues. Science 248:1058-1060

Sisodia SS, Koo EH, Beyreuther K, Unterbeck A, Price DL (1990) Evidence that beta-amyloid protein in Alzheimer's disease in not derived by normal processing. Science 248:492-495

Stokes CE, Hawthorne JN (1987) Reduced phosphoinositide concentrations in anterior temporal cortex of Alzheimer-diseased brains. J Neurochem 48:1018-1021

Ponte P, Gonzalez-DeWhitt P, Schilling J, Miller J, Hsu D, Greenberg B, Davis K, Wallace W, Lieberburg I, Fuller F, Cordell B (1988) A new A4 amyloid mRNA contains a domain homologous to serine protease inhibitors. Nature 331:525-527

Tanzi RE, Bird ED, Latt SA, Neve RL (1987) The amyloid beta protein gene is not duplicated in brains from patients with Alzheimer's disease. Science 238:666-669

Tanzi RE, McClatchey AI, Lamperti ED, Villa-Komaroff L, Gusella JF, Neve RL (1988) Protease inhibitor domain encoded by an amyloid protein precursor mRNA assaciated with Alzheimer's disease. Nature 331:528-530

Terry RD, Peck A, DeTeresa R, Schechter R, Horoupian DS (1981) Some morphometric aspects of the brain in senile dementia of the Alzheimer type. Ann Neurol 10:184-192

Wang R, Meschia JF, Cotter RJ, Sisodia SS (1991) Secretion of the beta/A4 amyloid precursor protein. Identification of a cleavage site in cultured mammalian cells. J Biol Chem 266:16960-16964

Weidemann A, Koenig G, Bunke D, Fischer P, Salbaum JM, Masters CL and Beyreuther K (1989) Identification, biogenesis, and localization of precursors of Alzheimer's disease A4 amyloid protein. Cell 57:115-126.

MECHANISTIC AND FUNCTIONAL ASPECTS OF OSCILLATORY CALCIUM SIGNALLING

Andrew P. Thomas, Thomas A. Rooney and Dominique C. Renard.
Department of Pathology and Cell Biology
Thomas Jefferson University
1020 Locust Street
Philadelphia, PA 19107, USA.

Introduction

Mobilization of calcium from intracellular stores and the extracellular medium to yield an increase in cytosolic free Ca2+ ($[Ca2+]_i$) is one of the most common forms of signal transduction utilized by extracellular stimuli in the control of cell function. It is now clear that receptor-induced increases in $[Ca2+]_i$ are generated, in part, by an elevation in the level of the second messenger inositol 1,4,5-trisphosphate ($Ins(1,4,5)P_3$) which is produced by a phospholipase C (PLC)-mediated hydrolysis of phosphatidylinositol 4,5-bisphosphate (Berridge and Irvine, 1984). $Ins(1,4,5)P_3$ is released into the cytosol where it interacts with an intracellular receptor which functions as a release channel for lumenal Ca^{2+} (Berridge and Irvine, 1989; Joseph and Williamson, 1989). However, intracellular Ca^{2+} stores are not uniformly sensitive to $Ins(1,4,5)P_3$, as demonstrated in permeabilized cell studies where $Ins(1,4,5)P_3$ can only release a fraction (30-50%) of the calcium accumulated by non-mitochondrial stores (Berridge and Irvine, 1984; Williamson et al., 1985; Joseph and Williamson, 1989). There is some debate as to the intracellular location of the $Ins(1,4,5)P_3$-sensitive Ca^{2+} store. In Purkinje cells antibodies to the receptor have revealed localized concentrations on the nuclear envelope and parts of the endoplasmic reticulum (E.R.) (Ross et al., 1989). Alternatively, small vesicular membrane structures enriched with calsequestrin and calcium pumps, but distinct from ER, termed calcisomes have been proposed to play a role in $Ins(1,4,5)P_3$-induced Ca^{2+} release (Volpe et al., 1988). The $Ins(1,4,5)P_3$ receptor has been purified and reconstituted (Ferris et al., 1989). In addition, the gene has been cloned and sequenced, demonstrating that it has partial homology with the ryanodine receptor (Mignery et al., 1989; Furuichi et al., 1989).

The majority of studies of receptor-mediated effects on $[Ca^{2+}]_i$ have been carried out in cell suspensions. In studies of this type receptor activation has been shown to produce relatively sustained increases in $[Ca^{2+}]_i$, which parallel the time course of the mono- or biphasic

NATO ASI Series, Vol. H 70
Phospholipids and Signal Transmission
Edited by R. Massarelli, L. A. Horrocks,
J. N. Kanfer, and K. Löffelholz
© Springer-Verlag Berlin Heidelberg 1993

Ins(1,4,5)P$_3$ increase induced by the agonist. The development of techniques to measure $[Ca^{2+}]_i$ at the single cell level has revealed a hitherto unsuspected degree of complexity in the control and organization of these Ca^{2+} signals. Receptor-mediated increases in $[Ca^{2+}]_i$ can display both temporal and spatial organization, which is often manifested in the form of $[Ca^{2+}]_i$ oscillations and waves (Berridge and Galione, 1988; Berridge and Irvine, 1989; Jacob, 1990a; Thomas et al., 1991). Oscillations of $[Ca^{2+}]_i$ with a range of patterns and shapes can be generated by Ca^{2+}-mobilizing agonists at the single cell level, but most of the kinetic parameters are relatively similar in the majority of cell types. For example, the amplitudes of most $[Ca^{2+}]_i$ oscillations are within the range of 200-1000 nM with transient spikes usually having higher amplitudes than sinusoidal oscillations (Jacob, 1990a). In addition, a well-defined latent period usually precedes the initial $[Ca^{2+}]_i$ increase. Individual cells in a single preparation often show marked intercellular variation in the latency, frequency, amplitude and dose-dependence of $[Ca^{2+}]_i$ oscillations (Prentki et al., 1988; Rooney et al., 1989; Millard et al., 1988). In many cell types the cell to cell variations in latent period and subsequent frequency results in asynchronous $[Ca^{2+}]_i$ oscillations which lead to the apparent non-oscillatory behavior observed in the measurements at the cell population level.

Frequency-modulated $[Ca^{2+}]_i$ oscillations in single hepatocytes

Woods et al., (1986) first demonstrated that hormone-induced $[Ca^{2+}]_i$ increases in hepatocytes are more complex than would be expected for a simple mechanism in which hormone binding to its receptor leads to an elevation of Ins(1,4,5)P$_3$ levels with a proportional release of Ca^{2+} into the cytosol. Using aequorin-injected hepatocytes it was shown that rather than giving a $[Ca^{2+}]_i$ rise followed by a sustained plateau, the $[Ca^{2+}]_i$ changes in individual cells stimulated with phenylephrine or vasopressin were composed of a series of discrete spikes in the continued presence of the agonist (Woods et al., 1986, 1987). With increasing concentrations of agonist there was an increase in the frequency of the oscillations, but the amplitude of the $[Ca^{2+}]_i$ spikes was found to be independent of agonists dose. At high concentrations of more potent hormones, such as vasopressin, the $[Ca^{2+}]_i$ responses occurred as a sustained $[Ca^{2+}]_i$ rise (Woods et al., 1987). Similar findings have also been obtained using digital imaging fluorescence microscopy of hepatocytes loaded with fura-2 (Rooney et al., 1989; Kawanishi et al., 1989). Figure 1 shows the $[Ca^{2+}]_i$ response of a single fura-2-loaded hepatocyte stimulated with increasing doses of vasopressin, with a washout period interposed between each addition to ensure complete recovery of cell responsiveness. Over the lower dose range (1 nM and below), which

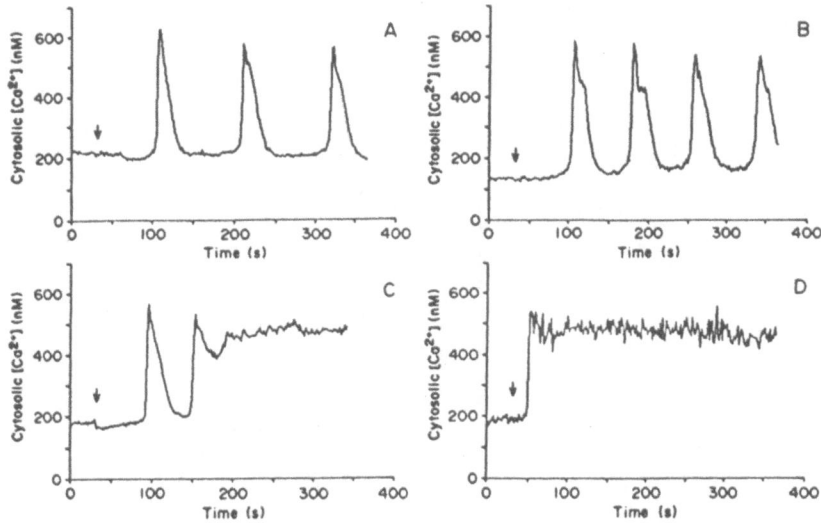

Figure 1. Dose response to vasopressin for $[Ca^{2+}]_i$ oscillations in a single hepatocyte. Vasopressin concentrations were A, 0.5 nM; b, 1.0 nM; C, 3.0 nM; D, 5.0 nM. Reprinted with permission from Rooney et al., 1989, where further experimental details can be found.

equates to the physiological range, vasopressin caused oscillatory $[Ca^{2+}]_i$ increases. Higher doses of vasopressin gave sustained or mixed $[Ca^{2+}]_i$ increases. However, regardless of the type of $[Ca^{2+}]_i$ response or agonist dose, the amplitude of $[Ca^{2+}]_i$ changes was the same.

Figure 2 shows dose response data from a large number of cells for the frequency of $[Ca^{2+}]_i$ oscillations in response to vasopressin and the α-adrenergic agonist phenylephrine. In contrast to the situation with vasopressin, most cells treated with phenylephrine do not yield sustained $[Ca^{2+}]_i$ increases even at high doses. This is presumably because phenylephrine is a weaker agonist than vasopressin in these 24 h cultured hepatocytes (Rooney et al., 1989). The dose-dependent increase in oscillation frequency is accompanied by a dose-dependent decrease in the latent period prior to the first $[Ca^{2+}]_i$ spike. Rooney et al. (1989) found that there was a correlation between latent period and the subsequent oscillation period over a range of agonist doses and different oscillation frequencies. This observation led to the suggestion that the same processes were occurring during the initial latent period as during the inter-spike period and that these processes were the parameters that

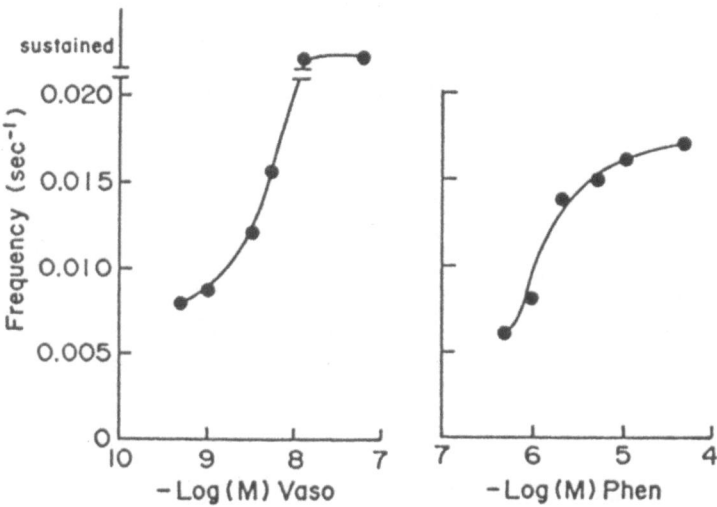

Figure 2.   Dose dependence of oscillation frequency for hepatocytes treated with vasopressin (left panel) and phenylephrine (right panel). Data represent mean values from 50-70 cells treated with each dose of agonist.

were regulated by agonist dose. The constancy of $[Ca^{2+}]_i$ spike amplitude over a range of agonist doses and with various hormones of differing maximal efficacy in the liver, has formed the basis for the proposal that Ca^{2+}-mediated effects in hepatocytes might be encoded in a frequency-modulated rather than an amplitude-modulated manner (Woods et al., 1986, 1987; Berridge et al., 1988).

In common with the amplitude of the $[Ca^{2+}]_i$ oscillations, the kinetic properties (rate of rise and fall) of the hormone-induced $[Ca^{2+}]_i$ transients in hepatocytes are independent of stimulus strength (Woods et al., 1986, 1987; Rooney et al., 1989). Even at agonist doses sufficiently high to give sustained rather than oscillatory $[Ca^{2+}]_i$ increases, the initial rate of $[Ca^{2+}]_i$ rise is also indistinguishable from the rates of rise of oscillatory $[Ca^{2+}]_i$ changes given by lower agonist doses. Furthermore, with a variety of different agonists and even direct activation of the G protein-linked PLC with AlF_4^-, we have found that the amplitude and rate of rise of the subsequent $[Ca^{2+}]_i$ oscillations are the same. Thus, it appears that once a $[Ca^{2+}]_i$ transient has been initiated, the amount and rate of Ca^{2+} mobilization are independent of the original stimulus. The only parameters modulated by agonist dose are the initial latent period prior to the first $[Ca^{2+}]_i$ increase and the frequency at which the $[Ca^{2+}]_i$ spikes occur.   Interestingly, although $[Ca^{2+}]_i$ oscillations in hepatocytes can occur in the absence of extracellular Ca^{2+}, it is these same parameters of latency and frequency that are modulated by the extracellular Ca^{2+} concentration (Rooney et al., 1989, 1990; Kawanishi et al., 1989). The kinetics of the individual $[Ca^{2+}]_i$ spikes are unaffected even by the complete removal of extracellular Ca^{2+}. These data suggest that the basic $[Ca^{2+}]_i$ oscillation mechanism relies only on Ca^{2+} release and reuptake by intracellular stores, but the availability of Ca^{2+} from outside of the cell is an important determinant of the ultimate signal strength.

Although the rates of rise of $[Ca^{2+}]_i$ oscillations in hepatocytes appear to be independent of the nature of the stimulus, there are significant variations in the falling phase (Woods et al., 1987; Rooney et al., 1989). For example, phenylephrine-induced $[Ca^{2+}]_i$ transients are always found to be of significantly shorter duration than those induced by angiotensin II, while vasopressin gives $[Ca^{2+}]_i$ spikes of intermediate width. These differences in $[Ca^{2+}]_i$ spike duration can be ascribed to differences in the rates at which the released Ca^{2+} is reaccumulated into the intracellular stores. Interestingly, although the rates of decline of the $[Ca^{2+}]_i$ transients are different for different agonists, they do not appear to vary in a dose-dependent manner, at least not over the range in which oscillation frequency is regulated by agonist dose. Kinetic data obtained using both aequorin and fura-2 for the rates of rise and fall of $[Ca^{2+}]_i$ in response to various hormones can be found in Woods et al. (1987) and Rooney et al. (1989, 1991).

Evidence has been obtained to suggest that protein kinase C (PKC) may be involved in regulating the rate of Ca^{2+} reuptake during the falling phase of the $[Ca^{2+}]_i$ oscillations, since the PKC inhibitors staurosporine and sphingosine prolong the $[Ca^{2+}]_i$ transients in response to the majority of agonists by reducing the rate of the falling phase (Sanchez-Bueno et al., 1990). By contrast, phorbol esters have no apparent effect on the shape of the $[Ca^{2+}]_i$ spikes but reduce the frequency of the oscillations (Woods et al., 1987), as would be expected from their ability to inhibit the PLC activation by hormones in liver. It has been suggested that differences in $[Ca^{2+}]_i$ oscillation shape may result from differential susceptibility of different receptors to this feedback inhibition (Woods et al., 1986, 1987; Sanchez-Bueno et al., 1990). Alternatively, there may be differences in the activation of PKC by various receptors, resulting in differential effects on intracellular Ca^{2+} pumps (Rooney et al., 1989).

Spatial organization of $[Ca^{2+}]_i$ oscillations in hepatocytes

Recent studies have shown that in addition to temporal organization in the form of $[Ca^{2+}]_i$ oscillations, there is a further level of organization of $[Ca^{2+}]_i$ at the subcellular level in individual hepatocytes (at least once these cells have been allowed to redevelop some degree of functional polarity in primary culture) (Rooney et al., 1990). In these hepatocytes, agonist-induced $[Ca^{2+}]_i$ oscillations do not occur synchronously throughout the cell, but originate from a specific locus adjacent to a region of the cell membrane and then propagate through the cell in the form of a wave of $[Ca^{2+}]_i$ increase (Rooney et al., 1990). Each oscillation in a series originates from the same subcellular locus and when a cell is stimulated sequentially with different agonists acting through distinct receptors, the site of $[Ca^{2+}]_i$ wave initiation is common to all agonists. $[Ca^{2+}]_i$ waves begin as a localized $[Ca^{2+}]_i$ increase in one region of the cell that may last for 1-3 seconds; the increased $[Ca^{2+}]_i$ then propagates through the rest of the cell over a period of a few seconds (depending on cell size). The falling phase of the $[Ca^{2+}]_i$ oscillations does not occur as a wave but is synchronous over the whole cell. Examples of images of changes in $[Ca^{2+}]_i$ that demonstrate such waves can be found Rooney et al. (1990) and Thomas et al. (1991).

We have examined the kinetics of $[Ca^{2+}]_i$ wave propagation by determining the time course of $[Ca^{2+}]_i$ changes at specific subregions of individual cells selected to be in line with the direction of wave propagation. Figure 3 shows a series of such $[Ca^{2+}]_i$ plots for three individual oscillations following phenylephrine treatment of a hepatocyte (Figs. 3B, 3C, 3D). The whole cell response to the agonist is shown in Figure 3A on a more condensed time base. The amplitude and rate of $[Ca^{2+}]_i$ increase is the same for each region of the cell, with the $[Ca^{2+}]_i$ changes offset by a period of time that is proportional to the distance of that subcellular region from the locus of $[Ca^{2+}]_i$ wave initiation. The fact that the rate and amplitude of $[Ca^{2+}]_i$ increase is the same throughout the cell argues against a simple diffusion mechanism for the spread of the $[Ca^{2+}]_i$ waves and suggests instead that a self-propagative mechanism may be involved. Plots of the type shown in Figure 3 have been used to calculate the rate at which the $[Ca^{2+}]_i$ waves propagate through cells by determining the time taken to reach half height at points of known distance from the locus of origin. As with the kinetics of $[Ca^{2+}]_i$ spikes measured over the whole cell, the rates of $[Ca^{2+}]_i$ wave propagation were found to be independent of agonist dose or agonist type (Rooney et al., 1990). Rates of $[Ca^{2+}]_i$ wave propagation with a variety of agonists measured in hepatocytes were in the range 20-25 $\mu M \cdot s^{-1}$. This is similar to the rates for $[Ca^{2+}]_i$ waves observed in egg cells (Gilkey et al., 1978; Miyazaki et al., 1986), astrocytes (Cornell-Bell, 1990) and endothelial cells (Jacob, 1990b), while those reported for cardiac myocytes (Takamatsu, 1990) are slightly faster. In both egg cells and cardiac myocytes, evidence has been

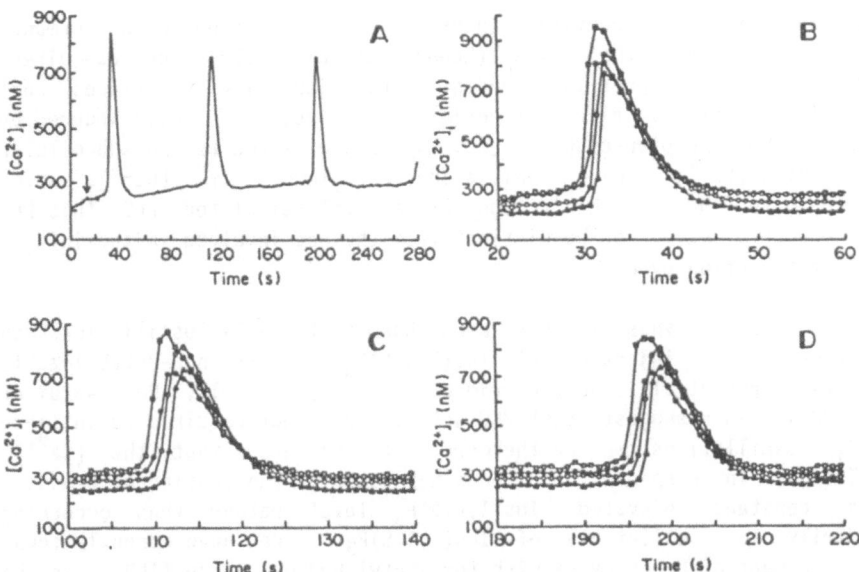

Figure 3. Oscillations of $[Ca^{2+}]_i$ measured at subregions within the same cell during a series of $[Ca^{2+}]_i$ waves. Panel A shows $[Ca^{2+}]_i$ measured over a single hepatocyte and panels B-D show the $[Ca^{2+}]_i$ changes during each of the $[Ca^{2+}]_i$ spikes at discrete intracellular regions selected to be in line with the direction of wave propagation. Note that the data in panels B-D are plotted on an expanded time base. Reprinted with permission from Rooney et al., 1990.

obtained to suggest that $[Ca^{2+}]_i$ waves spread by a self-propagating mechanism which is most readily explained by Ca^{2+}-induced Ca^{2+}-release (CICR) (Takamatsu, 1990; Berridge and Irvine, 1989). A similar mechanism could explain the $[Ca^{2+}]_i$ waves in hepatocytes (Rooney et al., 1990; Thomas et al., 1991). However, it is also possible that a Ca^{2+}-dependent activation of PLC could yield similar self-propagating waves of $[Ca^{2+}]_i$.

$[Ca^{2+}]_i$ oscillations independent of receptor activation

As noted above, $[Ca^{2+}]_i$ oscillations can be initiated by AlF_4^- which binds to the GDP-bound form of G-proteins, mimicking the gamma phosphate of GTP and leading to direct activation of PLC without a requirement for receptor activation. Moreover, the $[Ca^{2+}]_i$ oscillations induced by AlF_4^- also occur as $[Ca^{2+}]_i$ waves originating from an identical locus to that observed with hormonal stimulation of the same cells (Rooney et al., 1990). Because direct activation of G protein-dependent PLC with AlF_4^- initiates Ca^{2+} waves with identical sites of origin and kinetics to those induced by hormones, the receptor-coupled PLC may be localized to the subcellular region of $[Ca^{2+}]_i$ wave initiation. Thus, if $[Ca^{2+}]_i$ waves self-propagate through the action of a Ca^{2+}-stimulated PLC, this PLC would probably have to be distinct from the one involved in the initial signal transduction events.

In several cell types it has been shown that $[Ca^{2+}]_i$ oscillations can be generated by injection of $Ins(1,4,5)P_3$ or its non-metabolizable analogues (Oron et al., 1985; Berridge, 1988; Payne et al., 1988; Wakui et al., 1989). This suggests that PLC activation is not required to initiate $[Ca^{2+}]_i$ oscillations. Furthermore, it indicates that the $[Ca^{2+}]_i$ oscillations in response to hormones may occur in the continuous presence of a constant elevated $Ins(1,4,5)P_3$ level rather than occurring secondarily to oscillations of $Ins(1,4,5)P_3$. We have recently found that treatment of hepatocytes with tert-butyl hydroperoxide (TBHP), in the presence of an antioxidant that blocks the lipid peroxidation effects of this agent, induces $[Ca^{2+}]_i$ oscillations with similar kinetic properties to those initiated by hormones (Sakaida et al., 1991; Rooney et al., 1991). TBHP has been shown by others to inhibit intracellular and plasma membrane Ca^{2+} pumps. However, our studies have elucidated an additional effect of this agent mediated by its ability to oxidize hepatic glutathione levels. The elevated levels of GSSG modify the $Ins(1,4,5)P_3$ receptor in a way that apparently makes it more sensitive to $Ins(1,4,5)P_3$ (Rooney et al., 1991). Thus, it appears that in the intact cell TBHP treatment leads to a sensitization to basal $Ins(1,4,5)P_3$ levels such that sufficient Ca^{2+} is released to trigger oscillations. This mechanism is also presumably independent of oscillatory $Ins(1,4,5)P_3$ changes or PLC activation secondary to the $[Ca^{2+}]_i$ increases, since phenylephrine concentrations giving similar $[Ca^{2+}]_i$ oscillation frequencies to TBHP produce a substantial $Ins(1,4,5)P_3$ increase while TBHP has no measurable effect on this second messenger (Rooney et al., 1991).

We have also found that TBHP induces $[Ca^{2+}]_i$ waves in primary cultured hepatocytes from a subcellular locus that is identical to that observed for hormonal stimulation of the same cells. The oscillatory $[Ca^{2+}]_i$ waves activated by TBHP have similar kinetic properties to those produced by Ca^{2+}-mobilizing hormones and AlF_4^-. The finding that TBHP can activate $[Ca^{2+}]_i$ waves independent of PLC activation suggests that an asymmetric distribution of the $Ins(1,4,5)P_3$-sensitive Ca^{2+} pool or a standing gradient of basal $Ins(1,4,5)P_3$ levels may underlie the locus of $[Ca^{2+}]_i$ wave initiation in cultured hepatocytes. This most likely reflects a polarization of the signal transduction apparatus in a manner that mirrors the polarization of hormone receptors to the sinusoidal domain of the hepatocyte plasma membrane in the intact liver (see Rooney and Thomas, 1991; Thomas et al., 1991).

Proposed mechanisms for $[Ca^{2+}]_i$ oscillations

Several models have been proposed to explain the $[Ca^{2+}]_i$ oscillations generated by inositol phospholipid-linked hormones (Berridge and Galione, 1988; Berridge and Irvine, 1989; Jacob, 1990a). These can be classified into two groups according to whether they predict constant or oscillating levels of $Ins(1,4,5)P_3$. Models that incorporate $Ins(1,4,5)P_3$ oscillations in the generation of $[Ca^{2+}]_i$ transients have been suggested that involve either feedback inhibition of $Ins(1,4,5)P_3$ formation through DG-dependent PKC activation (Woods et al., 1987; Sanchez-Bueno et al., 1990), or a cooperative release of Ca^{2+} by $Ins(1,4,5)P_3$ in combination with feedforward activation of $Ins(1,4,5)P_3$ formation by a Ca^{2+}-activated PLC (Meyer and Stryer, 1988). $[Ca^{2+}]_i$ oscillations in the presence of constant $Ins(1,4,5)P_3$ levels have been suggested to occur either by Ca^{2+}-dependent feedback inhibition of $Ins(1,4,5)P_3$-induced Ca^{2+} release (Payne et al., 1988; Parker and Ivorra, 1990) or feed-forward activation of Ca^{2+} release from an $Ins(1,4,5)P_3$-insensitive Ca^{2+} pool by a CICR mechanism (Berridge and Galione, 1988; Rooney et al., 1989, 1990).

The data showing that $[Ca^{2+}]_i$ oscillations can occur in response to injection of non-metabolizable $Ins(1,4,5)P_3$ analogues and our data using TBHP to activate $[Ca^{2+}]_i$ oscillations independent of $Ins(1,4,5)P_3$ formation do not appear to be consistent with models involving oscillatory $Ins(1,4,5)P_3$ levels. Since PLC activation is apparently not required by these agents, it is also difficult to see how PKC activation could play a role in terminating the $[Ca^{2+}]_i$ transients, as suggested by the PKC feedback inhibition model. Furthermore, if the Ca^{2+} wave initiation site in cultured hepatocytes is determined by localized receptors and/or other elements of the signal transduction apparatus, it seems unlikely that mechanisms of $[Ca^{2+}]_i$ oscillations controlled at the receptor level could generate self-propagating $[Ca^{2+}]_i$ waves.

In the model proposed by Meyer and Stryer (1988), $[Ca^{2+}]_i$ oscillations are dependent on the cooperative release of Ca^{2+} by $Ins(1,4,5)P_3$ and subsequent feedforward activation of PLC by Ca^{2+}, resulting in a surge of $Ins(1,4,5)P_3$ formation. The falling phase of each $[Ca^{2+}]_i$ oscillation is dependent on Ca^{2+} sequestration by an $Ins(1,4,5)P_3$-insensitive Ca^{2+} pool, followed by a lowering of $Ins(1,4,5)P_3$ and a recovery period during which Ca^{2+} is reaccumulated by the Ins(1,4,5)P$_3$-sensitive Ca^{2+} pool. A Ca^{2+}-activated PLC model can explain the constant kinetics of the [Ca^{2+}]$_i$ transients at various agonist doses and, if the Ins(1,4,5)P$_3$-sensitive Ca^{2+} pools and a Ca^{2+}-sensitive PLC were distributed throughout the cell, such a mechanism could also generate self-propagating waves of [Ca^{2+}]$_i$ from a fixed origin. However, it does not appear that hormonal stimulation of Ins(1,4,5)P$_3$ formation in liver is dependent on an elevation of [Ca^{2+}]$_i$ above basal levels (Taylor and Exton, 1987; Renard et al., 1987). In addition, as noted above, our data relating to TBHP-induced [Ca^{2+}]$_i$ oscillations and waves do not appear to be consistent with mechanisms based on oscillating Ins(1,4,5)P$_3$ levels.

One model that has been proposed to explain the occurrence of [Ca^{2+}]$_i$ oscillations in the absence of transient Ins(1,4,5)P$_3$ production is based on the finding that Ca^{2+} exerts a negative feedback effect on Ins(1,4,5)P$_3$-induced Ca^{2+} release (Payne et al., 1988; Parker and Ivora, 1990). Although this model offers a potentially important mechanism for the termination of [Ca^{2+}]$_i$ spikes, it is not clear how it would give rise to the prolonged periods of near basal [Ca^{2+}]$_i$ levels which occur between [Ca^{2+}]$_i$ transients. An alternative mechanism that does not rely on Ins(1,4,5)P$_3$ oscillations is based on CICR (Berridge and Gallione, 1988; Berridge and Irvine, 1989; Rooney et al., 1989; Thomas et al., 1991). According to this model, Ca^{2+} is released continuously by Ins(1,4,5)P$_3$ and sequestered by Ins(1,4,5)P$_3$-insensitive Ca^{2+} pools until these reach capacity, at which point a spike of CICR is initiated. Ins(1,4,5)P$_3$ remains at a constant level dependent on agonist dose, which therefore controls the rate of Ca^{2+} transfer between the two Ca^{2+} pools. After each [Ca^{2+}]$_i$ spike, the released Ca^{2+} is accumulated by both Ins(1,4,5)P$_3$-sensitive and Ins(1,4,5)P$_3$-insensitive Ca^{2+} pools and the time taken for the CICR pools to refill to capacity sets the period between spikes (frequency), which is determined primarily by the level of Ins(1,4,5)P$_3$. Thus, the oscillation frequency and initial latent period are modulated by agonist dose, but the amplitude and kinetics of each [Ca^{2+}]$_i$ spike are entirely a property of the CICR pools, independent of Ins(1,4,5)P$_3$ concentration and agonist dose. This is consistent with the observations in hepatocytes (Woods et al., 1986, 1987; Rooney et al., 1989) and endothelial cells (Jacob, 1990b), where agonist dose regulates the latency and frequency of [Ca^{2+}]$_i$ oscillations without affecting oscillation amplitude or shape. The CICR mechanism is also well suited to explain the occurrence of self-propagating waves of [Ca^{2+}]$_i$, since the initiating Ca^{2+} released by Ins(1,4,5)P$_3$ could be localized to one subcellular region while the CICR pools would be distributed throughout the cell.

Functional significance of oscillations and $[Ca^{2+}]_i$ waves

A number of recent reviews have considered the implications and possible significance of $[Ca^{2+}]_i$ oscillations for cell function (Berridge et al., 1988; Berridge and Galione, 1988; Jacob, 1990a; Thomas et al., 1991). Much of the machinery of cellular Ca^{2+} homeostasis is designed to maintain a low resting level of $[Ca^{2+}]_i$. Therefore, limiting agonist-induced $[Ca^{2+}]_i$ increases to discrete spikes would reduce the amount of energy expended by homeostatic mechanisms opposing the effects of the agonist. Oscillatory $[Ca^{2+}]_i$ signalling also offers additional levels of control over the signalling information conveyed to the cell, including the ability to cause differential activation of Ca^{2+}-dependent processes based on the shape of the oscillation. For example, rapidly-activated processes will tend to follow all of the $[Ca^{2+}]_i$ changes while slowly-activated processes will give damped responses and may only respond to $[Ca^{2+}]_i$ transients of longer duration. A potentially important function of oscillatory $[Ca^{2+}]_i$ signalling is that frequency modulation can maintain better fidelity to the original signal at lower signal strength than can be achieved with amplitude-modulation (Berridge and Gallione, 1988; Berridge et al., 1988).

The functional significance of $[Ca^{2+}]_i$ waves has received rather less attention than the $[Ca^{2+}]_i$ oscillation phenomena. $[Ca^{2+}]_i$ waves could be an inevitable consequence of the oscillation mechanism in cells where the signal transduction elements are localized to one region. However, the capability to propagate the $[Ca^{2+}]_i$ signal from the site of stimulus detection (receptor) could be a key element of the regulation of cell function. Hormone receptors and the associated signal transduction apparatus are localized to the sinusoidal membrane domain in hepatocytes (Watanabe et al., 1986; Maurice et al., 1988), while many of the Ca^{2+}-responsive enzymes are distributed throughout the cell. The $[Ca^{2+}]_i$ waves in these cells may serve the function of ensuring that all parts of the hepatocyte experiences a maximal amplitude of signal during each $[Ca^{2+}]_i$ oscillation, despite the localized nature of the signal initiation site (Rooney et al., 1990). Thus, the presence of a self-propagating mechanism for distributing the $[Ca^{2+}]_i$ signal into the interior of the cell could be essential for normal cell function and full hormonal responsiveness.

Diffusive mechanisms of distributing the $[Ca^{2+}]_i$ signal from the plasma membrane site of second messenger generation to distal parts of the cell would result in a signal of diminishing amplitude and lower rate of $[Ca^{2+}]_i$ rise with increasing distance from the receptor domain. This problem can be overcome with mechanisms that incorporate self-propagation of Ca^{2+} release, such as CICR. Figure 4 demonstrates these two types of behavior using a computer simulation of predicted $[Ca^{2+}]_i$ changes at

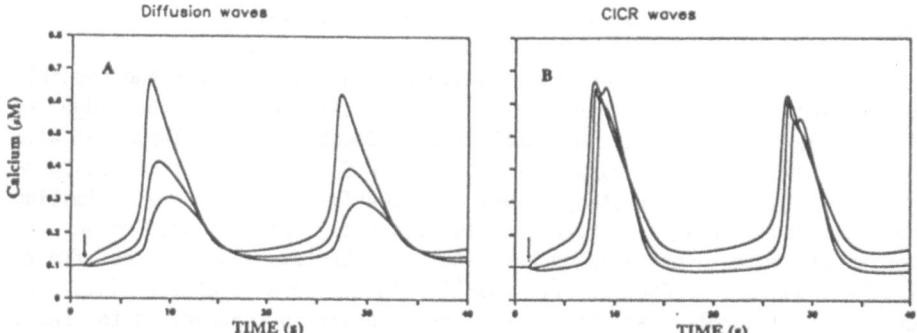

Figure 4. Simulation of $[Ca^{2+}]_i$ wave propagation by diffusive (left panel) and self-propagating (right panel) mechanisms. Reprinted with permission from Thomas et al., 1991, where full details of the simulation and parameters used can be found.

different points within a single cell (plotted in a similar manner to the real data of Fig. 3). The hormone receptors are modelled as being localized to a single compartment and $[Ca^{2+}]_i$ signals propagate either by diffusion (left panel) or CICR (right panel). More details can be found in Thomas et al., (1991). In the context of frequency-modulated Ca^{2+} signalling (Berridge et al., 1988; Berridge and Galione, 1988), where highly cooperative Ca^{2+} binding proteins such as calmodulin act almost as digital Ca^{2+} sensors, a low amplitude signal may result in no signal being detected at all in regions of the cell that are relatively distant from the receptors. Self-propagating $[Ca^{2+}]_i$ waves may also be important in the transmission of $[Ca^{2+}]_i$ waves from cell to cell (e.g. Cornell-Bell et al., 1990). The ability to transmit $[Ca^{2+}]_i$ waves between cells would allow synchronization of the function of multicellular units or even whole organs.

Conclusions

A wide range of hormones and other agonists that bind to cell surface receptors affect intracellular events as a result of an elevation of $[Ca^{2+}]_i$ mediated by the second messenger $Ins(1,4,5)P_3$. In the liver this signal transduction system plays a key role in the activation of glycogenolysis and a number of other metabolic pathways by norepinephrine, vasopressin and several other hormones. Measurements carried out in cell populations have indicated that there is a direct relationship between $Ins(1,4,5)P_3$ levels and $[Ca^{2+}]_i$. However, measurements of

$[Ca^{2+}]_i$ in single hepatocytes using digital imaging fluorescence microscopy have revealed that hormone-induced $[Ca^{2+}]_i$ changes have a complex spatial and temporal organization that is not apparent in cell population studies. The $[Ca^{2+}]_i$ responses in single hepatocytes consist of a series of discrete spikes which repeat at a constant interval and are separated by periods during which $[Ca^{2+}]_i$ returns close to the basal level. The frequency of these $[Ca^{2+}]_i$ oscillations increases with increasing agonist concentration, but the rate of rise and fall of the $[Ca^{2+}]_i$ spikes is independent of agonist dose. In hepatocytes maintained in primary culture the $[Ca^{2+}]_i$ oscillations do not occur synchronously throughout the cell, but originate from a specific plasma membrane locus and spread through the cell as a wave. Oscillatory $[Ca^{2+}]_i$ waves which initiated at the same subcellular locus can also be produced by agents that bypass the receptor and stimulate Ca^{2+} release from the $Ins(1,4,5)P_3$-sensitive Ca^{2+} pool directly. Based on the kinetics and spatial organization of the $[Ca^{2+}]_i$ oscillations in hepatocytes, it appears that each $[Ca^{2+}]_i$ spike is initiated by $Ins(1,4,5)P_3$ generation at a localized site of signal transduction and then distributed throughout the cell, with undiminished amplitude, by a self-propagative mechanism.

Acknowledgments

This work was supported by United States Public Health Service Grant DK38422. T. Rooney is the recipient of a Juvenile Diabetes Foundation International Research Fellowship.

References

Berridge MJ (1988) Inositol trisphosphate-induced membrane potential oscillations in Xenopus oocytes. J Physiol (Lond) 403:589-599

Berridge MJ, Galione A (1988) Cytosolic calcium oscillators. FASEB J 2:3074-3082

Berridge MJ, Irvine RF (1984) Inositol trisphosphate, a novel second messenger in cellular signal transduction. Nature (Lond) 312:315-321

Berridge MJ, Irvine RF (1989) Inositol phosphates and cell signalling. Nature (Lond) 341:197-205

Berridge MJ, Cobbold PH, Cuthbertson KSR (1988) Spatial and temporal aspects of cell signalling. Philos Trans R Soc Lond B Biol Sci 320:325-343

Cornell-Bell AH, Finkbeiner SM, Cooper M, Smith SJ (1990) Glutamate induces calcium waves in cultured astrocytes: Long-range glial signaling. Science 247:470-473

Ferris CD, Huganir RL, Supattapone S and Snyder SH (1989) Purified inositol 1,4,5-trisphosphate receptor mediates calcium flux in reconstituted lipid vesicles. Nature (Lond) 342:87-89

Furuichi T, Yoshikawa S, Miyawaki A, Wada K, Maeda N, Mikoshiba K (1989) Primary structure and functional expression of the inositol 1,4,5-trisphosphate-binding protein P_{400}. Nature (Lond) 342:32-38

Gilkey JC, Jaffe LF, Ridgway EB, Reynolds GT (1978) A free calcium wave traverses the activating egg of the medaka *Oryzias latipes*. J Cell Biol 76:448-466

Jacob R (1990a) Calcium oscillations in electrically non-excitable cells. Biochim Biophys Acta 1052:427-438

Jacob R (1990b) Imaging cytoplasmic free calcium in histamine stimulated endothelial cells and in fMet-Leu-Phe stimulated neutrophils. Cell Calcium 11:241-249

Joseph SK, Williamson JR (1989) Inositol polyphosphates and intracellular calcium release. Arch Biochem Biophys 273:1-15

Kawanishi T, Blank LM, Harootunian AT, Smith MT and Tsien RY (1989) Ca^{2+} oscillations induced by hormonal stimulation of individual fura-2-loaded hepatocytes. J Biol Chem 264:12859-12866

Maurice M, Rodgier E, Cassio D, Feldmann G (1988) Formation of plasma membrane domains in rat hepatocytes and hepatoma cell lines in culture. J Cell Sci 90:79-92

Meyer T, Stryer L (1988) Molecular model for receptor-stimulated calcium spiking. Proc Natl Acad Sci U.S.A. 85:5051-5055

Mignery GA, Sudhof TC, Takei K, Camilli P (1989) Putative receptor for inositol 1,4,5-trisphosphate similar to ryanodine receptor. Nature (Lond) 342:192-195

Millard PJ, Gross D, Webb WW, Fewtrell C (1988) Imaging asynchronous changes in intracellular Ca^{2+} in individual stimulated tumor mast cells. Proc Natl Acad Sci USA 85:1854-1858

Miyazaki S-I, Hashimoto N, Yoshimoto Y, Kishimoto T, Igusa Y, Hiramoto Y (1986) Temporal and spatial dynamics of the periodic increase in intracellular free calcium at fertilization of golden hamster eggs. Develop Biol 118:259-267

Oron Y, Dascal N, Nadler E, Lupu M (1985) Inositol 1,4,5-trisphosphate mimics muscarinic responses in Xenopus oocytes. Nature (Lond). 313:141-143

Parker I, Ivorra I (1990) Inhibition by Ca^{2+} of inositol trisphosphate-mediated Ca^{2+} liberation: A possible mechanism for oscillatory release of Ca^{2+}. Proc Natl Acad Sci USA 87:260-264

Payne R, Walz B, Levy S, Fein A (1988) The localization of calcium release by inositol trisphosphate in Limulus photoreceptors and its control by negative feedback. Philos Trans R Soc Lond B Biol Sci 320:359-379

Prentki M, Glennon MC, Thomas AP, Morris RL, Matschinsky FM, Corkey BE (1988) Cell-specific patterns of oscillating free Ca^{2+} in carbamylcholine-stimulated insulinoma cells. J Biol Chem 263:11044-11047

Renard D, Poggioli J, Berthon B, Claret M (1987) How far does phospholipase C activity depend on the cell calcium concentration? A study in intact cells. Biochem J 243:391-398

Rooney TA, Sass E, Thomas AP (1989) Characterization of cytosolic calcium oscillations induced by phenylephrine and vasopressin in single fura-2-loaded hepatocytes. J Biol Chem 264:17131-17141

Rooney TA, Sass E, Thomas AP (1990) Agonist-induced cytosolic calcium oscillations originate from a specific locus in single hepatocytes. J Biol Chem 265:10792-10796

Rooney TA, Thomas AP (1991) Organization of intracellular calcium signals generated by inositol lipid-dependent hormones. Pharmac Ther 49:223-237

Rooney TA, Renard DC, Sass E, Thomas AP (1991) Oscillatory cytosolic calcium waves independent of stimulated inositol 1,4,5-trisphosphate formation in hepatocytes. J Biol Chem (In press)

Ross CA, Meldolesi J, Milner TA, Satoh T, Supattapone S, Snyder SH (1989) Inositol 1,4,5-trisphosphate receptor localized to endoplasmic reticulum in cerebellar Purkinje neurons. Nature (Lond) 339:468-470

Sakaida I, Thomas AP, Farber JL (1991) Increases in cytosolic calcium ion concentration can be dissociated from the killing of cultured hepatocytes by tert-butyl hydroperoxide. J Biol Chem 266:717-722

Sanchez-Bueno A, Dixon CJ, Woods NM, Cuthbertson KSR, Cobbold PH (1990) Inhibitors of protein kinase C prolong the falling phase of each free-calcium transient in a hormone-stimulated hepatocyte. Biochem J 268:627-632

Takamatsu T, Weir WG (1990) Calcium waves in mammalian heart: quantification of origin, magnitude, waveform, and velocity. FASEB J 4:1519-1525

Taylor SJ, Exton JH (1987) Guanine nucleotide and hormone regulation of polyphosphoinositide phospholipase C activity of rat liver plasma membranes. Biochem J 248:791-799

Thomas AP, Renard DC, Rooney TA (1991) Spatial and temporal organization of calcium signalling in hepatocytes. Cell Calcium 12:111-126

Volpe P, Krause K-H, Hashimoto S, Zorzato F, Pozzan T, Meldolesi J, Lew DP (1988) "Calciosome," a cytoplasmic organelle: the inositol 1,4,5-trisphosphate-sensitive Ca^{2+} store of nonmuscle cells? Proc Natl Acad Sci USA 85:1091-1095

Wakui M, Potter BVL, Peterson OH (1989) Pulsatile intracellular calcium release does not depend on fluctuations in inositol trisphosphate concentration. Nature (Lond) 339:317-320

Watanabe J, Kanamura S, Kanai K (1986) Plasma membrane regionalization in cultured mouse hepatocytes. Anat Rec 214:1-7

Williamson JR, Cooper RH, Joseph SK, Thomas AP (1985) Inositol trisphosphate and diacylglycerol as intracellular second messengers in liver. Am J Physiol 248:C203-C216

Woods NM, Cuthbertson KSR, Cobbold PH (1986) Repetitive transient rises in cytoplasmic free calcium in hormone-stimulated hepatocytes. Nature (Lond) 319:600-602

Woods NM, Cuthbertson KSR, Cobbold PH (1987) Agonist-induced oscillations in cytoplasmic free calcium concentration in single rat hepatocytes. Cell Calcium 8:79-100

PHOSPHOLIPID HYDROLYSIS AND $[Ca^{2+}]_i$ CONTROL

J. Meldolesi, G. Gatti, M. Magni, E. Clementi, D. Zacchetti,
and *H. Scheer

Department of Pharmacology
CNR Center of Cytopharmacology
S. Raffaele Scientific Institute
University of Milano
Via Olgettina, 60
20132 Milano, Italy

The existence of a direct link between receptor-activated phospholipid hydrolysis and changes of cellular Ca^{2+} homeostasis was hypothesized quite some time ago but remained unclear, being based largely on anecdotal evidence, until the discovery almost a decade ago of the specific Ca^{2+} release activity of inositol 1,4,5-trisphosphate (IP_3) (reviewed by Berridge and Irvine, 1989). Initially, the reaction responsible for IP_3 generation, i.e., the hydrolysis of phosphatidylinositol 4,5-bisphosphate (PIP_2) by specific phospholipase(s) C (PLC), was thought to be the only metabolic process by which $[Ca^{2+}]_i$ was regulated. However, the field soon became more complex. In 1984-85, activation of the receptors coupled to PPI hydrolysis was shown to stimulate not only Ca^{2+} release from internal stores but also Ca^{2+} influx via channels in the plasmalemma, later named second messenger-operated channels (Meldolesi and Pozzan, 1987). The multiplicity and complex regulation .(not only by second messengers, but also via G proteins and the physiological state of intracellular Ca^{2+} stores) of these channels is discussed elsewhere (Meldolesi et al., 1991).

Around the same time, the other second messenger generated by PIP_2 hydrolysis, diacylglycerol (DAG), began to be appreciated as a feed-back regulator of the receptor-triggered response, working via the activation of protein kinase C (PKC) and the inhibitory effect of phosphorylation by PKC on receptor and G protein function (see Vicentini et al., 1985 for our contribution to the field, and Nishizuka, 1986 for a review). Moreover, DAG was found to modulate also the activity of voltage-gated Ca^{2+} channels (Di Virgilio et al., 1986; review: Tsien et al., 1988). Finally, during the past two years, DAG has been shown to originate not only from PIP_2, but also from phosphatidylcholine (PC). Hydrolysis of the latter

* Dept. of Pharmacology, University of Montreal, Canada. H.S. is a MRC-CNR Canada-Italy Exchange Scientist

NATO ASI Series, Vol. H 70
Phospholipids and Signal Transmission
Edited by R. Massarelli, L. A. Horrocks,
J. N. Kanfer, and K. Löffelholz
© Springer-Verlag Berlin Heidelberg 1993

phospholipid by either phospholipase C or D is believed to be stimulated by multiple mechanisms: direct interaction of the enzymes with G protein-coupled receptors, activation of PKC and as yet unspecified Ca^{2+}-dependent mechanism(s) (see Exton, 1990). The last two processes are thought to be of particular importance because they apparently maintain a high DAG-generating activity for long periods of time.

The following series of events can, in fact, be envisaged to occur following receptor activation: first, PIP_2 hydrolysis is triggered by direct receptor-PLC (specific for the phosphoinositide) interaction, with a subsequent burst in the formation of both DAG and IP_3 and a peak $[Ca^{2+}]_i$ increase. PKC is thereby activated (by DAG and Ca^{2+}) and, concomitantly, second-messenger-operated channels are opened, resulting in a persistent, although moderate, elevation of $[Ca^{2+}]_i$. In many cells these events are accompanied by plasma membrane hyperpolarization (due to activation of Ca^{2+}-dependent K^+ channels: Fasolato et al., 1988; Pandiella et al., 1989a), in other cells by a depolarization (e.g., by blockade of the M current). In the latter case voltage-gated Ca^{2+} channels can participate in the $[Ca^{2+}]_i$ increase. When stimulation is of moderate intensity, $[Ca^{2+}]_i$ begins to oscillate rhythmically or increases in its oscillation frequency (see Malgaroli et al., 1990). These activities (sustained by intracellular Ca^{2+} stores) usually persist so long as the receptor remains activated (review: Berridge and Galione, 1989). High PKC and $[Ca^{2+}]_i$ activate the PC-specific phospholipases (C and D), thus DAG levels increase, giving rise to a delayed, often large and persistent, plateau (for further details see Exton, 1990). Concomitantly, other phospholipases (C isoforms addressed to all types of phosphoinositides and A_2) are activated by the high $[Ca^{2+}]_i$. Thus, the relationship between phospholipid hydrolysis and $[Ca^{2+}]_i$ is not only complex but also mutual, i.e., each of the two can contribute to increase the other. Desensitization at the receptor and extra-receptor (for example, the channels) levels ultimately controls the evolution of the process and thus contributes to termination of the process.

So far, we have depicted a general scenario for our theme. In the rest of this presentation we will focus specifically on three particular aspects that our group has been investigating during the past few years.

Regulation of PIP$_2$ hydrolysis by receptors: only stimulation or also inhibition?

No doubt exists that PIP$_2$ hydrolysis is stimulated by a variety of receptors that interact directly with the PLC involved. The enzyme and the mechanisms of interaction change depending on the general structure of the receptors. Receptors organized according to the seven-membrane-spanning-domain structure interact via specific G proteins, and PLC is probably of the β family (see Taylor et al., 1991); those endowed with tyrosine kinase activity phosphorylate the PLC-γ1 enzyme, which thus becomes able to interact with PIP$_2$ even in the presence of the cytoskeletal protein, profilin (Meldolesi and Magni, 1991; for PLC classification see Rhee et al., 1989; for PLC-γ1 activation, Goldschmidt-Clermont et al., 1991). A question however remains open: are other receptors able to interact with PLC in an inhibitory fashion? By analogy with the situation with adenylate cyclase, this is often admitted a priori, with no attention to experimental data. The receptors suspected to share this activity are the same known to inhibit adenylate cyclase and, in addition, to modulate the function of numerous K$^+$ and Ca^{2+} channels, in all cases via G protein interaction. Our initial studies in this field demonstrated that, in pituitary lactotroph cells, the inhibition by dopamine (working via inhibitory D$_2$ receptors) of the phosphoinositide hydrolysis stimulated by TRH was delayed, nonspecific for PIP$_2$ and Ca^{2+}-dependent. In other words, the effect of dopamine consisted (largely, if not exclusively) in the inhibition of Ca^{2+} influx, and therefore the inhibition of phosphoinositide hydrolysis was indirect and Ca^{2+}-mediated (Vallar et al., 1989). Recently, we have addressed the same question in PC12 cells using clonidine to activate the inhibitory α_2 receptor and carbachol and bradykinin to stimulate PIP$_2$ hydrolysis via muscarinic and B$_2$ receptors. In this case, stimulation of the inhibitory α_2 receptor remained without any effect on the hydrolysis of both PIP$_2$ and other phosphoinositides, even though adenylate cyclase was inhibited and the muscarinic Ca^{2+}-release response was consistently reduced (Gatti et al., in preparation). We conclude that although a direct, negative regulation of PLC at the level of inhibitory receptors cannot be excluded yet, it is certainly neither as general nor as significant as the inhibitory adenylate cyclase-receptor linkage. These results cast serious doubts not only on the existence of the process, but, even if the process does exist, on its physiological role. In fact, control of [Ca^{2+}]$_i$ and DAG production seems to be regulated by inhibitory receptors working via channels and cAMP generation. Thus, the direct negative control of PIP$_2$ hydrolysis, suggested by others, can only play a minor role, if any.

Phospholipid hydrolysis stimulates cell growth

That phospholipid hydrolysis stimulates cell growth is largely based on experiments demonstrating that 1) the well-known family of tumor promoters, the phorbol esters, mimic the PKC activation property of DAG and 2) mitogenic responses are induced by the activation of various PIP_2 hydrolysis-coupled receptors, such as those for bombesin and angiotensin II. Moreover, various growth factor receptors (for EGF, PDGF, FGF and others, see Pandiella et al., 1989b; Meldolesi and Magni, 1991) are capable of stimulating PIP_2 hydrolysis by the mechanism summarized in the preceding section. This reaction could thus contribute to their physiological effects. However, it must be pointed out that PIP_2 hydrolysis is transient (over in a few min), whereas commitment to proliferation requires relatively long periods (most often several hr) of stimulation. Thus, it seems very likely that the lipid metabolism important for cell growth is not that stimulated directly at the receptor level by PIP_2 hydrolysis, but rather the prolonged PC hydrolysis response sustained by Ca^{2+} and PKC activation (see above and Exton, 1990; Meldolesi and Magni, 1991).

In this respect, two important observations have been made recently by our group (Magni et al., 1991). First, the activation of Ca^{2+} influx via second messenger-activated channels is particularly persistent with EGF compared to other receptor agonists; second, pharmacological blockade of these same channels by the imidazole drug SC38249 results in a considerable (> 50%) inhibition of the mitogenic activity of both EGF and serum. Clearly, mechanisms other than lipid metabolism (e.g., tyrosine phosphorylation of the raf-1 oncogene or other important substrates) play important roles in cell growth. With stimuli that increase neither Ca^{2+} nor lipid metabolism, these alternative mechanisms are probably responsible for the entire proliferative effect; when lipid metabolism is stimulated, the two types of signals are expected to cooperate synergistically.

In addition to PKC, a site at which lipid metabolites could operate is in the regulation of the ras protooncogene protein. In fact, the specific GTPase activator protein, GAP, is inhibited by DAG and other lipids, whereas the corresponding inhibitory protein is activated. Lipid metabolites appear therefore to stabilize the ras protein in its GTP-bound, active state (see Tsai et al., 1990). Interestingly, GAP is also a substrate for receptor tyrosine phosphorylation. Therefore this protein appears to be under a dual and possibly synergistic control of lipids and tyrosine phosphorylation. In the near future other examples of this kind might emerge so that, hopefully, the growth control mechanisms and their regulation will ultimately be clarified.

How many Ca^{2+} stores exist in eukaryotic cells?

If, indeed, as we have already discussed extensively, Ca^{2+} and lipid metabolism are intimately and mutually correlated, any scientist interested in the regulatory role of lipids needs to know about the origin of $[Ca^{2+}]_i$ changes in cells. Numerous Ca^{2+} channels, operated by voltage, receptors and second messengers (Meldolesi and Pozzan, 1987; Tsien et al., 1988; Meldolesi et al., 1991), exist at the plasma membrane, and in addition, various Ca^{2+} stores function within the cytoplasm. The Ca^{2+} from these stores can activate Ca^{2+}-dependent processes without increasing Ca^{2+} influx (Meldolesi et al., 1990). We have shown recently using the neurosecretory cell PC12 that most of the the Ca^{2+} segregated within intracellular stores is neither rapidly exchanged nor controlled by receptor activation. This Ca^{2+} is apparently distributed among various organelles, with secretion granules accounting for a significant fraction (~ 25% of the total), and mitochondria for only a very small percentage (Fasolato et al., 1991). For rapidly exchanging Ca^{2+}, two separate pools have been described in various cell types, one sensitive to IP_3, the other to drugs active on the muscle sarcoplasmic reticulum (caffeine and ryanodine, see Meldolesi et al., 1990). However, in our work on a group of clones recently isolated from PC12 cells, we found sensitivity to the latter drugs to be expressed only by some clones, but not by all of them. Moreover, working on a caffeine-sensitive clone, we found that the Ca^{2+} storage and release activities controlled by the drug cannot be dissociated from the activities operated by IP_3. In other words, the experimental evidence indicates the existence of a single rapidly exchanging Ca^{2+} store residing in a single organelle that apparently expresses two types of intracellular channels, the IP_3 and the ryanodine receptors (see Zacchetti et al., 1991). This observation opens a number of interesting problems in both molecular and cell biology. In particular: what are the mechanisms that address the specific components of rapidly exchanging Ca^{2+} stores to their final destination? How is it that in various types of cells, these destinations are different? These and many other related questions are going to make our scientific life busy during the next few years.

References

Berridge, MJ, Irvine RF (1989) Inositol phosphates and cell signalling. Nature 341: 197-205

Berridge, MJ, Galione A (1988) Cytosolic calcium oscillators. FASEB J 2:3074-3082

Di Virgilio F, Pozzan T, Wollheim CB, Vicentini LM, Meldolesi J (1986) Tumor promoter phorbol myristate acetate inhibits Ca^{2+} influx through voltage gated Ca^{2+} channels in two secretory cell lines, PC12 and Rinm5F. J Biol Chem 261:32-36

Exton, JH (1990) Signalling through phosphatidylcholine breakdown. J Biol Chem 265:1-4

Fasolato C, Pandiella A, Meldolesi J, Pozzan T (1988) Generation of inositolphosphates, cytosolic Ca^{2+} and ionic fluxes in PC12 cells treated with bradykinin. J Biol Chem 263:17350-17359

Fasolato C, Zottini M, Clementi E, Zacchetti D, Meldolesi J, Pozzan T (1991) Intracellular Ca^{2+} pools in PC12 cells. J Biol Chem 266:20159-20167

Goldschmidt-Clermont PJ, Kim JW, Machesky LM, Rhee SG, Pollard TD (1991) Regulation of phospholipase C-γ1 by profilin and tyrosine phosphorylation. Science 251:1231-1233

Magni M, Meldolesi J, Pandiella A (1991) Ionic events induced by epidermal growth factor: evidence that hyperpolarization and stimulated cation influx play a role in the stimulation of cell growth. J Biol Chem 266: 6329-6335

Malgaroli A, Fesce R, Meldolesi J (1990) Spontaneous $[Ca^{2+}]_i$ fluctuations in rat chromaffin cells do not require inositol 1,4,5-trisphosphate elevations but are generated by a caffeine- and ryanodine-sensitive intracellular Ca^{2+} store. J Biol Chem 265:3005-3008

Meldolesi J, Pozzan T (1987) Pathways of Ca^{2+} influx at the plasma membrane: voltage-, receptor- and second messenger-operated channels. Exp Cell Res 171: 271-283

Meldolesi J, Magni M (1991) Lipid metabolites and growth factor action. Trends Pharmacol Sci 12:362-364

Meldolesi J, Madeddu L, Pozzan T (1990). Intracellular Ca^{2+} storage organelles in non muscle cells: heterogeneity and functional assignment. Biochim Biophys Acta 1055:130-140

Meldolesi J, Clementi E, Fasolato C, Zacchetti D, Pozzan T (1991) Stimulation of Ca^{2+} influx following receptor activation: facts, hypotheses and missing links. Trends Pharmacol Sci 12:289-292

Nishizuka Y (1986) Studies and perspectives of protein kinase C. Science 233:305-312

Pandiella A, Magni M, Lovisolo D, Meldolesi J (1989) The effects of epidermal growth factor on membrane potential. J Biol Chem 264:12914-12921

Rhee SG, Suh P-G, Ryu S-H, Lee SY (1989) Studies of inositol phospholipid-specific phospholipase C. Science 244: 546-550

Taylor SJ, Chai HZ, Rhee SG, Exton JH (1991) Activation of the β1 isozyme of phospholipase C by α subunits of the Gq class of G proteins. Nature 350:516-518

Tsai M-H, Yu C-L, Stacey DW (1991) A cytoplasmic protein inhibits the GTPase activity of H-Ras in a phospholipid-dependent manner. Science 250:982-985

Tsien RW, Lipscombe D, Madison DV, Bley KR, Fox AP (1988) Multiple types of neuronal calcium channels and their selective modulation. Trends Neurosci 11:431-438

Vallar L, Vicentini LM, Meldolesi J. (1988) Inhibition of inositol phosphate production is a late, Ca^{2+}-dependent effect of D_2 dopaminergic receptor in rat lactotroph cells. J Biol Chem 263:10127-10134

Vicentini LM, Di Virgilio F, Ambrosini A, Pozzan T, Meldolesi J (1985) Tumor promoter phorbol 12-myristate, 13-acetate inhibits phosphoinositide hydrolysis and cytosolic Ca^{2+} rise induced by the activation of muscarinic receptors in PC12 cells. Biochem Biophys Res Comm 127:310-317

Zacchetti D, Clementi E, Fasolato C, Lorenzon P, Zottini M, Grohovaz F, Fumagalli G, Pozzan T, Meldolesi J (1991) Intracellular Ca^{2+} pools in PC12 cells. A unique, rapidly-exchanging pool is sensitive to both inositol 1,4,5-trisphosphate and caffeine-ryanodine. J Biol Chem 266:20152-20158

PHOSPHOINOSITIDE HYDROLYSIS INDICATES FUNCTIONAL RECEPTORS IN ASTROCYTES AND IN NEOPLASTIC CELLS FROM THE HUMAN CNS

S. Murphy, G. Bruner and M.L. Simmons
Department of Pharmacology
University of Iowa College of Medicine
Iowa City, IA 52242
USA

Astrocytes, the majority cell type in the CNS, have the potential to play a dynamic role in intercellular communication. Situated adjacent to neurons and to cells of the microvasculature, these glial cells display a variety of surface receptors (Murphy and Pearce, 1987). Astrocytes synthesize and release peptide mediators, such as met-enkephalin, somatostatin, endothelin-1 and -3 (Shinoda et al., 1989; MacCumber et al., 1990; Ehrenreich et al., 1991), and eicosanoids including prostaglandins E_2, $F_{2\alpha}$, D_2, thromboxane A_2, and the leukotrienes LTB_4 and LTC_4 (Murphy et al., 1988). Regulation of the synthesis and release of many of these agents from astrocytes is not completely understood. We have focussed on the mechanism by which prostanoid production in astrocytes is regulated by purinergic (P_{2y}) receptor agonists. Most recently, we have demonstrated the evoked release of an astrocyte-derived (vaso)relaxing factor (ADRF) which is not a prostanoid but a nitrosyl compound with properties similar to nitric oxide. Thus, activation of specific receptors induces the synthesis and release of products from astrocytes with the potential to directly affect not only neuronal, but also vascular cell function.

Receptor-coupling

Binding sites for a range of neurotransmitter ligands have been described in primary cultures of astrocytes (Murphy and Pearce, 1987). Evidence from freshly isolated cells (Salm and McCarthy, 1989) suggests that at least a subset of these sites on astrocytes are retained in the mature CNS. In addition, there are interesting regional and developmental variations in receptor expression (Wilkin et al., 1990). To reveal whether these astrocyte binding sites are functional receptors as opposed to mere acceptors, we have looked for evidence of receptor coupling to cyclases, phospholipases (PLC, PLA_2, PLD) and calcium channels. A range of receptors are positively linked to adenylate cyclase (β-adrenergic,

NATO ASI Series, Vol. H 70
Phospholipids and Signal Transmission
Edited by R. Massarelli, L. A. Horrocks,
J. N. Kanfer, and K. Löffelholz
© Springer-Verlag Berlin Heidelberg 1993

TABLE 1. Agents which Induce PPI Hydrolysis and Influence Calcium Flux in Astrocytes

Agonist	Receptor	IP1	IP2	IP3	Ca2+ Increase
Norepi.	α_1	+	+	+	+
Carbachol	M_1	+	+	+	+
Glutamate	Q_p	+	+	+	+
Serotonin	$5HT_2$	+	+	+	+
Histamine	H_1	+	?	?	?
ATP	P_{2y}	+	+	+	+
Endothelin	ET-1,3	+	+	+	+
Bradykinin	B_2	+	+	+	+
A23187		+	+	NE	+
Thapsigargin		NE	NE	NE	+

NE, no effect

VIP, ACTH, MSH, secretin, glucagon, H_2-histamine, dopamine, 5HT, A_2-adenosine, and the prostaglandins E_1, E_2 and I_2). Receptors such as the muscarinic cholinergic M2, α_2-adrenergic, A_1-adenosine and melatonin are negatively coupled to adenylate cyclase. We have found (Table 1) that a number of astrocyte receptors are linked via G protein(s) to PLC and the subsequent hydrolysis of polyphosphoinositides (PPI), including α_1-adrenergic, muscarinic, histamine, 5HT, P_{2y}-purinergic, glutamate (Q_p metabotropic receptor), oxytocin, vasopressin, bradykinin and ACTH (Pearce and Murphy, 1988; Wilkin and Cholewinski, 1988).

Activation of some of these receptors evokes complex changes in intracellular calcium. Clearly, these changes include components from intra- and extracellular sources of calcium, which may contribute to that part of agonist-induced PPI hydrolysis which is calcium-dependent. Figure 1A shows a typical agonist-evoked change in intracellular calcium, in this case a response to carbachol. The rapid rise is due to IP_3-induced calcium mobilization, while the sustained 'tail' results from calcium influx. Thapsigargin, an inhibitor of calcium ATPase with no effects on PPI hydrolysis, causes a leak of calcium from IP_3-sensitive and insensitive stores. The elevated and sustained calcium level results from the inability of the cell to re-sequester calcium (Figure 1B).

Receptors coupled to PLC show predictable down-regulation. We compared the actions of agonists and phorbol esters to determine whether their effects were analogous or homologous (Pearce et al., 1988). Preincubation of [^{3}H]inositol labeled cultures with phorbol myristate acetate results in a time and concentration-dependent decrease in the accumulation of

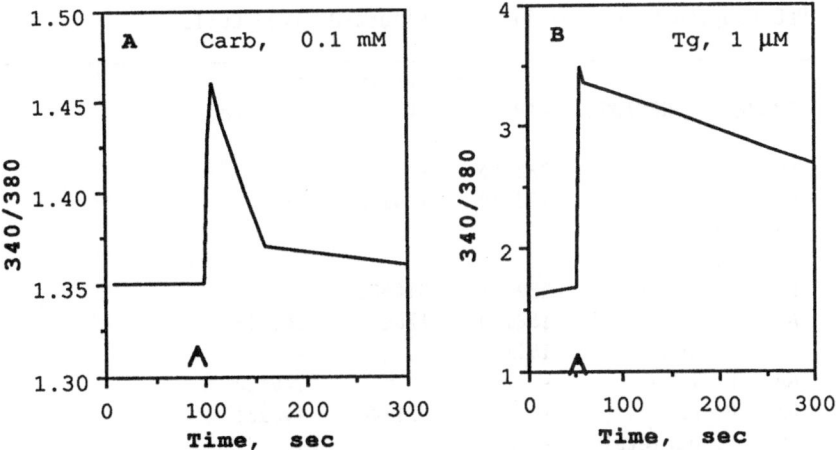

Figure 1. Effects of (A) carbachol and (B) thapsigargin on intracellular calcium in fura-loaded astrocytes.

inositol phosphates evoked by muscarinic receptor stimulation. Much longer preincubation periods with higher concentrations are required to elicit the same effect with the muscarinic agonist carbachol. These effects are not additive and in both cases, the response to agonist recovers within two days. Both pretreatments are effective in reversing the carbachol-evoked mobilization of calcium. Receptor-binding studies reveal that both pretreatments result in a loss of muscarinic receptors and also cause the uncoupling of a G protein from the receptor. However, whereas phorbol-pretreatment causes heterologous desensitization, the carbachol pretreatment results only in homologous downregulation.

To confirm that PLC-coupled receptors are retained on astrocytes in the mature brain, we have investigated PPI turnover in cells derived from human glioma biopsy material (Murphy et al., 1990b). The fact that these cells originate from glioma material and that they can be maintained in culture suggests that they are indeed transformed. The results were surprising (Table 2). Cells from low grade astrocytomas displayed PPI hydrolysis in response to ATP and norepinephrine, but muscarinic receptors coupled to PLC were absent. In contrast, cells from the highly malignant glioblastomas expressed muscarinic but not purinergic responses. Intermediate grade anaplastic astrocytomas expressed a variety of responses. This apparent correlation of grade with receptor expression is not confounded by patient age or by CNS region of origin. In addition, the A172 cell line derived from a human glioblastoma express an abundance of muscarinic receptors coupled to PPI hydrolysis. This distribution of receptors might reflect the de-differentiated state of astrocytes comprising the glioblastoma, and the more differentiated state of cells in

TABLE 2. PLC-Coupled Receptors in Human Glioma-Derived Cells

| TUMOR | GRADE | REGION | TOTAL IPs (% Basal) | | |
			ATP	NE	Carbachol
89-362	LG	Op. chiasma	232±46	-	-
89-373	LG	R.temp	192±38	174±32	-
89-543	LG	Pilocytic	-	-	-
89-653	LG	L. frontal	-	-	-
90-422	LG	Temporal	229±17	314±22	-
89-236	AA	Temp, medial	152±18	176±32	140±13
89-454	AA	R. temp	151±18	-	-
89-517	AA	L. temp	-	-	176±28
89-437	GM	R. cbm	-	271±40	443±31
89-441	GM	L. parietal	-	-	-
89-484	GM	L. parietal	-	-	215±20
90-170	GM	R. cbm	-	-	185±01
90-641	GM	Frontal lobe	-	-	-
90-637	GM	L. frontal	152±14	199±15	-
90-642	GM	L. temp.	-	502±108	-
A172	GM	Cell line	-	-	1753±454

LG, low grade astrocytoma; AA, anaplastic astrocytoma; GM, glioblastoma multiforme. Total [^{3}H]inositol phosphates (IPs) were determined in [^{3}H]inositol-labeled cells exposed to agonists (100 μM) for 30 min. Values are means (% basal) ± sem, n=3.
- indicates not significantly different from basal.

lower grade tumors. If so, our observation supports the idea that purinergic and α_1-adrenergic receptors are retained on mature cells. The functional significance of the retention of muscarinic receptors on undifferentiated astrocytes, or their re-expression in de-differentiated cells, raises intriguing questions about their role in cell cycle progression.

Role of PPI hydrolysis in receptor-mediated eicosanoid production

Activation of receptors linked to the hydrolysis of PPI generates DAG. The fate of DAG interested us because it could be further metabolized via DAG lipase to AA, and thus form a substrate for eicosanoid production. In initial experiments we found that astrocytes synthesize and release a

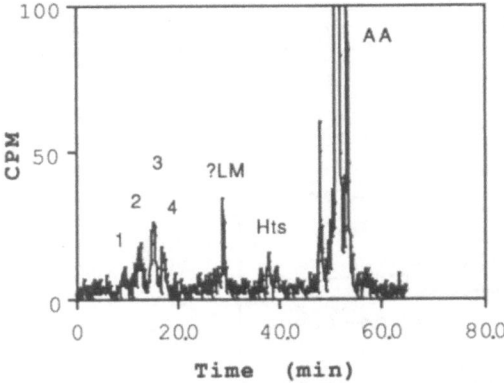

Figure 2. Evoked release of arachidonate metabolites from astrocytes. Cell cultures were labeled with [³H] AA for 18 hours and then incubated for 60 min in the presence of 5 μM A23187. Conditioned medium was extracted and eicosanoids separated by HPLC. 1=TXB₂, 2=PGF₂α, 3=PGE₂, 4=PGD₂, ?LM=an unknown lipoxygenase metabolite, Hts=HETES.

number of prostanoids. In response to the calcium ionophore A23187 the release of PGs E₂, D₂, F₂α and also TXA₂ can be detected (Figure 2).

It is interesting that the bulk (80%) of arachidonic acid (AA) is not metabolized but is released unchanged into the medium. On presentation to astrocytes, AA causes the hydrolysis of PPI and the accumulation of IPs (including IP₃). This effect is not secondary to AA metabolism, as it is unaltered in the presence of cyclooxygenase and lipoxygenase inhibitors (Murphy and Welk, 1989). This observation raises the possibility that agonist-evoked release of unmetabolized AA from cells could be significant, not only in terms of substrate supply for further metabolism but also in cell-cell signaling. In addition, we routinely see large amounts of a lipoxygenase product (LM in Figure 2) released from these astrocytes which has so far eluded identification. As this lipoxygenase product is released in such amounts, it may represent a novel mediator from astrocytes.

To discover the origin of the AA for prostanoid production, we attempted to evoke AA and prostanoid release with agonists that stimulated PLC. In no case were we able to see release, suggesting that the AA for prostanoid production is not supplied by the polyphosphoinositides. This conclusion was supported by experiments in which PLA₂ was blocked by mepacrine. Exposure of the cells now to A23187 did not affect the hydrolysis of inositol-containing phospholipids but effectively abolished the liberation of AA and prostanoids (Pearce et al., 1987).

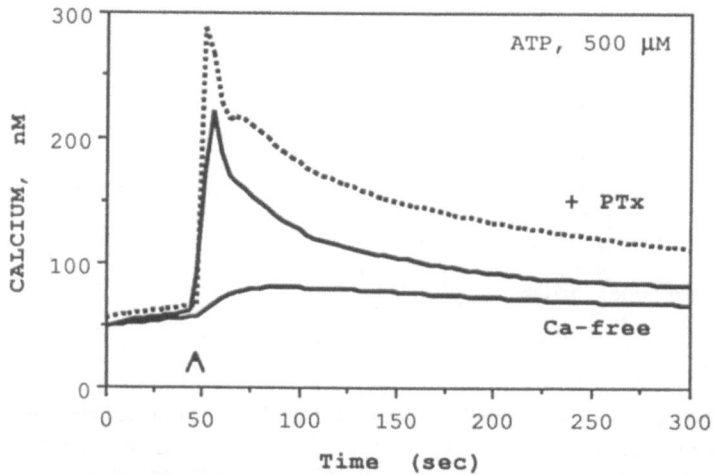

Figure 3. Changes in intracellular calcium in response to ATP. Astrocytes on coverslips were loaded with fura-2 and stimulated at 20 °C. Cells pretreated with pertussis toxin (100 ng/ml) are indicated ± PTx. Cells stimulated in the absence of extracellular calcium are indicated Ca-free.

Recently, we have discovered that ATP and ADP, acting at P_2-purinergic receptors, are potent stimuli for thromboxane (TX) release (Pearce et al., 1989). To verify our proposal that ATP mobilizes AA by activating a P_2 receptor we investigated the efficacy of a range of ATP analogs. The rank order of potency suggests involvement of a P_{2y} receptor (Bruner and Murphy, 1990a). In astrocytes, this receptor is coupled to PLC but not to PLD (Bruner and Murphy, 1990b). The mechanism for ATP activation of PLA_2 could be through mobilization of intracellular calcium (via IP_3), activation of calcium influx (through a channel), or direct coupling of the enzyme to the P_{2y} receptor. Direct coupling was attractive because this could explain why purinergic agonists stimulate eicosanoid production while other receptor-mediated agonists do not. To investigate the possibility that PLA_2 activation is not dependent on prior PLC stimulation, we used pertussis toxin (PTx) in an attempt to distinguish between these two pathways. Inositol phosphate accumulation was used as a measure of PLC stimulation, and TX production was used as a measure of PLA_2 activation. PTx inhibits both TX and IP formation in astrocytes, but the degree of sensitivity is markedly different. PTx inhibits TX production at concentrations (IC_{50} = 75pg/ml) which leave IP_3 formation intact.

This finding was supported by measurements of ATP-induced changes in intracellular calcium (Figure 3). ATP causes a large rise in cell calcium which is comprised of influx and mobilization from intracellular stores. These components are unaffected by PTx-treatment, suggesting that calcium fluxes are not required for AA mobilization and TX synthesis.

Our evidence suggests, therefore, that neither the generation of IP3 by ATP, nor calcium influx are required for PLA2 activation. In addition, we propose that the purinergic receptor in astrocytes is linked not only to PLC, but also to PLA2 and to a calcium channel. Whether this is one receptor coupled to the two lipases via different G proteins remains to be seen. Whichever, the function of the P2y receptor coupled to PLC is unresolved.

PPI hydrolysis and the activation of the synthase for a guanylyl cyclase activating factor

Evidence is accumulating to suggest that cells in the brain, apart from the microvasculature, release a nitrosyl compound which is a guanylyl cyclase-activating factor (GAF) derived from L-arginine (Garthwaite, 1991). The action of this GAF is blocked by hemoglobin and methylene blue (an inactivator of soluble guanylyl cyclase), potentiated by superoxide dismutase, and its production is inhibited by arginine analogs such as N-monomethyl arginine (NMA). As yet, there is no consensus as to the identity of this GAF, but it is either NO or a NO-containing compound such as a nitrosothiol.

Using a combination of NO-chemiluminescence detection and vascular ring bioassay, we have found that astrocytes grown on microcarrier beads and stimulated with various agonists release a vasorelaxant nitrosyl compound (Murphy et al., 1990a, 1991). This astrocyte-derived relaxing factor (ADRF) is inactivated by hemoglobin, production is blocked by NMA, inhibition by this arginine analog is reversed with L-arginine and, after sodium iodide reflux, there is detectable NO in the effluent from stimulated cells (Figure 4).

The agents which are effective in stimulating ADRF release are proven agonists at PLC-coupled receptors (α_1-adrenergic, B2, Qp). This implies that PPI hydrolysis and the mobilization of calcium and/or activation of PKC might be responsible for activating the ADRF-synthase in these cells. However, two pieces of evidence argue against this. First, astrocytes do not produce ADRF in response to carbachol, ATP, serotonin or histamine, all of which cause PPI hydrolysis. Secondly, exposure to thapsigargin and the subsequent rise in intracellular calcium (Figure 1) is not a sufficient stimulus for ADRF production (see Figure 4).

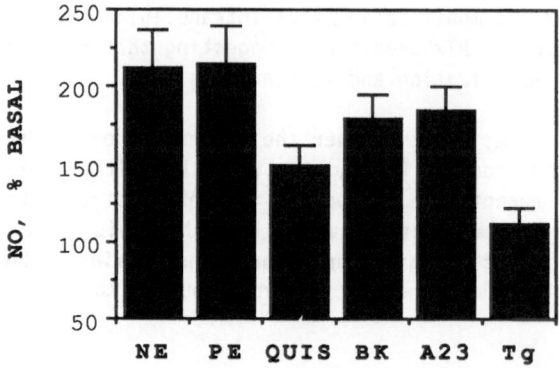

Figure 4. Chemiluminescence detection of NO in medium conditioned by astrocytes after stimulation with norepinephrine (NE, 100 μM), phenylephrine (PE, 100 μM), quisqualate (QUIS, 100 μM), bradykinin (BK, 1 μM), A23187 (A23, 5 μM) and thapsigargin (Tg, 0.1 μM) for 30 min at 37 °C. Values are means (% basal) ± sem of 5-10 experiments.

As an alternative to chemiluminescence, and to determine whether ADRF is a GAF, we have employed a sensitive target cell assay which exploits the abundant expression of soluble guanylyl cyclase in RFL-6 cells (a rat lung fibroblast line). These cells were first introduced (Forstermann et al., 1990) to detect GAF in subcellular fractions derived from brain and endothelial cells. RFL-6 cells do not themselves produce a GAF, unlike other potential target cells such as neurons, vascular endothelium and smooth muscle. Either conditioned medium from stimulated donor (astroglial) cells can be rapidly applied to the RFL-6 (target) cells, donor cells are plated directly onto target cells, or the two cell types are grown on separate substrates but within the same well. Superoxide dismutase is added to preserve ADRF, together with a phosphodiesterase inhibitor to conserve cGMP. Cells are scraped and cGMP is extracted and acetylated before radioimmune assay. Basal cGMP production in RFL-6 cells is low but soluble guanylyl cyclase activity, relected in their response to sodium nitroprusside, is very high. These cells do not themselves respond to the agents which evoke ADRF release, at least in terms of cGMP production.

Rapid transfer of medium conditioned by astrocytes exposed to NE (100 μM) for 60 seconds results in a two-fold increase in cGMP in RFL-6 cells. As an alternative, we have grown astrocytes and RFL-6 cells on separate

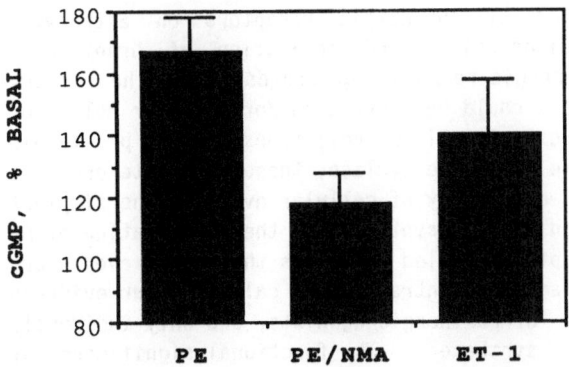

Figure 5. Elevation of cGMP in RFL-6 cells by phenylephrine (PE, 100 μM) and endothelin-1 (ET-1, 0.1 μM) stimulation of adjacent astrocytes. N-monomethyl arginine (NMA, 10 μM) was added prior to PE. Results represent the mean ± sem from three to seven experiments.

substrates but sharing the same medium. When phenylephrine (100 μM) is applied there is a similar two-fold increase in cGMP in the target cells. If NMA is added as a competitive inhibitor, the effects of PE on cGMP are inhibited by 75%. Using this assay for ADRF we can also detect small stimulatory effects of endothelin-1, which is known to activate PPI hydrolysis in astrocytes (Figure 5).

A number of recent reports claim separation and purification from brain of a soluble GAF synthase (Bredt and Snyder, 1990; Schmidt et al., 1991). This synthase (EC 1.14.23.-) converts L-arginine to citrulline with a concomitant synthesis of NO. The synthesis of NO can be blocked with NMA, indicating that the enzyme is similar to that in endothelial cells and in macrophages. GAF synthase requires calmodulin, is inactive at resting free calcium, fully active at 400 nM calcium, and NADPH enhances NO formation. The enzyme in cerebellum is a homodimer of 155 kDa subunits. The subcellular compartmentation of ADRF synthase, and its regulation by agonists and co-factors, are now being actively pursued in our laboratory.

Astroglial cells have intimate anatomical associations with the pial surface, with the microvasculature and with adjacent neurons. Clearly, from both in vivo and in vitro evidence, astrocytes display an array of receptors for signal molecules which could originate from neuronal or vascular cells. While, as yet, we know little of the functions of these receptors in vivo, activation of purinergic receptors on astrocytes promotes the release of eicosanoids, and activation of bradykinin, quisqualate, ET-1 and α1-adrenergic receptors generates ADRF. The targets for these astrocyte products could be neurons and/or vascular cells, or they may function as autacoids. As all CNS cell types express prostanoid receptors and contain soluble guanylate cyclase, these astrocyte products could initiate or modulate a variety of cellular events. The simplest explanation for the activation of ADRF synthase, or the mobilization of AA by agonists acting at receptors coupled to PLC is that these events are mediated by the resulting changes in intracellular calcium. Our evidence suggests that this is not so for purinergic agonists, and only indirectly supports the case for ADRF synthase. The functional significance of receptors which hydrolyze PPI in astroglial cells remains enigmatic.

Acknowledgments

We wish to thank Greg Welk, Sherry Kardos and Erik Miller for their contribution to these studies, and Michael Hart and Steven Moore (Division of Neuropathology) for supplying discarded human biopsy material and help with the HPLC analysis. This work is supported by NIH grants NS24621, NS29226, RRO5372, and ACS grant IN122K.

REFERENCES

Bredt DS, Snyder SH (1990) Isolation of nitric oxide synthetase, a calmodulin-requiring enzyme. Proc Natl Acad Sci USA 87:682-685
Bruner G, Murphy S (1990a) ATP-evoked arachidonic acid mobilization in astrocytes is via a P2Y-purinergic receptor. J Neurochem 55:1569-1575
Bruner G, Murphy S (1990b) Regulation of phospholipase D in astroglial cells by calcium-activated protein kinase C. Mol Cell Neurosci 1:146-150
Ehrenreich H, Kehrl JH, Anderson RW, Rieckmann P, Vitkovic L, Coligan JE, fauci AS (1991) A vasoactive peptide, endothelin-3, is produced by and specifically binds to primary astrocytes. Brain Res 538:54-58
Forstermann U, Gorsky LD, Pollock JS, Ishii K, Schmidt HHHW, Heller M, murad F (1990) Hormone-induced biosynthesis of endothelium-derived relaxing factor/nitric oxide-like material in N1E-115 neuroblastoma cells requires calcium and calmodulin. Mol Pharmacol 38:7-13
Garthwaite J (1991) Glutamate, nitric oxide and cell-cell signalling in the nervous system. Trends Neurosci 14:60-67

MacCumber MW, Ross CA, Snyder SH (1990) Endothelin in brain: Receptors, mitogenesis, and biosynthesis in glial cells. Proc Natl Acad Sci USA 87:2359-2363

Murphy S, Pearce B (1987) Functional receptors for neurotransmitters on astroglial cells. Neuroscience 22:381-394

Murphy S, Pearce B, Jeremy J, Dandona P (1988) Astrocytes as eicosanoid producing cells. Glia 1:241-245

Murphy S, Welk G (1989) Arachidonic acid evokes inositol phospholipid hydrolysis in astrocytes. FEBS Lett 257:68-70

Murphy S, Minor RL, Welk G, Harrison DG (1990a) Evidence for an astrocyte-derived vasorelaxing factor with properties similar to nitric oxide. J Neurochem 55:349-351

Murphy S, Welk G, Thwin SS (1990b) Stimulation of thromboxane release from primary cell cultures derived from human astrocytic glioma biopsies. Glia 3:241-245

Murphy S, Minor RL, Welk G, Harrison DG (1991) CNS astroglial cells release nitrogen oxides with vasorelaxant properties. J Cardiovasc Pharmacol 17 (Suppl. 3):S265-S268

Pearce B, Murphy S (1988) Neurotransmitter receptors coupled to inositol phospholipid turnover and calcium flux: consequences for astrocyte function. In: Kimelberg H (ed) Glial Cell Receptors, Raven Press, New York, p 197-221

Pearce B, Morrow C, Murphy S (1988) Characteristics of phorbol ester- and agonist-induced downregulation of astrocyte receptors coupled to inositol phospholipid metabolism. J Neurochem 50:936-944

Pearce B, Jeremy J, Morrow C, Murphy S, Dandona P (1987) Inositol phospholipids are probably not the source of arachidonic acid for eicosanoid synthesis in astrocytes. FEBS Lett 211:73-77

Pearce B, Murphy S, Morrow C, Jeremy J, Dandona P (1989) ATP-evoked calcium mobilisation and prostanoid release from astrocytes: P2-purinergic receptors linked to phosphoinositide hydrolysis. J Neurochem 52:971-977

Salm AK, McCarthy KD (1989) Expression of β-adrenergic receptors by astrocytes ioslated from adult rat cortex. Glia 2:346-352

Schmidt HHHW, Pollock JS, Nakane M, Gorsky LD, Forstermann U, Murad F (1991) Purification of a soluble isoform of guanylyl cyclase-activating-factor synthase. Proc Natl Acad Sci USA 88:365-369

Shinoda H, Marini AM, Cosi C, Schwartz JP (1989) Brain region and gene specificity for neuropeptide gene expression in cultured astrocytes. Science 245:415-417

Wilkin GR, Cholewinski A (1988) Peptide receptors on astrocytes. In: Kimelberg H (ed) Glial Cell Receptors, Raven Press, New York, p 223-2422

Wilkin GP, Marriott DR, Cholewinski AJ (1990) Astrocyte heterogeneity. Trends Neurosci 13:43-46

THE G-PROTEINS REGULATING PHOSPHOINOSITIDE BREAKDOWN

J. H. Exton, S. J. Taylor and J. L. Blank
Howard Hughes Medical Institute
Department of Molecular Physiology and Biophysics
Vanderbilt University School of Medicine
831 Light Hall
Nashville, TN 37232 USA

Introduction

The stimulation of phosphatidylinositol 4,5-bisphosphate (PIP_2) hydrolysis is a widespread cellular response to many hormones, growth factors and neurotransmitters (Berridge 1987). It is catalyzed by a phospholipase C (PLC) and yields two signaling molecules: inositol 1,4,5-trisphosphate (IP_3) which releases Ca^{2+} from stores in the endoplasmic reticulum, and 1,2-diacylglycerol (DAG) which activates protein kinase C. The growth factors activate the PLC through the tyrosine kinase activity of their receptors (Kriz et al 1990), whereas the other agonists act through guanine nucleotide-binding regulating proteins (G-proteins). Despite the recognition several years ago that G-proteins were involved in the regulation of PLC, some of these G-proteins have only recently been identified (Taylor et al 1990, Smrcka et al 1991, Blank et al 1991). These G-proteins are insensitive to pertussis toxin, but it is clear that toxin-insensitive G-proteins are also involved in the regulation of PLC in some tissues (Exton 1988).

Purification of G-proteins that regulate PIP_2 phospholipase C

The first G-protein that was shown to activate PLC was purified from bovine liver plasma membranes treated with guanosine 5'-(3-0-thio) triphosphate (GTPγS) and other GTP analogues (Taylor et al 1990). This treatment yielded a protein that maintained its ability to activate partially purified liver PLC through a series of chromatographic steps (heparin-Sepharose, Q-Sepharose, Sephacryl S-300, octyl-Sepharose and Mono Q). The heparin-Sepharose step was important since it separated the G-protein from the PLC. At the Mono Q step, the PLC activator protein was still contaminated with G_i α-subunits, but these were eliminated after ADP-ribosylation with pertussis toxin and rechromatography on Mono Q (Taylor et al 1990). As shown in Fig. 1, a typical final preparation contained 42 and 43 kDa proteins that were recognized by an antipeptide antiserum (8645)

NATO ASI Series, Vol. H 70
Phospholipids and Signal Transmission
Edited by R. Massarelli, L. A. Horrocks,
J. N. Kanfer, and K. Löffelholz
© Springer-Verlag Berlin Heidelberg 1993

raised to a sequence common to all G-protein α-subunits, but not by an antipeptide antiserum (588) raised to a sequence found in the α-subunits of G_s, G_i, G_o and G_t (Taylor et al 1990, 1991). A 35kDa protein was also present that was recognized by an antipeptide antiserum raised to a sequence in the β-subunit of G_t.

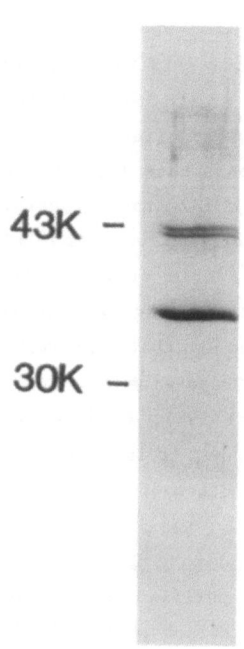

Figure 1 Silver-stained gel of G protein activator of PIP_2 phospholipase.

In a separate study, the G-proteins that activate PLC were also purified from bovine liver membranes (Blank et al 1991). The activity of the G-proteins was assayed by reconstituting them with partially purified PLC from bovine liver or with highly purified PLC from bovine brain, in the presence and absence of 100 μM GTPγS. The chromatographic steps were similar to those used for the activated α-subunit, except that the sequence was heparin-Sepharose, DEAE-Sephacel, octyl-Sepharose, hydroxylapatite, Mono Q and Sephacryl S-300. The final preparation was resolved from G_i and G_s and also consisted of 42 and 43 kDa proteins that were recognized by the "α-common" antiserum (8645). These proteins were present in approximately equal amounts. A 35kDa protein recognized by β-subunit antibodies was also present at about the level of the combined α-subunits. Thus the two approaches yielded the same two G-proteins either in the form of holomers or free α-subunits.

Characterization of G-proteins that regulate PIP_2 phospholipase C

As expected from studies of agonist regulation of PLC in hepatocytes (Lynch et al 1986), the G-proteins isolated from liver were not subject to ADP ribosylation by pertussis toxin (Blank et al 1991). When present in liver plasma membranes or purified in the heterotrimeric form, they were rapidly activated by poorly hydrolyzable analogues of GTP, but not other nucleotides. However, the concentration of GTPγS required for half-maximal activation was very high (4 μM). This implies that the affinity of the G-proteins for GTP and its analogues is much lower than observed with other members of this type of protein (Bourne et al 1991). Reconstitution of the G-proteins with bacterially expressed M1 muscarinic cholinergic receptors and carbachol causes a dramatic increase in GTPγS binding and GTPase activity, but this is not seen with the M2 muscarinic receptor (J. Blank, J. Exton and E. Ross, unpublished observations). Thus the G-proteins couple selectively to a receptor linked to phosphoinositide hormone, but not to one linked to adenylate cyclase.

The binding of GTP analogues to the G-protein required Mg^{2+} and was inhibited by high concentrations of a GDP analogue. The ability of the GTPγS activated G-protein to stimulate PLC was inhibited by excess $\beta\gamma$ subunits prepared from liver plasma membranes (Blank et al 1991). NaF activated the G-proteins, and the activation was enhanced by $AlCl_3$. All these properties are characteristic of heterotrimeric G-proteins in which $\beta\gamma$ subunits exert an inhibitory effect on α-subunits and AlF_4 induces activation of GDP-liganded α-subunits by substituting for the γ-phosphate of GTP.

Identification of G-proteins that regulate PIP$_2$ phospholipase C

The immunological studies described above indicated that the G-proteins were not members of a known class of these proteins. However, a new class of G-protein, termed G_q, has recently been identified by molecular cloning (Strathmann & Simon 1990). A series of cDNAs encoding novel α-subunits lacking the site for ADP-ribosylation by pertussis toxin have been identified in a mouse brain cDNA library by polymerase chain reaction. Members include α_q, α_{11} and α_{14}, and exhibit less than 50% amino acid sequence identity with other α-subunits (Strathmann & Simon 1990). Independently, Pang and Sternweis (1990) have identified unique α-subunits in mammalian brain using affinity chromatography on $\beta\gamma$ chromatography. These have weights of approximately 42kDa and are not subject to ADP-ribosylation by pertussis toxin. They co-chromatograph with the 42kDa α-subunit from bovine liver. Tryptic peptides derived from these subunits show identity to deduced sequences of α_q and α_{11} (Pang & Sternweis 1990). Furthermore, these workers have raised antipeptide antisera (WO82, WO83) to unique sequences in α_q. Both of these antisera recognize the 42kDa α-subunit purified from bovine liver, but only WO83 recognizes the 43kDa α-subunit (Taylor et al 1991). It is therefore concluded that the 42kDa protein is similar or identical to α_q, whereas the 43kDa is another member of the G_q class. Sequence data should identify the exact nature of both proteins.

Interaction of G-proteins with PIP$_2$ phospholipase C

The studies of Taylor et al (1990) identified the 42kDa α-subunit as a regulator of PLC. Subsequently, Smrcka et al (1991) found that α_q purified from bovine brain (Pang & Sternweis 1990) activated PLC in the presence of AlF$^-_4$, but not GTP analogues. They attributed the lack of response to GTP analogues to the low affinity of α_q for these nucleotides. In both the studies of Taylor et al (1990) and Smrcka et al (1991), the PLCs with which the α-subunits were reconstituted were partially purified from bovine liver or brain. More recently the isozyme specificity of the PLC activated by the 42kDa α-subunit from liver has been examined using the β_1, γ_1 and δ_1 isozymes purified from bovine brain (Taylor et al 1991). The results (Fig. 2) show unequivocally that only the β_1 isozyme is activated. A similar conclusion has been reached using partially purified PLCβ_1 from bovine brain and liver (Smrcka et al 1991, K. Shaw & J. H. Exton, unpublished observations).

Further support for the PLC isozyme specificity of the G-protein activation is provided by the results of adding isozyme-specific monoclonal and polyclonal antibodies to liver plasma membranes incubated with the activated 42kDa α-subunit or GTPγS (Taylor et al 1991). Only antibodies to the β_1 isozyme inhibited the activation of the androgenous PLC. These results confirm the report by Carter et al (1990) that antibodies raised against a PLC-β isozyme inhibited GTPγS stimulation of PLC in rabbit brain membranes, although other antibodies were not tested in that study.

The 42 and 43 kDa G_q-related α-subunits isolated from bovine liver have been partially resolved chromatographically and it has been demonstrated that both activate PLCβ_1 with approximately equal effectiveness (S. J. Taylor and J. H. Exton, unpublished observations). Interestingly, the PLC isozyme that is the target of the growth factor receptor tyrosine kinase is the γ_1 isozyme, and neither the β_1 or δ_1 isozyme is phosphorylated (Kriz et al 1990).

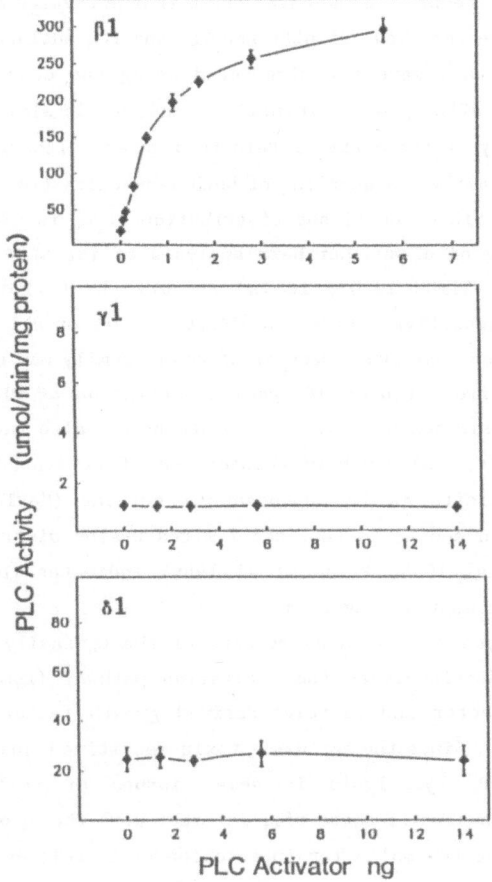

Figure 2. Isozyme specificity of PIP$_2$ phospholipase (PLC) activated by the 42kDa α-subunit from liver (PLC activator).

Discussion

There is now much evidence that the pertussis toxin-insensitive G-proteins that activate PLC are members of the G$_q$ family and have the properties of heterotrimeric G-proteins. In the basal state (not activated by agonist-occupied receptors) they have very low affinity for GTP and its analogues. This accounts for the fact that they were not discovered using the conventional $[^{35}\gamma]$GTPγS binding assay. The 42kDa protein identified in liver is similar or identical to α_q, but the identity of the 43kDa protein is unknown, although it appears to be a member of the G_q family. Sequencing of both α-subunits from liver is currently being undertaken. Since the tissue distribution of G_q is widespread, as shown by Western blotting of α_q and Northern analysis of its mRNA (Pang & Sternweis 1990, Strathmann & Simon 1990), it is probably the G-protein that mediates pertussis toxin-insensitive activation of PLC in most or all tissues. However, it must be recognized that other members of the G_q family may also be ubiquitous.

It is clear from studies of agonist activation of PLC in a variety of tissues that pertussis toxin-sensitive G-proteins are also involved. The nature of these is obscure. Although microinjection of activated α_o subunits into Xenopus oocytes elicits an IP$_3$-dependent Cl current (Moriarty et al 1990), implying that G$_o$ can couple to PLC, the limited tissue distribution of this G-protein (Mumby et al 1988, Price et al 1989) indicates that another toxin-sensitive G-protein must be involved.

The PLC isozyme activated by members of the G_q family is clearly the β_1 isozyme and this distinguishes the activation pathway from that utilized by epidermal growth factor and platelet-derived growth factor, in which the γ_1 isozyme is involved. Since the pertussis toxin-sensitive G-proteins that mediate agonist-induced PIP$_2$ hydrolysis in some tissues or cells have not been identified, it is unknown if these will interact with the β_1 or another isozyme.

The details of the molecular interaction of G_q-related proteins with the relevant agonist receptors and with PLCβ_1 are unknown, although there is no evidence indicating that they differ mechanistically from those for signal transduction involving G_s. The widespread involvement of G-proteins, apparently of the G_q class, in many responses to hormones and neurotransmitters raises the possibility that mutations and modifications of these proteins could conceivably underlie some disease processes, including oncogenesis.

113

References

Berridge MJ (1987) Inositol trisphosphate and diacylglycerol: two interacting second messengers. Ann Rev Biochem 56:159-193

Blank JL, Ross AH, Exton JH (1991) Purification and characterization of two G-proteins which activate the β_1 isozyme of phosphoinositide-specific phospholipase C. Identifiction as members of the G_q class. J Biol Chem In press

Bourne HR, Sanders DA, McCormick F (1991) The GTPase superfamily: conserved structure and molecular mechanism. Nature 349:117-127

Kriz R, Lin L-L, Sultzman L, Ellis C, Heldin C-H, Pawson T, Knopf J (1990) Phospholipase C isozymes: structural and functional similarities. In: Protooncogenes in cell development. Wiley, Chichester (Ciba Foundation Symposium 150) p 112-127

Lynch CJ, Prpic V, Blackmore PF, Exton JH (1986) Effect of islet-activating pertussis toxin on the binding characteristics of Ca^{2+}-mobilizing hormones and on agonist activation of phosphorylase in hepatocytes. Mol Pharmacol 29:196-203

Mumby S, Pang I-H, Gilman AG, Sternweis PC (1988) Chromatographic resolution and immunological identification of the α_{40} and α_{41} subuits of guanine nucleotide-binding regulatory proteins from bovine brain. J Biol Chem 263:2020-2026

Pang I-H, Sternweis PC (1990) Purification of unique α-subunits of GTP-binding regulatory proteins (G proteins) by affinity chromatography with immobilized $\beta\gamma$ subunits. J Biol Chem 265:18707-18712

Price SR, Tsai S-C, Adamik R, Angus CW, Serventi IM, Tsuchiya M, Moss J, Vaughan M (1989) Expression of $G_{o\alpha}$ mRNA and protein in bovine tissues. Biochemistry 28:3803-3807

Smrcka AV, Hepler JR, Brown KO, Sternweis PC (1991) Regulation of a polyphosphoinositide-specific phospholipase C activity by purified G_q. Science 251:804-807

Strathmann M, Simon MI (1990) G protein diversity: a distinct class of α-subunits is present in vertebrates and invertebrates. Proc Natl Acad Sci USA 87:9113-9117

Taylor SJ, Chae HZ, Rhee SG, Exton JH (1991) Activation of the β1 isozyme of phospholipase C by purified α subunits of the G_q class of G proteins. Nature In press

Taylor SJ, Smith JA, Exton JH (1990) Purification from bovine liver membranes of a guanine nucleotide-dependent activator of phosphoinositide specific phospholiase C. Immunologic identification as a novel G-protein α-subunit. J Biol Chem 265:17150-17156

ACTIONS AND INTERACTIONS OF CHOLINERGIC AND EXCITATORY AMINOACID RECEPTORS ON PHOSPHOINOSITIDE SIGNALS, EXCITOTOXICITY AND NEUROPLASTICITY

Pawels Kurian, Fulton T. Crews, L. Judson Chandler and Norbert J. Pontzer
Department of Pharmacology and Therapeutics, University of Florida College of Medicine, Gainesville, FL 32610 - 0267, USA.

The excitatory neurotransmitters acetylcholine and glutamate are involved in neuronal plasticity, which is thought to be an essential component of learning and memory. The loss of cognitive ability in Alzheimer's disease and in age-associated memory impairment has been suggested to be secondary to a loss of central nervous system cholinergic transmission. Drugs which specifically disrupt cholinergic transmission have profound effects on learning and memory. A loss of cholinergic neurons clearly occurs early in the course of Alzheimer's disease when memory loss is the only prominent symptom. In studies using experimental models of Alzheimer's disease, lesioning cholinergic neurons also disrupts the ability of animals to learn. Glutamate has also been implicated in memory processes. Drugs that block glutamate receptors, particularly the N-methyl-D-aspartate (NMDA) receptor subtype, can produce cognitive deficits. An *in vitro* model of synaptic plasticity, long-term potentiation (LTP), is thought to be mediated in part through NMDA receptors. These studies suggest that both cholinergic and glutamatergic signals play an important role in memory processes and cognitive function.

We have investigated receptor transduction systems activated by acetylcholine and glutamate in order to determine which ones may be relevant to synaptic plasticity associated with learning and memory. A variety of receptors have been shown to stimulate the hydrolysis of phosphoinositides (PI) to produce inositol polyphosphate and diacylglycerol second messengers. Early studies of phosphoinositide hydrolysis in brain slices in the presence of lithium showed that norepinephrine, carbachol, serotonin, glutamate, calcium ionophores and potassium-induced depolarization could all stimulate PI hydrolysis (Fig. 1). The ability of calcium ionophores and depolarization to stimulate PI hydrolysis in brain slices led to the suggestion that calcium influx can activate a phospholipase responsible for PI hydrolysis. *In vitro* experiments with membranes prepared from rat cerebral cortex showed that GTP as well

NATO ASI Series, Vol. H 70
Phospholipids and Signal Transmission
Edited by R. Massarelli, L. A. Horrocks,
J. N. Kanfer, and K. Löffelholz
© Springer-Verlag Berlin Heidelberg 1993

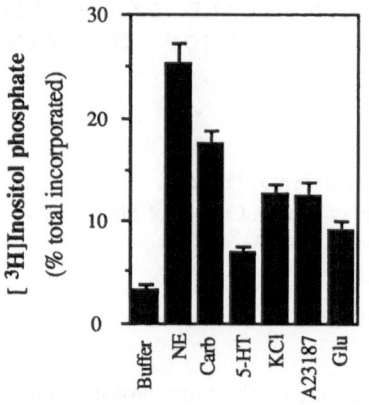

Fig. 1. *Stimulation of PI hydrolysis in the presence of Li+.* [3H]Inositol was incorporated into cerebral cortical slices as described by Gonzales and Crews (1984). Slices were stimulated in the presence of 10 mM Li+ for 60 min and the accumulation of [3H]inositol phosphates expressed as the percentage of total incorporated [3H]inositol. Shown is the mean plus the S.E.M. from 3-6 experiments each done in triplicate.

as calcium can stimulate phosphoinositide hydrolysis (Gonzales and Crews, 1988; Gonzales and Crews, 1985). Furthermore, nanomolar changes in intracellular calcium levels cause parallel changes in PI hydrolysis (Chandler and Crews, 1990a; Chandler, 1990b). These studies and others indicate that both receptor and depolarization induced calcium influx leads to phosphoinositide hydrolysis. Taken together, these findings suggest that Phosphoinositide specific phospholipase(s) coupled to PI hydrolysis may be activated by either receptor activation of by a GTP binding protein or stimulation of calcium flux.

Since lithium is known to disrupt the phosphoinositide cycle by inhibiting inositol phosphatases, we developed a method to measure the individual inositol polyphosphates in the absence of lithium. In these studies, carbachol (1 mM) stimulated an almost two-fold increase in the formation of [3H]InsP$_3$ and about a 15 fold increase in the formation of [3H]Ins(1,3,4,5)P$_4$ (Fig. 2). HPLC separation of [3H]InsP$_3$ formed during stimulation indicated that two isomers were present (Ins(1,4,5)P$_3$ the active isomer for release of intracellular calcium, and Ins(1,3,4)P$_3$ an inactive Ins(1,3,4,5)P$_4$ metabolite). Approximately equal amounts of the two [3H]InsP$_3$ isomers were found after 5 minutes of carbachol stimulation. This indicates that a significant fraction of the Ins(1,4,5)P$_3$ isomer is phosphorylated by the 3-kinase to Ins(1,3,4,5)P$_4$, which is subsequently dephosphorylated by a 5-phosphatase to Ins(1,3,4)P$_3$. Comparisons of the effects of various agonists on [3H]Ins(1,3,4,5)P$_4$ and [3H]InsP$_3$ formation in the absence of lithium, indicate that only carbachol, a cholinergic agonist, and quisqualate, a glutamatergic agonist, significantly increase [3H]InsP$_3$ and [3H]Ins(1,3,4,5)P$_4$ levels (Fig. 2). The lack of response of the other agonists may be due to different mechanisms of coupling to phosphoinositide hydrolysis as has been suggested previously for norepinephrine (Crews *et al.,* 1989; Crews *et al.,* 1988; Pontzer and Crews, 1990).

Carbachol stimulated phosphoinositide hydrolysis is known to occur primarily through muscarinic receptors of the M$_1$ subtype (Gonzales and Crews, 1984). The small efficacy of

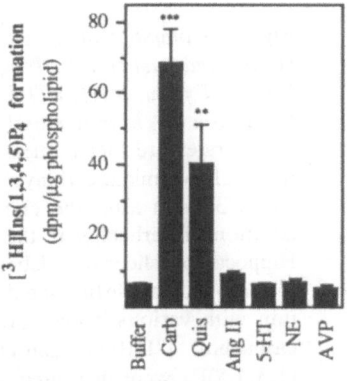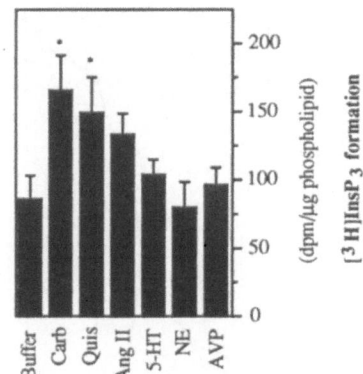

Fig. 2. *Stimulation of Ins(1,3,4,5)P₄ and InsP₃ formation by various neurotransmitters in the absence of lithium.* Cerebral cortical slices were labeled with [³H]inositol for 1 hour followed by stimulation with the various agonists for 5 min. The individual inositol phosphates were separated on Dowex columns as described by Chandler *et al.* (1991).

oxotremorine for stimulation of [³H]Ins(1,3,4,5)P₄ formation suggests that carbachol stimulated Ins(1,3,4,5)P₄ formation is also through M_1 receptors (Fig. 3). The differences between the pronounced increases in phosphoinositide hydrolysis by norepinephrine and serotonin in the presence of lithium versus their lack of efficacy in the absence of lithium are not clear. The effect of lithium may be to amplify weak signals which are not large enough to measure in the absence of lithium.

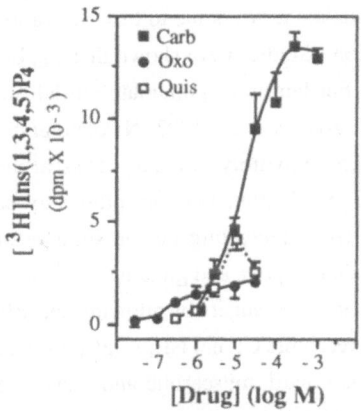

Fig. 3. *Stimulation of [³H]Ins-(1,3,4,5)P₄ formation by various concentrations of carbachol, oxotremorine and quisqualate.* Hippocampal slices were incubated with [³H]inositol for 60 min and then stimulated with various concentrations of the agonist for 30 min.

Although the stimulation of Ins(1,3,4,5)P₄ formation produces a large signal with carbachol and glutamate, receptors which are not coupled to Ins(1,3,4,5)P₄ formation may not appear to produce significant stimulation in the absence of lithium (Kurian *et al.*, 1992). We have previously proposed that at least a portion of the NE signal is mediated by calcium influx (Chandler and Crews, 1989). Thus, these finding suggest that receptors are differentially coupled to Phosphoinositide hydrolysis by differences in efficacy and mechanism of coupling.

To further investigate differences in coupling of receptors to phosphoinositide hydrolysis, membrane preparations of rat cerebral cortex were incubated

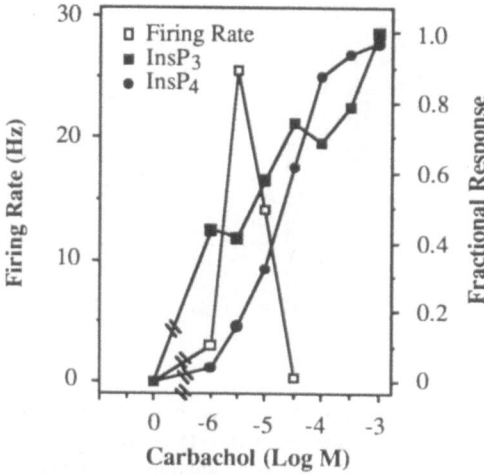

Fig. 4. *Comparison of carbachol stimulated neuronal firing with [³H]InsP₃ and [³H]Ins(1,3,4,5)P₄ formation in hippocampal slices.* Firing rates are means measured in the CA1 pyramidal cell layer of four slices 30 min after the cumulative addition of carbachol in the buffer. Hippocampal slices were labeled with [³H]inositol and stimulated for 30 min with various concentrations of carbachol. [³H]InsP₃ and [³H]Ins-(1,3,4,5)P₄ were measured after 30 min of carbachol stimulation.

with [³H]phosphoinositides in the presence of various compounds. Consistent with previous studies, GTPγS and calcium were observed to stimulate PtdIns(4,5)P₂ hydrolysis. However, in the presence of GTPγS, only carbachol further increased PtdIns(4,5)P₂ breakdown. Glutamate, quisqualate and norepinephrine were without effect. These findings suggest that muscarinic receptors activated by carbachol are coupled to phospholipase C activation through a guanine nucleotide binding protein (G-protein). Since both G_o and G_q have been reported to activate PLC (Smrcka *et al.*, 1991), glutamate and quisqualate may couple to PLC activation through G-proteins not active in our membrane preparation. Studies have shown that G_o, but not G_q, is sensitive to inhibition by pertussis toxin. Furthermore, glutamate stimulated PtdIns(4,5)P₂ hydrolysis is sensitive to pertussis toxin (Sugiyama *et al.*, 1987; Nicoletti *et al.*, 1988) whereas muscarinic receptor stimulated PtdIns(4,5)P₂ hydrolysis is not. Muscarinic receptors may thus be linked to PI hydrolysis through G_q while glutamate receptors may be linked to PI hydrolysis through both G_o and calcium flux, depending on the subtype of glutamate receptor. The differential coupling of muscarinic receptors and glutamate receptors may have functional implications for the formation of second messengers by allowing specific interactions between receptors. Muscarinic and glutamate receptors but not NE, AngII, 5-HT or AVP stimulate the formulation of Ins(1,3,4,5)P₄. In this regard, muscarinic and glutamate receptors appear to be particularly important for memory and learning.

A comparison of carbachol stimulated InsP₃ and Ins(1,3,4,5)P₄ formation with neuronal firing rates in hippocampal slices suggests an important function for Ins(1,3,4,5)P₄ in the regulation of electrophysiological responses (Fig. 4). After exposure to low concentrations of carbachol we observe the well documented increase in firing rate. However, after exposure to higher

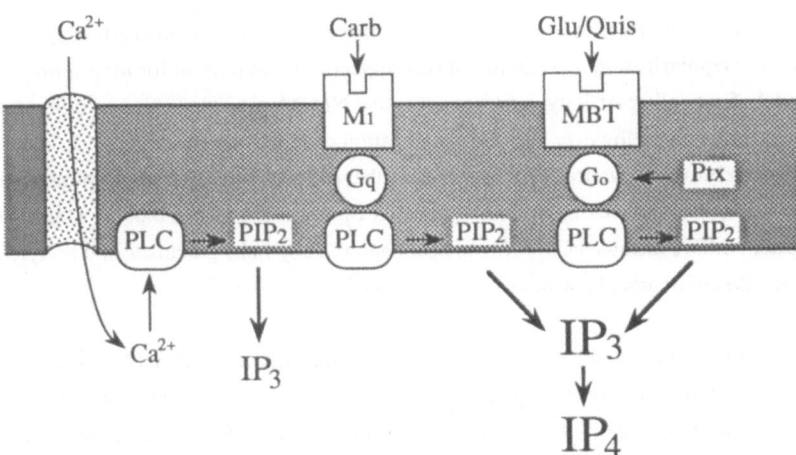

Fig. 5. *Schematic model depicting separate pathways for activating phospholipase C (PLC) coupled to PI hydrolysis.* It is hypothesized that the metabotropic (MBT) glutamate receptor subtype couples to PLC activation through a pertussis toxin (Ptx) sensitive guanine nucleotide binding protein (G_o), whereas the M_1 subtype of muscarinic receptors couple to PLC activation through a pertussis toxin insensitive guanine nucleotide binding protein (G_q). Both of these receptors stimulate the formation of Ins(1,4,5)P3 (IP3) and Ins(1,3,4,5)P4 (IP4). A third pathway for activating PLC is through stimulation on calcium influx which can then directly activate PLC. This pathway appears to only form Ins(1,4,5)P3.

concentrations of carbachol, cell firing actually decreases. This abatement of firing only occurs at concentrations of carbachol that also increase Ins(1,3,4,5)P_4 formation. Pharmacological studies suggest that abatement of firing is linked to activation of M_1 muscarinic and the subsequent hydrolysis of PtdIns(4,5)P_2 (Pontzer and Crews, 1990). In preliminary intracellular studies, abatement of firing was associated with a large sustained membrane depolarization. This membrane depolarization may be sufficient to change other voltage dependent processes, including removing the Mg^{2+} block from NMDA receptors.

The interaction of cholinergic and glutamatergic receptors may occur through the formation of Ins(1,3,4,5)P_4, which appears to be particularly increased by these transmitters. Ins(1,3,4,5)P_4 has been shown to release intracellular Ca^{2+}, to initiate refilling of intracellular Ca^{2+} stores, and to open a cation channel in neuronal membranes. Muscarinic stimulated Ins(1,3,4,5)P_4 formation may produce a large sustained depolarization through these or other mechanisms. The implications of a large sustained depolarization caused by muscarinic receptors are many fold. One receptor thought to be involved in neuronal excitation leading to plasticity as well as excitotoxicity is the NMDA receptor.

Activation of the NMDA receptor opens a non-specific cation channel allowing a large calcium influx. Depolarization and calcium influx appear to be important for long-term potentiation, a model of synaptic plasticity, whereas excessive NMDA mediated calcium influx may also lead to neurotoxicity. However, NMDA gated cation channels are blocked by Mg^{2+} at normal neuronal levels of polarization. We hypothesize that a muscarinic cholinergic stimulated increase in $Ins(1,3,4,5)P_4$ induces a depolarization that removes the Mg^{2+} block. This allows a much greater NMDA mediated Ca^{2+} flux to occur and thereby links glutamate and acetylcholine in the production of neuronal plasticity and excitotoxicity.

In summary, $PtdIns(4,5)P_2$ hydrolysis occurs through a number of mechanisms. Certain receptors appear to predominantly act through G-proteins whereas other receptors may predominantly stimulate hydrolysis by mobilizing calcium. Although many receptors increase levels of $Ins(1,4,5)P_3$, glutamate and cholinergic muscarinic receptors stimulate $Ins(1,3,4,5)P_4$ formation in large quantities. $Ins(1,3,4,5)P_4$ has specific binding sites in brain and appears to act as a second messenger mediating different signals than those mediated by $Ins(1,4,5)P_3$. It is hypothesized that muscarinic receptors modulate NMDA receptor by causing a depolarization that is sufficient to remove the Mg^{2+} block and thereby allow neuronal plasticity that is essential for memory and learning.

ACKNOWLEDGMENTS: This research was supported by grants from the National Institute of Health, Grants # AA 06069 and AG 06660.

REFERENCES

Chandler LJ, Crews FT (1989) Calcium stimulated versus G-protein mediated phosphoinositide hydrolysis in brain. In: Freysz L, Hawthorne JN, Toffano G (eds) Neurochemical Aspects of Phospholipid Metabolism. Liviana Press, Padova, Italy, p 135-140

Chandler LJ, Crews FT (1990) Calcium versus G-protein mediated phosphoinositide hydrolysis in rat cerebral cortical synaptoneurosomes. J Neurochem 55:1022-1030

Chandler LJ, Crews FT (1990) Calcium versus G-protein activated phosphoinositide hydrolysis in synaptoneurosomes from young and old rats. In: Cotman C, Khachaturian Z, Pettegrew J (eds) Annals NY Acad Sci Calcium, Membranes, Aging, and Alzheimer's Disease. 568:187-192

Chandler LJ, Kurian P, Crews FT (1991) Effects of ethanol on inositol 1,3,4,5-tetrakisphosphate metabolism by rat brain homogenates. Alcoholism: Clin & Exp Res 15:136–140

Crews FT, Gonzales RA, Raulli R, McElhaney R, Pontzer N, Raizada MK (1988) Interaction of calcium with receptor stimulated phosphoinositide hydrolysis in brain and liver. Ann New York Acad Sci 522:88–95

Gonzales RA, Crews FT (1984) Characterization of the cholinergic stimulation of phosphoinositide hydrolysis in rat brain slices. J Neurosci 4:3120–3127

· Gonzales RA, Feldstein JB, Crews FT, Raizada MK (1985) Receptor mediated inositide hydrolysis is a neuronal response: Comparison of primary neuronal and glial cultures. Brain Res 345:350–355

Gonzales RA, Crews FT (1988) Guanine nucleotide and calcium stimulated inositol phospholipid hydrolysis in brain membranes. J Neurochem 50:1522–1528

Kurian P, Narang N, Crews FT (1992) Decreased carbachol stimulated inositol 1,3,4,5-tetrakisphosphate formation in senescent rat cerebral cortical slices. Neurobiol Aging (in press)

Nicolletti F, Wroblewski JT, Fadda E, Costa E (1988) Pertussis toxin inhibits signal transduction at a specific metabolotropic glutamate receptor in primary cultures of cerebellar granule cells. Neuropharmacol 27:551–556

Pontzer NJ, Crews FT (1990) Desensitization of muscarinic stimulated hippocampal cell firing is related to PI hydrolysis and inhibited by lithium. J Pharmacol Exp Therap 253:921–929

Raulli R, Danysz W, Wroblewski JT (1991) Pretreatment of cerebellar granule cells with concanavalin A potentiates quisqualate-stimulated phosphoinositide hydrolysis. J Neurochem 56:2116–2124

Samrcka AV, Hepler JR, Brown KO, Sternweis PC (1991) Regulation of polyphosphoinositide-specific phospholipase C activity by purified G_q. Science 251:804–807

Sugiyama H, Ito I, Hirono C (1987) A new type of glutamate receptor linked to inositol phospholipid metabolism. Nature 325:531–533

THE PHOSPHOLIPID ENVIRONMENT OF ACTIVATED SYNAPTIC MEMBRANE RECEPTORS MAY PROVIDE BOTH INTRACELLULARLY AND RETROGRADELY ACTING SIGNALS FOR THE REGULATION OF NEURO(MUSCULAR) TRANSMISSION.

E. Heilbronn and L. Järlebark
Unit of Neurochemistry and Neurotoxicology
Stockholm University
S-106 91 Stockholm
Sweden

Summary: The activation of two receptors of skeletal muscle and myotube in culture, the nicotinic acetylcholine receptor (nAChR) and the ATP-activated P_2-purinergic receptor (P_2R) resulted, in both cases, in increased intracellular levels of diacylglycerol (DAG). In the case of the receptor-ion channel macromolecule the intracellular DAG increases were seen after activation of nAChR by a cholinergic ligand and blocked by the nAChR inhibitors α-bungarotoxin or d-tubocurarine; they were dependent on the presence of external Ca^{2+}, which points to the action of a phospholipase A_2, present in the membrane and activated directly, probably via a G-protein, by nAChR. In the second case the P_2R activates a G-protein-phospholipase C system which results in phosphoinositide turnover and a simultaneous increase in inositol phosphates and DAG, followed by intracellular Ca^{2+} movement and influx of Ca^{2+}. It is discussed if DAG increases, when occurring close to the sarcolemma, might result in lipoxygenase products moving into the synapse and acting as "retrograde" signals. A preliminary experiment with arachidonic acid and a mouse phrenic nerve-diaphragm preparation was performed and showed no changes in

NATO ASI Series, Vol. H 70
Phospholipids and Signal Transmission
Edited by R. Massarelli, L. A. Horrocks,
J. N. Kanfer, and K. Löffelholz
© Springer-Verlag Berlin Heidelberg 1993

MEPPs, while higher AA concentrations may have decreased the EPPs' amplitudes.

Introduction

Neurotransmitters/modulators at a synapse either induce intracellular biosynthesis of second messengers by activating membrane receptor - G-protein - enzyme cascades or they open ligand-gated ion channels. The second messengers produced are part of regulatory systems largely affecting ion channels. Both ways, the arrival of the extracellular signal is transduced into the final, characteristic response of the cell in question. It has also recently been recognized that an activated neuroreceptor may directly interact with a G-protein in the membrane and this way modulate ion channels. Some intracellular messengers such as the unsaturated arachidonic fatty acid (AA), released from cell membrane phosphatidylinositol or phosphatidylcholine by an activated phospholipase A_2 or biosynthesized from the product of an activated receptor - G-protein - phospholipase C-induced turnover of phosphoinositides to inositolphosphates and diacylglycerol (DAG), a rather common second messenger and its lipoxygenase-metabolites, seem, at a synapse, to act as regulators/modulators of presynaptic transmitter release. Recent and earlier (Häggblad et al, 1987) results obtained in our laboratory have shown that drug-induced changes in the properties and the composition of the membrane phospholipids surrounding the nicotinic acetylcholine receptor (nAChR) of the skeletal muscle model, the electrocyte of Torpedo marmorata, change the properties of the receptor. We therefore asked the question if the activation of this receptor by cholinergic agonists in itself induces changes in the phospholipid composition of its closeby membrane environment and, if so, would this result in the formation of lipid-derived messengers which are used by the cell both in postsynaptic modulation and, after passage into the synapse, as retrograde modulators of neuromuscular transmission at the presynapse? A modulatory action at the postsynapse might involve the activity of the receptor and/or that of ion channels while an extracellular action at the presynapse might modulate ion-channels, transmitter/precursor uptake systems or autoreceptors. Alternatively, perhaps modulation could even take place after penetration of the retrogradely moving substance into the presynapse. Presynaptic modulation of transmitter release has recently been described.

Recent experiments and results

In work on modulators of neuromuscular transmission and on excitation-contraction coupling in skeletal muscle of chick, mouse and rat, using myotubes in culture as models, we have studied two different sarcolemmal receptor systems, that of the nAChR and that of the receptor for the extracellularly acting modulator adenosine triphosphate (ATP), a subtype of the P_2-purinergic receptors (P_2R) which activates a G protein - phospholipase C system (Häggblad and Heilbronn, 1987; Häggblad and Heilbronn, 1988; Eriksson and Heilbronn, 1989; Häggblad et al, 1990) and resembles closely but not entirely, the $P_{2y}R$ subtype (Burnstock and Buckley, 1985). In both cases we found that receptor activation increases intracellular levels of DAG. In the second case, using $[2-^3H]myo$-inositol-loaded cells, ATP and some of its nonhydrolyzable derivatives were shown to increase intracellular inositol phosphate (IP_1, IP_2, IP_3) levels concomitantly with those of DAG. Thus phophoinositide turnover occurred, mediated by a G-protein and phospholipase C (PL-C). Second to the phosphoinositide (PI) turnover and thus the formation of IP_3, and using fura 2-loaded cells, the cytosolic Ca^{2+} level of the myotubes was found to rise. In most cells this increase was caused by a two-step mechanism (Häggblad et al, 1990; Eriksson and

Heilbronn, submitted). A rapid and transient fluorescence peak was seen and found to be due to Ca^{2+} release from intracellular stores (the presence of extracellular Ca^{2+} was shown not to be necessary), presumably from the main Ca^{2+} store of the myotube, the sarcoplasmic reticulum (SR). This peak was immediately followed by a more sustained one due to influx of Ca^{2+} from the outside of the cell (Fig 1). No influx was seen in the absence of extracellular Ca^{2+} and influx rapidly disappeared upon extracellular addition of a dihydro-pyridinium type of calcium channel blocker, or was prevented by it.

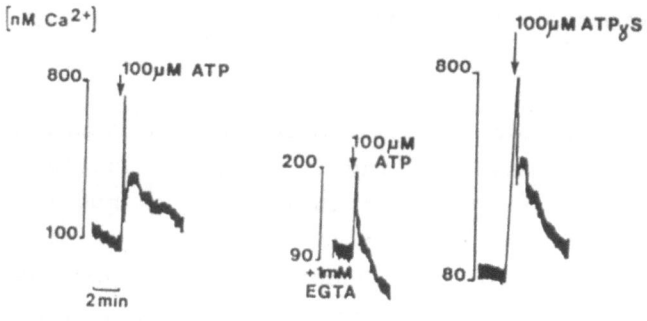

Fig 1. Calcium transients in fura 2-loaded chick myotubes following external ATP; effects of EGTA and ATPγS, adenosine (3-thio) triphosphate.

In another series of experiments we analyzed the intracellular levels of DAG after carbachol-induced activation of the myotube nAChR and also after depolarization of the myotube membrane by raising extracellular $[K^+]$ to 40 mM. The activation of nAChR opens its ion channel, allowing the passage of monovalent cations and Ca^{2+}. This starts membrane depolarization and opens further, voltage-dependent, cation channels. Finally, a sarcolemmal action potential is created and spreads into the interior of the cell via the tubular system. These events are prevented by the action of a number of nAChR antagonists, among them α-bungarotoxin (αBgt) and higher concentrations of d-tubocurarine (10-100 μM). Until recently, formation of membrane phospholipid-derived second messengers was not considered to be an immediate consequence of nAChR activation. However, Adamo et al (1985) suggested that agonist-induced activation of nAChR might induce a PI turnover in myotubes. In our experiments using [2-³H]glycerol-loaded cells, we found that the DAG levels of myotubes in culture increased when their nAChR is activated by carbachol (Fig 2). This increase could be prevented with either αBgt or inhibitory concentrations of d-tubocurarine. Subsequently, we found that depolarization of [2-³H]glycerol-loaded cells by high [K⁺] also caused increases in [DAG] (Heilbronn et al, submitted).

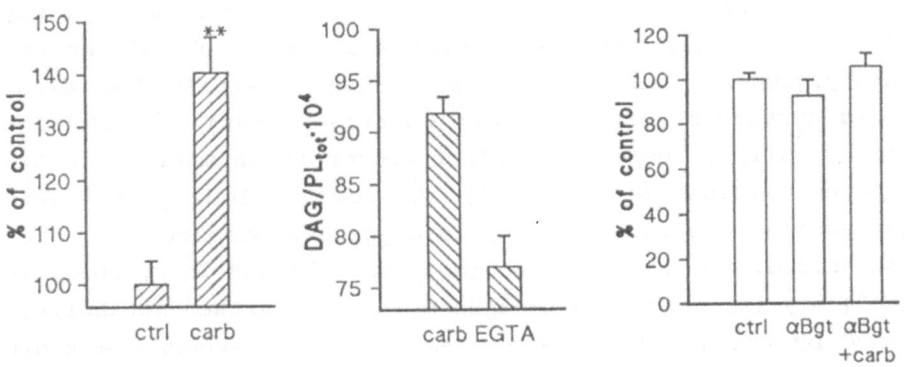

Fig 2.
a) Carbachol-activation of nAChR induces increases in intracellular DAG levels in [³H]glycerol-incubated chick myotubes (embryonal day 11).
b) Extracellular Ca²⁺ is needed.
c) αBgt prevents carbachol-induced DAG increases.

External Ca²⁺ was shown to be necessary for the [DAG] increases, an observation suggesting that activation of a phospholipase A₂ was their cause.

These findings brought back the original question. Intracellular DAG levels may increase along several different ways
which points to DAG as a key substance in synaptic modulation.
Continuing with nAChR, the everlasting model, it is accepted
that a DAG/Ca^{2+} activated protein kinase C contributes to its
control by phosphorylating the receptor. Such phosphorylation
may be a step towards receptor desensitization (Huganir et al,
1986; Huganir and Greengard, 1987; review: Huganir and
Greengard, 1990), a way to modulate neurotransmission by
weakening or preventing the effects of the released transmitter at its receptor. DAG may contribute to the modulation
of other proteins by activating their phosphorylation. As well
known, however, intracellular DAG is also metabolized.
Stimulation of PI turnover is expected to result in the
production of AA-enriched DAG, which in turn is thought to be
metabolized through one of two major routes: it may be
recycled to PI after conversion to phosphatidic acid or it may
release free AA by the sequential actions of DAG-lipase and
monoacylglycerol-lipase, thus providing substrate for eicosanoid synthesis (Irvine, 1982; Berridge, 1984). A modulator
of phospholipid, especially PI, turnover is in many cells the
Ca^{2+}-level (Ziboh et al, 1984). At low (<0.1mM) [Ca^{2+}] more
lipoxygenase products than cyclo-oxygenase products are biosynthesized, i e more eicosanoids. Thus changes in cytoplasmic
[Ca^{2+}] may influence the amounts of AA and further metabolites
formed by the lipoxygenase system. These compounds are known
to be able to pass over membranes. One of the metabolites, 12-
hydroperoxy-5,8,10,14-eicosatetraenoic acid (12-HPETE), as
well as AA have been suggested to act as messengers mediating
pre-synaptic neurotransmitter release (Piomelli et al, 1987).
Lynch and Voss (1990) described AA induced increase of IP
metabolism and glutamate release. Piomelli et al (1989)
recently reported that activation of Ca^{2+}/calmodulin-dependent
protein kinase II (CaM-kinase II) in nerve terminals results
in enhanced transmitter release (reviews: Augustine et al,
1987; Piomelli and Greengard, 1990). Subsequently CaM-kinase
II (rat brain cortex) was found to be inhibited both by AA and
especially by 12-HPETE (IC$_{50}$=0,7 μM), acting as second

messengers and thus modulating synaptic function through the inhibition of CaM-kinase II-dependent presynaptic protein phosphorylation (Llinás et al, 1985; Piomelli et al, 1989). Specific inhibition of Ca^{2+}-calmodulin-dependent protein phosphorylation by AA was also demonstrated in intact synaptosomes isolated from rat forebrain. CaM-kinase II catalyzes the phosphorylation of several proteins, including synapsin I, a protein associated with synaptic vesicles and may this way participate in presynaptic modulation of neuro-transmitter release (Llinás et al, 1985).

In a first effort to study the effects of AA and metabolites, we applied AA to an isolated mouse phrenic nerve-diaphragm preparation in normal Ringer solution, whereafter miniature endplate potential (MEPP) and evoked endplate potential (EPP) amplitudes were recorded at 1 Hz stimulation. Experiments were performed in the presence of d-tubocurarine (2 μM) in order to avoid muscle twitching. (We thank Prof S Thesleff, Dept of Pharmacology, University of Lund, for helping us with this test.) Increasing amounts of AA (10-800 μM final conc) were added. No effects on MEPPs and very few giant MEPPs were seen. Higher concentrations of AA, however, may have decreased the EPP amplitudes. There were, however, some disturbing solvent effects. Further, the results of this preliminary experiment are hard to interpret. There is reason to suspect that postsynaptic effects occur; Ehrengruber and Zahler (1991) showed recently that on cultured chromaffin cells from bovine adrenal medulla externally added AA and other cis-unsaturated fatty acids partially inhibit cholinergic induced activation when the agonist is added after AA. Agonist-evoked Ca^{2+} influx was blocked, not that evoked by the voltage gated Ca^{2+}-channel. Preincubation with 5-, 12-, 15-HETE or 12-HPETE did not inhibit the receptor-dependent Ca^{2+} influx, thus their possible formation was not responsible. AA may thus, in a physiological response, contribute to muscle relaxation. A presynaptic effect could also be expected. Boksa et al (1988) found that high and low affinity choline uptake is inhibited by AA (10-150 μM) and depletion of [^{3}H]ACh content in rat cerebral cortical synaptosomes was observed, while endogenous [ACh] was not changed. Changes in membrane fluidity may have been responsible. ACh synthesis may have been inhibited.

Discussion and conclusion

The outer membrane of skeletal muscle-derived myotube in culture contains neuro- and neuromodulator receptors. We studied two of these and found that their activation increases the DAG content of the cell. This increase is achieved by 2 different mechanisms, involving respectively phospholipase C and, apparently, phospholipase A_2, depending upon the type of receptor. The phospholipase A_2 pathway is dependent upon influx of extracellular Ca^{2+} and is also induced when the outer membrane is depolarized with high external $[K^+]$ only. DAG is known to participate in a number of intracellular events, including regulation of physiologically important proteins by protein kinase C. It is also metabolized, by the action of DAG-lipase and monoacylglycerol lipase, to AA, thus providing a substrate for eicosanoid synthesis, or it may be converted to phosphatidic acid, and PI or release AA, able to pass over cell membranes and thus be suitable to act as cell-to-cell messengers. Some very preliminary tests suggest that AA added to a mouse phrenic nerve-diaphragm preparation impairs EPP amplitudes. If so, AA may: a) not be the final and most active messenger and b) the effects suggested by the experiment could either be pre- or postsynaptic. They might be explained with changes in ' receptor activity, as several authors have described changes in the nAChR ion channel function e g of chromaffin cells. However, at the neuromuscular junction nAChR occurs also as an autoreceptor. Indeed, nicotinic and muscarinic (mAChR) acetylcholine autoreceptors have been suggested to have opposite functions at the motornerve ending when controlling transmitter release. A further presynaptic receptor involved in the regulation of ACh release is an adenosine receptor. All these receptors may be targets for modulation. Inhibition of transmitter synthesis and reduced storage may be a further modulatory action of AA or lipoxygenase metabolites. Finally, the large changes in free Ca^{2+}-levels of skeletal muscle in rest or activity may influence the biosynthesis of the lipid messengers.

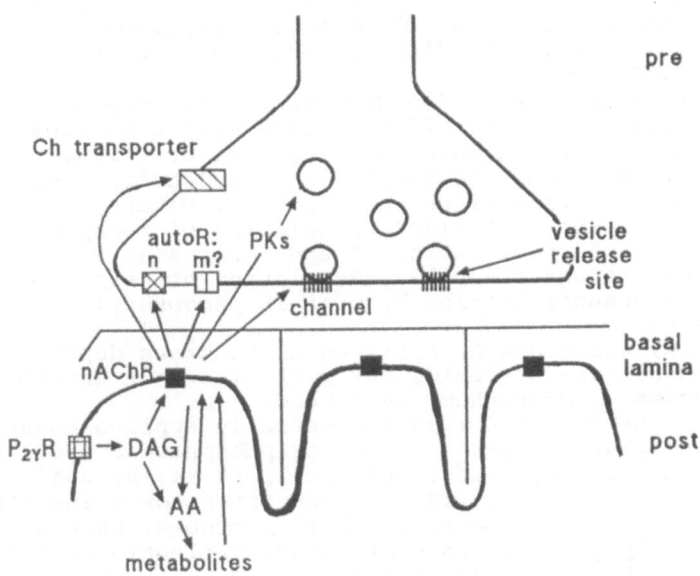

Fig 3. A model of a neuromuscular synapse showing possible targets for AA/eicosanoid modification.

References

Adamo S, Zani BM, Nervi C, Senni MI, Molinaro M, Eusebi F (1985) Acetylcholine stimulate phosphatidylinositol turnover at nicotinic receptors of cultured myotubes. FEBS Lett 190:161-164

Augustine GJ, Charlton MP, Smith SJ (1987) Calcium action in synaptic transmitter release. Annu Rev Neurosci 10:633-693

Berridge MJ (1984) Inositoltrisphosphate and diacylglycerol as second messengers. Biochem J 220:345-360

Boksa P, Mykita S, Collier B (1988) Arachidonic acid inhibits choline uptake and depletes acetylcholine content in rat cerebral cortical synaptosomes. J Neurochem 50:1309-1318

Burnstock G, Buckley N (1985) The classification of receptor for adenosine and adenine nucleotides. In: Methods used in Adenosine Research, Methods in Pharmacology, vol 6, ed DM Paton. Plenum Press, New York, pp 193-212

Ehrengruber MU, Zahler P (1991) Inhibition of the nicotinic ion channel by arachidonic acid and other unsaturated fatty acids in chromaffin cells from bovine adrenal medulla. Chimia 45:45-49

Eriksson H, Heilbronn E (1989) Extracellularly applied ATP alters the calcium flux through dihydropyridine-sensitive channels in cultured chick myotubes. Biochem Biophys Res Commun 159:878-885.

Eriksson H, Heilbronn H. The development of responses to extracellular ATP in cultured chick myotubes. Submitted.

Huganir RL, Delcour AH, Greengard P, Hess GP (1986) Phosphorylation of the nicotinic acetylcholine receptor regulates its rate of desensitization. Nature 321:774-776

Huganir RL, Greengard P (1987) Regulation of receptor function by protein phosphorylation. Trends Pharmacol Sci 8:472-477

Huganir RL, Greengard P (1990) Regulation of neurotransmitter receptor desensitization by protein phosphorylation. Neuron 5:555-567

Heilbronn E, Järlebark L, Eriksson H. Membrane depolarization-induced release of lipids may participate in transmission regulation. J Neurochem, submitted.

Häggblad J, Heilbronn E (1987) Externally applied adenosine-5'triphosphate causes inositol triphosphate accumulation in cultured chick myotubes. Neurosci Lett 74:199-204

Häggblad J, Heilbronn E (1988) P_2-purinoceptor stimulated phosphoinoitide turnover in chick myotubes. Calcium mobilization and the role of guanyl-nucleotide-binding proteins. FEBS Lett 235:133-136

Häggblad J, Eriksson H, Hedlund B, Heilbronn E (1987) Forskolin blocks carbachol-mediated ion-permeability of chick myotube nicotinic receptor and inhibits binding of ^{3}H-phencyclidine to Torpedo microsac nicotinic receptors. Naunyn-Schmiedeberg's Arch Pharmacol 336:381-386

Häggblad J, Eriksson H, Heilbronn E (1990) Cell surface ATP (P2y) purinoceptors trigger and modulate multiple calcium fluxes in skeletal muscle cells. In: Progress in Brain Research, Cholinergic Neurotransmisson: Functional and Clinical Aspects, vol 84, eds S-M Aquilonius, P-G Gillberg. Elsevier Science Publishers B V, pp 111-116

Irvine RF (1982) How is the level of free arachidonic acid controlled in mammalian cells? Biochem J 204:3-16

Llinás R, McGuiness TL, Leonard CS, Sugimoro M, Greengard P (1985) Intraterminal injection of synapsin I or calcium/calmodulin-dependent protein kinase II alters neurotransmitter release at the squid giant synapse. Proc Natl Acad Sci, USA 82:3035-3039

Lynch MA, Voss KL (1990) Arachidonic acid increases inositol phospholipid metabolism and glutamate release in synaptosomes prepared from hippocampal tissue. J Neurochem 55:215-221

Piomelli D, Greengard P (1990) Lipoxygenase metabolites of
  arachidonic acid in neuronal transmembrane signalling.
  Trends Pharmacol Sci 11:367-373
Piomelli D, Volterra A, Dale N, Siegelbaum SA, Kandel ER,
  Schwartz JH, Belardetti F (1987) Lipoxygenase metabolites
  of arachidonic acid as second messengers for presynaptic
  inhibition of Aplysia sensory cells. Nature 328:38-43
Piomelli D, Wang JKT, Sihra TS, Nairn AC, Czernik AJ,
  Greengard P (1989) Inhibition of Ca^{2+}/calmodulin-dependent
  protein kinase II by arachidonic acid and its metabolites.
  Proc Natl Acad Sci USA 86:8550-8554
Ziboh VA, Isseroff RR, Pandey R (1984) Phospholipid
  metabolism in calcium-regulated differentiation in
  cultured murine keratinocytes. Biochem Biophys Res
  Commun 122:1234-1240

THE ROLE OF ARACHIDONIC ACID AS A RETROGRADE MESSENGER IN LONG-TERM POTENTIATION.

M.A. Lynch, K.L. Voss, M.P. Clements and T.V.P. Bliss,
Division of Neurophysiology and Neuropharmacology,
National Institute for Medical Research,
Mill Hill,
London NW7 1AA, UK

In the past decade, a great deal of emphasis has been placed on establishing the mechanisms underlying long-term potentiation (LTP) in the hippocampus but despite intensive research, there is little agreement about the events leading to its establishment. However, it is agreed that there are at least two pharmacologically different phases of LTP, induction and maintenance, and in the past few years these two relatively distinct phases have, in general, been considered separately.

An understanding of the biophysical properties of the N-methyl-D-aspartate (NMDA) receptor has led to the fairly widely-held view that induction of LTP is achieved by simultaneous occupation of the postsynaptically-located NMDA receptor by glutamate and depolarization of the postsynaptic membrane (see Collingridge and Bliss, 1987). Entry of calcium through the NMDA-associated calcium channel results when the magnesium block is removed (Mayer et al., 1984; Nowak et al., 1984) and the subsequent increase in postsynaptic calcium then presumably sets in motion a sequence of events which lead to induction of LTP. The importance of postsynaptic calcium in the induction process was established by the observation that postsynaptic injection of the calcium chelator EGTA blocked induction of LTP in area CA1 (Lynch et al., 1983); this finding was later confirmed (Malenka et al., 1988). A short-term decremental potentiation of synaptic responses, induced by increasing calcium postsynaptically (Malenka et al., 1988) or treatment of hippocampal slices with NMDA (Kauer et al., 1988) appears to require only postsynaptic involvement but conjunction experiments have indicated that some presynaptic involvement is required for the induction of the longer-lasting potentiation of synaptic responses, associated with tetanus-induced LTP (Hvalby et al., 1986; Gustafsson and Wigstrom, 1986).

The events leading to maintenance of LTP are ill-understood and in spite of intensive investigation it has not yet been established whether presynaptic, postsynaptic, or changes in both areas are required. Morphological studies designed to investigate this question suggested that both presynaptic (Landfield et al., 1988) and postsynaptic (Desmond and

NATO ASI Series, Vol. H 70
Phospholipids and Signal Transmission
Edited by R. Massarelli, L. A. Horrocks,
J. N. Kanfer, and K. Löffelholz
© Springer-Verlag Berlin Heidelberg 1993

Levy, 1988) changes were associated with maintenance of LTP. A number of studies investigating receptor changes in LTP tend to support the view that postsynaptic changes played the more significant role (Muller et al., 1988; Davies et al., 1989). More recently quantal analysis studies which theoretically should provide a straightforward answer to the question of whether release of transmitter plays a role in maintenance of LTP has provided data which is subject to interpretative difficulties; on the one hand, data from two groups suggests that LTP is associated with an increase in transmitter release (Bekkers and Stevens, 1990; Malinow and Tsien, 1990) while data from a third group fails to confirm this (Foster and McNaughton, 1991).

The study of signal transduction mechanisms in LTP has attracted considerable attention in recent years and particular emphasis has been focussed on the role of protein kinase C (PKC), activity of which is required for the complete expression of LTP. Three observations have underlined its importance. (1) Both PKC activity (Akers et al., 1986) and phosphorylation of protein F1 (GAP43, B50; Lovinger et al., 1986), a presynaptically-located substrate for PKC (Gispen et al., 1985) are increased, in proportion to the degree of potentiation, following induction of LTP. It was shown that PKC activation did not occur at the time of induction of LTP, but rather some minutes afterwards, suggesting that the immediate induction event was PKC-independent (Routtenberg et al., 1985). We can deduce from these findings that the role of PKC is in the maintenance of LTP and that at least part of this PKC-dependent phase involves presynaptic modulation. (2) Inhibitors of PKC, albeit non-specific inhibitors, block LTP in dentate gyrus (Lovinger et al., 1987) and area CA1 (Reymann et al., 1988; Malinow et al., 1988, 1989). These reports agree that induction of LTP appears to be PKC-independent and that the effect of the inhibitors is to prevent the persistence of the synaptic response. More recently, intracellular injection of more selective inhibitors of PKC (PKC19-31) and calcium-calmodulin kinase 11 (CaMK11273-302), as well as bath application of H7, a non-specific kinase inhibitor, has indicated that postsynaptically located PKC or CaCaMK11 might play a role shortly after induction of LTP while presynaptic PKC plays a role in maintenance (Malinow et al., 1989). (3) Activators of PKC induce a form of potentiation which was relatively short-lived, persisting for less than an hour in most cases (Malenka et al., 1986; Gustafsson et al., 1988; Muller et al., 1988). Clearcut occlusion between tetanus-induced and phorbol ester-induced LTP has been reported by one group (Malenka et al., 1986) but this finding has been disputed (Gustafsson et al., 1988; Muller et al., 1988) suggesting that the two forms of potentiation are not identical. When all of these data are considered together it can be concluded that (a) PKC activation cannot be the sole mechanism underlying maintenance of LTP but rather contributes to the earlier part of it and (b) increased activity of PKC in the presynaptic terminals appears to be involved in the process.

During the past several years, efforts in this laboratory have concentrated on examining presynaptic changes accompanying the maintenance of LTP in perforant path-granule cell synapses. We have shown that LTP is associated with a relatively long-lasting increase in release into perfusate of newly-synthetized (Dolphin et al., 1983) and endogenous (Bliss et al., 1986) glutamate in the *in vivo* preparation. Although we have consistently reproduced our observation that endogenous glutamate release is increased in LTP (Bliss et al., 1986; Errington et al., 1987; Lynch et al., 1989a, 1991) another group has failed to replicate this funding (Aniksztejn et al., 1989). In further support of presynaptic involvement in maintenance of LTP, we have found that LTP is associated with an increase in K^+-induced, calcium-dependent release of prelabelled glutamate release in the *ex vivo* preparation (Lynch and Bliss, 1986; Lynch et al., 1989b). The association between LTP and the apparent increase in transmitter release was strengthened by the observation that when induction of LTP was blocked by commissural stimulation (Bliss et al., 1986; Lynch et al., 1989b), AP5 (Errington et al., 1987) or the lipoxygenase and phospholipase inhibitor, nordihydroguaiaretic acid (NDGA; Lynch et al., 1989), the increase in glutamate release was also blocked. We can conclude from this data that presynaptic changes are involved in the maintenance phase of LTP.

Earlier experiments suggested the involvement of increased polyphosphoinositide turnover in LTP (Bar et al., 1984; Clements et al., 1989), and since the two second messengers produced by increased phospholipase C (PLC) activity have been implicated in the release process, such a change might contribute to the LTP-associated increase in transmitter release. Inositol(1,4,5)trisphosphate ($InsP_3$) stimulates release of calcium from intracellular stores; transmitter release is calcium-dependent (Katz and Miledi, 1965) and $InsP_3$-induced increase in intraterminal calcium might therefore lead to increased transmitter release in LTP. The other second messenger, 1,2-diacylglycerol (DG) stimulates PKC, which has been implicated in the release process (Lynch and Bliss, 1986; Malenka et al., 1987) and therefore it seemed reasonable to propose that an LTP-associated increase in polyphosphoinositide turnover might underlie an LTP-associated increase in transmitter release. To test this hypothesis LTP was induced unilaterally in the dentate gyrus of urethane-anaesthetized rats, while the contralateral dentate received only test shocks and no high-frequency train and so served as the control. In parallel experiments, a high-frequency train to the commissural input to the dentate gyrus preceded the perforant path tetani and so induction of LTP was blocked (Douglas et al., 1983). Electrophysiological recordings were taken for 30 min before the tetani

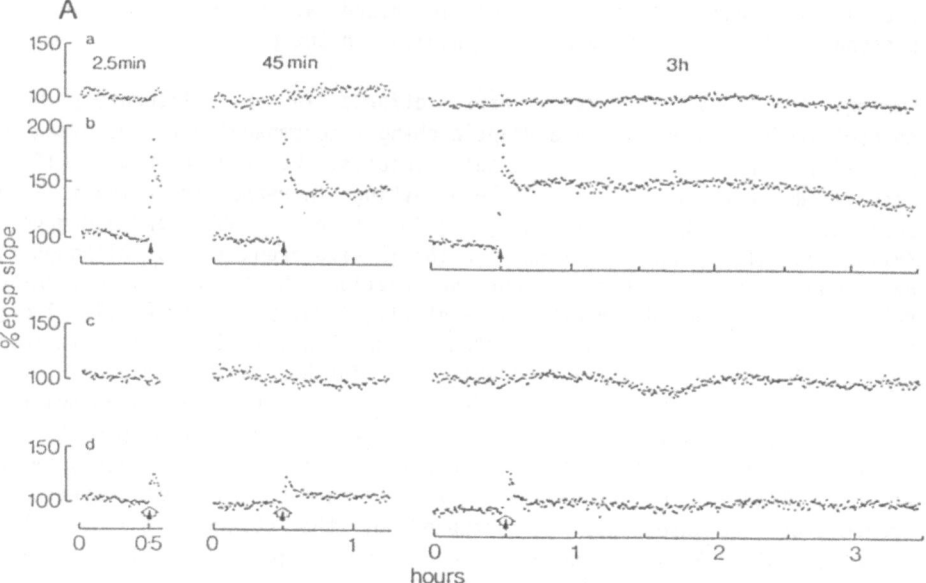

Figure 1. LTP was induced unilaterally in the dentate gyrus of urethane-anaesthetized rats by three high frequency trains of stimuli to the perforant path (a; 250 Hz for 162 msec, 30 sec interval). The contralateral dentate received only test shocks and no high-frequency train and so served as the control (b). In a second series of experiments, a high-frequency train to the commissural input to the dentate gyrus preceded the perforant path tetani to block induction of LTP (c; Douglas et al., 1983). In these experiments, the contralateral control dentate gyrus also received commissural stimulation but no perforant path tetanus (d). Electrophysiological recordings were taken for 30 min before the tetani and for either 2.5 min, 45 min or 3 h after the last high-frequency train to the perforant path in the experimental group (or the equivalent time in control experiments).

and for either 2.5 min, 45 min or 3 h after the high-frequency train to the perforant path.

The mean population epsp slope for the 9 animals is shown in Figure 1. An increase in epsp slope was observed only in the group receiving tetanic stimulation to the perforant path. Release of [^{3}H]glutamate (Figure 2) was significantly increased in slices obtained from tetanized tissue compared to control at each of the time intervals (* $p < 0.05$; Student's t-test for paired means). No change was observed when commissural

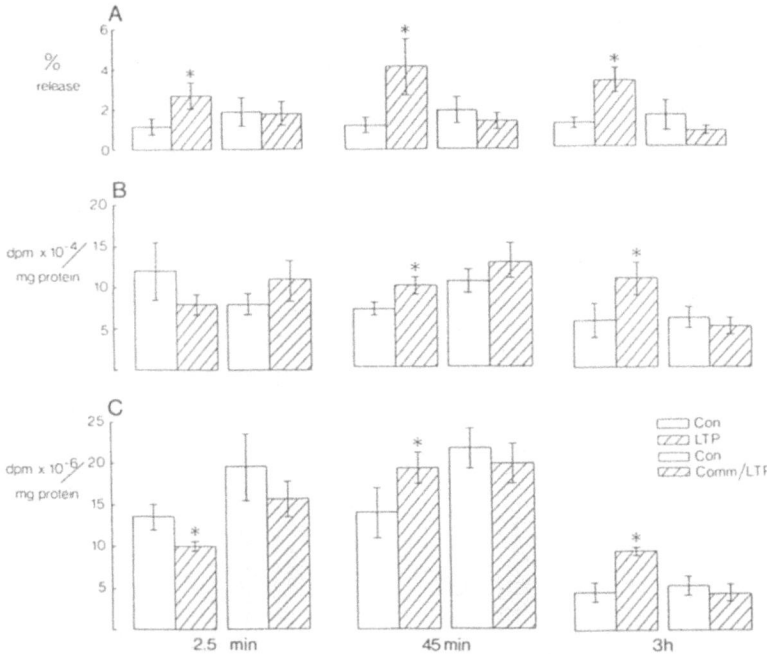

Figure 2. Tissue, either slices or the impure synaptosomal preparation P$_2$, was prepared at the end of the electrphysiological recordings described in Fig. 1. This figure shows changes in K$^+$-induced, calcium-dependent [^{3}H]glutamate release in slices (A) ^{3}H-inositol labelling of inositol phosphates in P$_2$ (B) and [^{14}C]arachidonic acid labelling of diacylglycerol in P$_2$ (C).

stimulation blocked induction of LTP. Polyphosphoinositide turnover was examined in the impure synaptosomal preparation P$_2$ obtained from the same tissue. [^{3}H]Inositol labelling of inositol phosphates (IP) and [^{14}C]arachidonic acid labelling of DG were examined as a measure of polyphosphoinositide turnover. A significant increase in labelling of both IPs and DG was observed in the group displaying LTP 45 min and 3 h after tetanization (* $p < 0.05$; Student's t-test for paired means). At 2.5 min after induction of LTP, there was a decrease in labelling which was significant in the case of DG (* $p < 0.05$). No significant changes were observed when LTP was blocked by commissural stimulation.

The data from these experiments indicate that, in this *ex vivo* preparation, the enhancement of glutamate release occurs as early as 2.5 min after induction of LTP and persists for as long as 3 hours. The findings also indicate that there is an increase in polyphosphoinositide

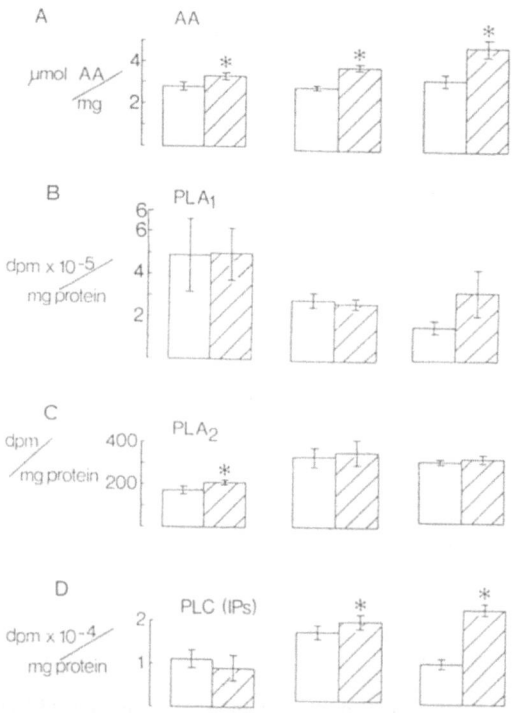

Figure 3.   LTP was induced *in vivo* in the dentate gyrus and followed for 2.5 min, 45 min or 3 hours as described in the legend to Figure 1. At the end of the experimental periods, slices of dentate gyrus were prepared and the arachidonic acid concentration associated with the membrane fraction of control and potentiated tissue was measured by HPLC. The results in A show that there was an enhanced concentration of free arachidonic acid associated with potentiated tissue compared to control at the three time intervals investigated (* p < 0.05; Student's t-test for paired means). PLA_1, PLA_2 and PLC activities were also examined. PLA_1 activity was similar in control and potentiated tissue at the three time intervals while there was a significant increase in activity of PLA_2 at 2.5 min after induction of LTP. In the case of PLC activity which was assessed in slices as [^{3}H]inositol labelling of inositol phosphates (IP), there was an increase 45 min and 3 hours after induction of LTP (p < 0.05; Student's t-test for paired means) but at 2.5 min there was no change.

turnover at the two later time intervals after induction of LTP but not at 2.5 min. These data lead us to propose that increased polyphosphoinositide turnover in synaptosomes is not responsible for the increase in transmitter release observed shortly after induction of LTP but it might underlie the later and more persistent change associated with maintenance of LTP. This is consistent with the findings that induction of LTP appears to be PKC-independent and that the role of PKC in expression of LTP is presynaptic (see above).

While our results suggests that maintenance of LTP requires presynaptic involvement, the trigger for induction of LTP is probably an increase in postsynaptic calcium concentration following NMDA receptor activation. We have therefore postulated the existence of a retrograde messenger which would be released from the postsynaptic site of induction to trigger presynaptic changes leading to an increase in glutamate release (Bliss et al., 1986). Four points suggested to us that arachidonic acid might be a candidate retrograde messenger. First, it is a small, lipid-soluble molecule which could traverse membranes without difficulty. Second, it was reported to act as an intracellular messenger in *Aplysia* (Piomelli et al., 1987). Third, it can be liberated from membrane phospholipids as a result of an increase in phospholipase activity. Since PLA_2 can be stimulated by an increase in calcium, which occurs postsynaptically at the time of induction of LTP, the machinery for liberation of arachidonic acid from a postsynaptic site was present. Fourth, an increase in presynaptic PKC activity in LTP has been reported (see above); fatty acids, including arachidonic acid, activate PKC (Linden et al., 1986), so a presynaptic target for arachidonic acid had already been described.

If arachidonic acid is a retrograde messenger in LTP, then first, any agent which interferes with its production and/or metabolism should interfere with induction of LTP; second, it should be released into the synaptic cleft; and third, it should induce a form of potentiation, which occludes with LTP. Therefore the first approach in testing the hypothesis was to investigate the effect of NDGA, which at low concentrations, inhibits activity of lipoxygenase but at higher concentrations also blocks activity of PLA_2 (Humes et al., 1983). NDGA (2×10^{-4}M) blocked induction of LTP, but whether this is due to its inhibitory effect on lipoxygenase or PLA_2 is unclear (Lynch et al., 1989a). More recently two reports confirmed the involvement of the arachidonate cascade in LTP (Okada et al., 1989; Massicotte et al., 1990); both reported that the PLA_2 inhibitor, p-bromophenacyl bromide, blocked induction of LTP. The second observation consistent with the retrograde messenger hypothesis was that arachidonic acid concentration in perfusate was increased after induction of LTP compared to before. In these experiments, a push-pull cannula straddling the molecular layer of the dentate gyrus in urethane-anaesthetized rats sampled perfusate for a 30-minute period before induction of LTP and for 90 minutes afterwards. While the concentration of arachidonic acid and its two 12-lipoxygenase metabolites, 12-hydroxyeicosatetraenoic acid and 12-hydro-peroxyeicosatetraenoic acid, were increased after induction of LTP, the concentrations of linolenic, linoleic, oleic, palmitic and stearic acids were unchanged (Lynch et al., 1991). The third finding consistent with the retrograde messenger hypothesis was that arachidonic acid induced a form of delayed, activity-induced potentiation which occluded with tetanus-induced LTP (Williams et al., 1989). None of the other fatty acids investigated mimicked the effect of arachidonic acid (Williams and Bliss, 1991). These three findings indicate that arachidonic acid, but not other fatty acids, plays some role in LTP but the delayed nature of the potentiation induced by arachidonic acid indicates that it cannot be the early retrograde factor responsible for triggering immediate changes in the presynaptic terminal which result in increased transmitter release. The identity of the early retrograde factor has yet to be established. One possible candidate is nitric oxide (NO), described as an NMDA-inducible, fast-acting messenger in cultured cerebellar cells (Garthwaite et al., 1988). Evidence suggests that there is little NO synthase, the enzyme responsible for NO release, in area CA1 (Snyder and Bredt, 1991). Despite this, it was recently reported that NO synthase inhibitors partially blocked LTP (Odell et al., 1991; Schuman and Madison 1991). The possibility that NO might play a role in dentate gyrus us currently being investigated.

Arachidonic acid is liberated from membrane phospholipids by the action of phospholipases. In order to investigate the enzymatic processes involved in arachidonic acid liberation in LTP, the *ex vivo* preparation was used. LTP was induced *in vivo* in one dentate gyrus of urethane-anaesthetized rats by tetanic stimulation of the perforant path (Lynch et al., 1989a); the contralateral dentate gyrus was used as the control. Electro-physiological changes were followed for 2.5 min, 45 min or 3 hours after the tetani and dissected dentate gyri were used to investigate the concentration of arachidonic acid and the activity of PLA_1, PLA_2 or PLC. The results obtained are shown in Figure 3. At each of the time intervals after induction of LTP, the concentration of arachidonic acid in membranes of slices prepared from potentiated tissue was significantly increased compared to control. The increase, which was small but nevertheless significant, was more pronounced 45 min and 3 hours after induction of LTP. This result parallels the finding that arachidonic acid induces a form of potentiation which develops over a period of about an hour (Williams et al., 1989). Figure 3 also shows that the activity of PLA_1 was similar in control and potentiated tissue at the three time intervals investigated. In contrast PLA_2 activity was significantly increased 2.5 min after induction of LTP but not at 45 min or 3 hours. In the case of PLC activity, which was assessed by investigating [^{3}H]inositol labelling of inositol phosphates, there was a significant increase in potentiated tissue compared to control at the two later time points but not at 2.5 min after tetanic stimulation. These data suggest that arachidonic acid is liberated from membrane phospholipids by the action of PLA_2 immediately after induction of LTP and by the action of PLC at later time intervals.

The findings from the *ex vivo* study are in agreement with data obtained from the *in vivo* experiment indicating that LTP was associated with a long-lasting increase in perfusate concentration of arachidonic acid (Lynch et al., 1991). Consistent with this data is the observation that induction of LTP is blocked by the PLA_2 inhibitor, p-bromophenacyl bromide (Okada et al., 1989; Massicotte et al., 1990). We suggest that the increase in concentration of postsynaptic calcium immediately after induction of LTP might stimulate activity of PLA_2 and thence liberation of arachidonic acid. Such a sequence of events, associated with NMDA receptor activation, has already been established, in part, in cultured striatal (Dumuis et al., 1988), cerebellar (Lazarewicz et al., 1990) and hippocampal (Sanfeliu et al., 1990) cells. Moreover, results from recent experiments in this laboratory indicate that NMDA receptor activation in dentate gyrus of hippocampus results in increased arachidonic acid release in a preparation enriched in postsynaptic membranes but not in a preparation enriched in either glia or synaptosomes (Clements and Lynch, unpublished).

We propose that the later increase in liberation of arachidonic acid is due to activation of the metabotropic quisqualate receptor and subsequent increase in PLC activity and formation of 1,2-DG. Data from cultured striatal (Dumuis et al., 1990) and hippocampal (Patel et al., 1990) tissue provides evidence consistent with this proposal. There is further support for this thesis; (1) AP3, an antagonist at the metabotropic quisqualate receptor, blocks induction and maintenance of LTP (Izumi et al., 1991), and (2) we have recently found that in dentate gyrus, ibotenate, which stimulates PLC activity (Nicoletti et al., 1986) by activation of the quisqualate receptor, triggers an increase in liberation of arachidonic acid (Clements and Lynch, unpublished).

At least three criteria are required of a retrograde messenger. First, release of the messenger should be from a postsynaptic site, second, the putative retrograde factor should appear in the synaptic cleft and third, the retrograde messenger should exert changes in the presynaptic terminal which will lead to an increase in glutamate release. To examine the first of these criteria, LTP was induced *in vivo* and control or potentiated tissue was used to obtain preparations enriched in synaptosomes, glia or postsynaptic densities. The free arachidonic acid concentration associated with membranes from each of these preparations was assessed by HPLC. The results showed that there was an increase in arachidonic acid concentration in potentiated tissue in the preparation enriched in postsynaptic densities but not glia or synaptosomes (Clements et al.,

1991). While these data are consistent with the hypothesis that the source of the increase in arachidonic acid is postsynaptic, they do not provide proof of the source and the possibility exists that arachidonic acid is released from another site at the time of induction of LTP and taken up into the postsynaptic density. The second criterion required of a retrograde messenger is that it is released into the synaptic cleft. Results from the *in vivo* experiments described above, which indicated that there was an increase in perfusate concentration of arachidonic acid (but not other fatty acids), provide evidence that the second criterion is fulfilled (Lynch et al., 1991). Finally, if arachidonic acid is a retrograde factor in LTP then it should induce changes in the presynaptic terminal which will lead to an increase in glutamate release. Using the *ex vivo* preparation, we have reported that arachidonic acid increases K^+-induced glutamate release (Lynch and Voss, 1990), perhaps directly or alternatively indirectly by stimulating synaptosomal PLC activity (Lynch and Voss, 1991). In support of the finding that arachidonic acid might exert its effect presynaptically by stimulating synaptosomal PLC activity, we have also reported that in synaptosomes prepared from dentate gyrus, both arachidonic acid and $InsP_3$ increase calcium concentration in a concentration-dependent but non-additive manner (Lynch and Voss, 1991). These data provide some evidence that the three basic criteria required of a retrograde messenger are met by arachidonic acid.

A scheme, in which arachidonic acid plays the role of delayed retrograde messenger in LTP, has emerged based on the findings described. At the time of induction of LTP, NMDA receptor activation results in an increase in calcium influx into the postsynaptic area with a subsequent increase in calcium concentration. The enhanced calcium concentration stimulates activity of PLA_2 which in turn liberates arachidonic acid from membrane phospholipids. Arachidonic acid is released into the synaptic cleft and stimulates the presynaptic terminal to release glutamate. The subsequent activation of the metabotropic quisqualate receptor stimulates PLC activity in the postsynaptic region and this, in concert with an increase in DG lipase activity, is responsible for the more persistent increase in arachidonic acid. The scheme depicts arachidonic acid as a trophic factor and suggests that its continued presence is required for a persistent increase in glutamate release. While the results obtained so far can be accommodated by this scheme, several outstanding questions remain to be answered. Among these questions are (1) arachidonic acid may not be the early retrograde messenger, so what is the identity of this molecule? (2) what is the precise nature of the interaction between arachidonic acid and the presynaptic terminal? (3) is the postsynaptic region the source of the released arachidonic acid? (4) is arachidonic acid itself or one of its metabolites responsible for the changes described? Finally, how persistent is the retrograde signal? How is it switched off? Does this switch signal a decline to baseline of glutamate release?

REFERENCES

Akers RF, Lovinger DM, Colley PA, Linden DJ, Routtenberg A (1986) Translocation of protein kinase C activity may mediate hippocampal long-term potentiation. Science 231:587-589

Aniksztejn L, Roisin MP, Amsellem R, Ben-Ari Y (1989) Long term potentiation in the hippocampus of the anaesthetized rat is not associated with a sustained enhanced release of endogenous excitatory amino acids. Neuroscience 28:387-392

Bar PR, Weigant F, Lopes da Silva FH, Gispen WH (1984) Tetanic stimulation affects the metabolism of phosphoinositides in hippocampal slices. Brain Res 321:381-385

Bekkers JM, Stevens CF (1990) Presynaptic mechanisms for long-term potentiation in the hippocampus. Nature 346:724-726

Bliss TVP, Douglas RM, Errington ML, Lynch, MA (1986) Correlation between long-term potentiation and release of endogenous amino acids from dentate gyrus of anaesthetized rats. J Physiol (Lond) 377:391-408

Bliss TVP, Lynch MA (1988) Long-term potentiation of synaptic transmission in the hippocampus; properties and mechanisms. In: Landfield PW and Deadwyler SA (eds) Long-term potentiation; from Biophysics to Behaviour. Alan R. Liss, New York, pp 3-77

Clements MP, Lynch ML, Bliss TVP (1989) The increase in phosphoinositide turnover associated with long-term potentiation may be mediated through a GTP binding protein. Neurosci Res Commun 3:11-19

Clements MP, Bliss TVP, Lynch ML (1991) Increase in arachidonic acid concentration in a postsynaptic membrane fraction following the induction of long-term potentiation in the dentate gyrus. Neurosci 45:379-389

Collingridge GL, Bliss TVP (1987) NMDA receptors - their role in long-term potentiation. Trends Neurosci 10:288-293

Davies SN, Lester RAJ, Reymann KG, Collingridge GL (1989) Temporally-distinct pre- and postsynaptic mechanisms maintain long-term potentiation. Nature 338:500-503

Desmond NL, Levy WB (1988) Anatomy of associative long-term synaptic modification. In: Landfield PW and Deadwyler SA (eds) Long-term potentiation; from Biophysics to Behaviour. Alan R. Liss, New York, pp. 265-305

Dolphin AC, Errington ML, Bliss TVP (1982) Long-term potentiation in perforant path *in vivo* is associated with increased glutamate release. Nature 297:496-498

Douglas RM, Goddard GV, Riives M (1982) Inhibitory modulation of long-term potentiation: evidence for a postsynaptic locus of control. Brain Res 240:259-272.

Dumuis A, Sebben M, Haynes L, Pin J-P, Bockaert J (1988) NMDA receptors activate the arachidonic acid cascade system in striatal neurons. Nature 336:68-70

Dumuis A, Pin J-P, Oomagari K, Sebben M, Bockaert J (1990) Arachidonic acid released from striatal neurons by joint stimulation of ionotropic and metabotropic quisqualate receptors. Nature 347:182-184

Errington ML, Lynch MA, Bliss TVP (1987) Long-term potentiation in the dentate gyrus: Induction and increased glutamate release are blocked by D(-)aminophosphonovalerate. Neuroscience 20:279-284

Foster TC, McNaughton BL (1991) Long-term enhancement of CA1 synaptic transmission is due to increased quantal size, not quantal content. Hippocampus 1:79-92

Gispen WH, Leunissen JLM, Oestreicher AB, Verkleij AJ, Zwiers H (1985) Presynaptic localization of B50 phosphoprotein: The ACTH-sensitive protein kinase substrate involved in rat brain polyphosphoinositide metabolism. Brain Res 328:381-385

Garthwaite J, Charles SL, Chess-Williams R (1988) Endothelium-derived relaxing factor on activation of NMDA receptors suggests role as intercellular messenger in the brain. Nature 336:385-387

Gustafsson B, Huang Y-Y, Wigstrom H (1988) Phorbol ester-induced synaptic potentiation differs from long-term potentiation in the guinea pig hippocampus *in vitro*. Neurosci Lett 85:77-81

Humes JL, Sadowski S, Galavage M, Goldenberg M, Subers E, Kuehl Jr FA, Bonney RJ (1983) Pharmacological effects of non-steroidal antiinflammatory agents on prostaglandin and leukotriene synthesis in mouse peritoneal macrophages. Biochem Pharmacol 32:2319-2322

Hvalby O, Lacaille J-C, Hu G-Y, Andersen P (1987) Postsynaptic long-term potentiation follows coupling of dendritic glutamate application and synaptic activation. Experientia 43:599-601

Izumi Y, Clifford DB, Zorumski CF (1991) 2-Amino-3-phosphonopropionate blocks the induction and maintenance of long-term potentiation in rat hippocampal slices. Neurosci Lett 122:187-190

Katz B, Miledi R (1965) The effect of calcium on acetylcholine release from motor nerve endings. Proc R Soc Lond (Biol) 161:496-503

Kauer JA, Malenka RC, Nicoll RA (1988) NMDA application potentiates synaptic transmission in the hippocampus. Nature 334:250-252

Landfield PW, Applegate MD, Pitler TA , Kerr DS (1988) Presynaptic mechanisms in short- and long-term potentiation: relevance to brain aging. In: Landfield PW and Deadwyler SA (eds) Long-term potentiation; from Biophysics to Behaviour. Alan R. Liss, New York, pp 377-408

Lazarewicz JW, Wrobleski JT, Costa E (1990) N-methyl-D-aspartate-sensitive receptors induce calcium-mediated arachidonic acid release in primary cultures of cerebellar granule cells. J Neurochem 55:1875-1881

Linden DJ, Murakami K, Routtenberg A (1986) A newly discovered protein kinase C activator (oleic acid) enhances long-term potentiation in the intact hippocampus. Brain Res 379:358-363

Lovinger DM, Colley PA, Akers RF, Nelson RB, Routtenberg A (1986) Direct relation of long-term synaptic potentiation to phosphorylation of membrane protein F1, a substrate for membrane protein kinase C. Brain Res 399:205-211

Lynch G, Larson J, Kelso S, Barrioneuvo G, Schottler F (1983) Intracellular injection of EGTA block induction of hippocampal long-term potentiation. Nature 305:719-721

Lynch MA, Bliss, TVP (1986) Long-term potentiation of synaptic transmission in the hippocampus: effect of calmodulin and oleoyl-acetyl-glycerol (OAG) on release of 3H-glutamate. Neurosci Lett 65:171-176

Lynch MA, Errington ML, Bliss TVP (1989a) Nordihydroguaiaretic acid blocks the synaptic component of long-term potentiation and the associated increases in glutamate and arachidonate: An *in vivo* study in the dentate gyrus of the rat. Neuroscience 30:693-701

Lynch MA, Errington ML, Bliss TVP (1989b) The increase in [^{3}H]-glutamate release associated with long-term potentiation in the dentate gyrus is blocked by commissural stimulation. Neurosci Lett 103:191-196

Lynch MA, Voss KL (1990) Arachidonic acid increases inositol phospholipid metabolism and glutamate release in synaptosomes prepared from hippocampal tissue. J Neurochem 55:215-221

Lynch MA, Voss KL (1991) Presynaptic changes in long-term potentiation; elevated synaptosomal calcium concentration and basal phosphoinositide turnover in dentate gyrus. J Neurochem 56:113-118

Lynch MA, Clements MP, Voss KL, Bramham CR, Bliss TVP (1991) Is arachidonic acid a retrograde messenger in long-term potentiation? Biochem Soc Trans (in press)

Malenka RC, Madison DV, Nicoll RA (1986) Potentiation of synaptic transmission in the hippocampus by phorbol esters. Nature 321:175-177

Malenka RC, Ayoub GS, Nicoll RA (1987) Phorbol esters enhance transmitter release in rat hippocampal slices. Brain Res 403:198-203

Malenka RC, Kauer JA, Zucker RS, Nicoll RA (1988) Postsynaptic calcium is sufficient for potentiation of hippocampal synaptic transmission. Science 242:81-84

Malinow R, Madison DV, Tsien RW (1988) Persistent protein kinase C activity underlying long-term potentiation. Nature 335:820-824

Malinow R, Schulman H, Tsien RW (1989) Inhibition of postsynaptic PKC or CaMK11 blocks induction but not expression of LTP. Science 45:862-866

Malinow R, Tsien RW (1990) Presynaptic enhancement shown by whole cell recordings of long-term potentiation in hippocampal slices. Nature 346:177-179

Massicotte G, Oliver MW, Lynch G, Baudry M (1990) Effect of bromophenacyl bromide, a phospholipase A_2 inhibitor, on the induction and maintenance of LTP in hippocampal slices. Brain Res 537:49-53

Mayer ML, Westbrook GL, Guthrie PB (1984) Voltage-dependent block by Mg^{2+} of NMDA responses in spinal cord neurones. Nature 309:261-263

Muller D, Joly M, Lynch G (1988) Contributions of quisqualate and NMDA receptors to the induction and expression of LTP. Science 242:1694-1697

Muller D, Turnbull J, Baudry M, Lynch G (1988) Phorbol-ester-induced synaptic facilitation is different than long-term potentiation. Proc Nat Acad Sci US 85:6997-7000

Nicoletti F, Meek JL, Iadarola MJ, Chuang DM, Costa E (1986) Coupling of inositol phospholipid metabolism with excitatory amino acid recognition sites in rat hippocampus. J Neurochem 46:40-46

Nowak L, Bregestovski P, Ascher P, Herbet A, Prochiantz A (1984) Magnesium gates glutamate-activated channels in mouse central neurones. Nature 307:462-465

O'Dell TJ, Hawkins RD, Kandel ER, Arancio O (1991) Tests on the roles of two diffusible substances in long-term potentiation: Evidence for nitric oxide as a possible early retrograde messenger. Proc Natl Acad Sci 88:11285-11289

Okada D, Yamagishi S, Sugiyama H (1989) Differential effects of phospholipase inhibitors in long-term potentiation in the rat hippocampal mossy fibres and Schaffer/commissural synapses. Neurosci Lett 100:141-146

Patel AJ, Sanfeliu C, Hunt A (1990) Development and regulation of excitatory amino acid receptors involved in the release of arachidonic acid in cultured hippocampal neural cells. Devel Brain Res 57:55-62

Piomelli D, Volterra A, Dale N, Siegelbaum SA, Kandel ER, Schwartz JH, Belardetti F (1987) Lipoxygenase metabolites of arachidonic acid are second messengers for presynaptic inhibition of *Aplysia* sensory cells. Nature 328:38-43

Reymann KG, Frey U, Jork R, Matties H (1988) Polymixin B, an inbitor of protein kinase C, prevents the maintenance of long-term potentiation in hippocampal CA1 neurones. Brain Res 440:305-314

Routtenberg A, Lovinger DM, Steward O (1985) Selective increase in phosphorylation of a 47-kD protein (F1) directly related to long-term potentiation. Behav Neural Biol 43:3-11

Sanfeliu C, Hunt A, Patel AJ (1990) Exposure of N-methyl-D-aspartate increases release of arachidonic acid in primary cultures of rat hippocampal neurones and not in astrocytes. Brain Res 526:241-248

Schuman EM, Madison DV (1991) A requirement for the intercellular messenger nitric oxide in long-term potentiation. Science 254:1503-1506

Snyder SH, Bredt DS (1991) Nitric oxide as a neuronal messenger. Trends Pharmacol Sci 12:125-128

Wigstrom H, Gustafsson B, Huang Y-Y, Abraham WC (1986) Hippocampal long-term potentiation is induced by pairing single afferent volleys with intracellularly injected depolarizing pulses. Acta Physiol Scand 126:317-319

Williams JH, Bliss TVP (1991) Arachidonate-induced potentiation of synaptic transmission in the rat hippocampus *in vitro* is not mimicked by other cis-unsaturated fatty acids. J Physiol 434:21P

Williams JH, Errington ML, Lynch MA, Bliss TVP (1989) Arachidonic acid induces a long-term activity-dependent enhancement of synaptic transmission in the hippocampus. Nature 341:739-742

RESISTING MEMORY STORAGE: ACTIVATING ENDOGENOUS PROTEIN KINEASE C INHIBITORS

Aryeh Routtenberg
Northwestern University, Cresap Neuroscience Laboratory,
2021 Sheridan Road, Evanston, Illinois 60208, U.S.A.

We have recently proposed a model of long-term potentiation (LTP), a physiological model of memory storage in the brain, that has as its major theme the view that signal transduction mechanisms on both sides of the synapse leads to the enhanced synaptic communication of LTP (Colley and Routtenberg, 1992). This involves a sequence of events requiring bidirectional communication, the first signal being presynaptic transmitter release. A modified version of this model is shown in Figure 1. In the present chapter I will first review the model and then discuss the likely control points in the sequence and the synergistic relation between messengers and retrograde signals.

The proposed sequence is briefly reviewed here; a detailed description is found in Colley and Routtenberg (1992). The first stage depicted in Figure 1 revolves around the influx of calcium into the presynaptic terminal. This has at least three consequences: mobilization of transmitter, activation of phospholipase A2 (PLA2) and protein kinase C (PKC). PKC requires either diacylglycerol (DAG) or cis-unsaturated fatty acid (CUFA) or both to be translocated and activated (see below).

The second stage involves activation of glutamate receptor subtypes which elevate calcium and along with DAG and CUFA

NATO ASI Series, Vol. H 70
Phospholipids and Signal Transmission
Edited by R. Massarelli, L. A. Horrocks,
J. N. Kanfer, and K. Löffelholz
© Springer-Verlag Berlin Heidelberg 1993

activate PKC post-synaptically (stages 3-5). At this point a
retrograde signal activates a receptor presynaptically that
generates a DAG signal which along with an elevated intraterminal calcium and CUFA translocates/activates PKC in the
presynapse (stage 6). This sets into motion an increased release
of transmitter and terminal growth, cytoskeletal reorganization
and gene expression at the cell body, all of which produce a
consequence which it is proposed here prolongs the activation
state of the terminal (stage 7a). Coordinated with these changes
would be a parallel set of alterations occurring in the
postsynaptic side involving a change in receptor sensitivity,
cytoskeletal reorganization and gene expression (stage 7b).
There are many more details in the Figure that will not be
discussed here; the reader can peruse the detailed review by
Colley and myself (1992).

Control points in the model: The regulation of plasticity

As one looks at Figure 1 from a systems point of view, one is
actually compelled, given that this sequence of events does not
possess alternative paths, to predict that blockade of any point
in the sequence aborts the process, blocking synaptic
enhancement. This raises the question, If this process is so
important why is it so vulnerable to inhibitors and why is it not
redundant so that such vulnerability would be reduced?

This model was constructed, in fact, by focussing on those experiments which blocked LTP with agents that block a specific enzyme or receptor. As illustrated by the solid dots on the arrows these blockade effects are found throughout the sequence of events. Thus, the process has no alternative pathways and so can be aborted by blockade at one point thereby preventing any enhancing effects on synaptic communication.

Could there be functional significance to the broad capability to block LTP or memory formation blockade? As suggested earlier (Routtenberg, 1972 - twenty years ago!) the storage of information is not a promiscuous process in which any and all inputs are automatically encoded in memory. Rather, it is a selective one with brain editors located in particular cell types (see Collier et al., 1987). To store non-selectively would severely contaminate the registers with useless information, interfering with the storage of those important events that are necessary for the survival of the organism. This may explain why there exist brain reward systems that can function to make evaluative "good-bad" judgments (Routtenberg, 1978).

This susceptibility to exogenous inhibitors, I submit, reflects an endogenous and necessary property of the system: there exist multiple control points in the system to prevent the spurious storage of irrelevant information. To accomplish this it is suggested that endogenous inhibitors are activated and can block the sequence at the same points in the process at which exogenous inhibitors have been shown to act. While this may be true at a variety of synapses, it will be particularly interesting to study those synapses which appear to be specialized for the editing function.

One such editing system was demonstrated by Collier et al., 1987. Activation of the hippocampal granule cell- mossy fiber system caused a complete forgetting of newly acquired information without affecting either old information already stored or the ability to store new information. Thus, this system prevented the storage of information that had occurred several min after the event had taken place. Perusal of Figure 1 suggests how information stored at time zero could be erased 10 min after it has occurred. In this instance, activation of the mossy fiber system could activate a protease to produce endogenous PKC inhibitors that would, like exogenous PKC inhibitors (Colley et al., 1990) eliminate the potentiated response up to 30 min after it was induced. This predicts that application of a protease inhibitor to CA3 cells would have a special effect on LTP relative to that of CA1 or granule cells.

This proposal first requires the existence of endogenous inhibitors of PKC. Indeed, there is a growing body of information which indicates that a variety of endogenous PKC inhibitors exist (e.g. Chan et al., 1986b; McDonald et al., 1987). It is interesting to consider the fact that the majority of these inhibitors are peptides that are likely to be derived from larger proteins. If so, then their generation would require a proteolytic step. This could explain why heroic doses of a protease inhibitor are required to obtain a blockade effect (Staubli et al., 1984). That is, it may be the case that leupeptin at physiological dosages would actually enhance memory formation, or LTP would also be enhanced by preventing the generation of PKC inhibitors. Given the fact that leupeptin is a rather broad spectrum protease inhibitor having neutral thiol and serine protease actions it may be predicted that enhancing actions on synaptic plasticity would be engendered by its use.

If endogenous PKC inhibitors play a role in regulating synaptic plasticity then, from the model, one would predict presence of PKC inhibitor on both sides of the synapse. Moreover, given that there are different PKC subtypes on both sides, different inhibitors could exist for a particular PKC subtype. That this may be the case is indicated by our recent finding (Sheu et al., 1991) that the S100 protein selectively inhibits the betaPKC phosphorylation of the neuron-specific PKC substrate, protein F1 (aka GAP43, B50, pp46, neuromodulin; see Benowitz and Routtenberg, 1987). One needs to be cautious here as we have recently discovered that PKC phosphorylation of endogenous substrates like F1 is preferential for the betaPKC subtype (Sheu et al., 1990). Thus, S100 may be an endogenous inhibitor not for a specific subtype but rather for physiologically-natural PKC-substrate interactions. It will be interesting, in this regard, to determine what the physiological substrates in brain for the gamma PKC subtype are and then to determine what the effect of S100 might be on this interaction.

From a cellular point of view it will be necessary to understand how the S100 protein which is derived from brain glial cells would influence phosphorylation of protein F1 which is present exclusively within the axonal process of neurons. It is possible that S100 inhibits PKC in glial cells and that the consequence of that inhibition then influences synaptic function. But this does not explain why S100 has a specific action on beta phosphorylation of protein F1. If this biochemical observation using purified enzyme, substrate and inhibitor peptide by Sheu et al. (1991) occurs in vivo, this suggests the need to search for a specific mechanism for neuron-glia interaction.

There is a potential structural substrate for this neuron-glia interaction. Ultrastructural studies of perforant path synapses on dendritic spines of the granule cells revealed a continuous network of protoplasmic glia that wrap around each and every synapse sometimes with thin slips of process no greater than 100 Angstrom units (Tarrant and Routtenberg, 1977). If dimerized S100 can cross the plasma membrane then the glia are in a

position to effectively control the development of synaptic
enhancement.

If the S100 protein has a selective effect on betaPKC then it
would be predicted from the model that its action would be on the
presynaptic side. A few years ago, Chan et al. (1986)
demonstrated that calmodulin was a powerful inhibitor of PKC
phosphorylation both of H1 histone and protein F1. At that time
no subtype specific information was available. It may now be
interesting if this endogenous molecule (calmodulin) which has
such widespread influence on a variety of functions also
regulates PKC activity in a subtype specific manner.

Synergism of PKC Regulation: Regulation by Convergence

Another mechanism for control is illustrated in Figure 1 on the presynaptic side which shows that betaPKC is regulated by calcium, CUFA and DAG. How might these work to activate PKC? A brief review of the signal transduction mechanism is in order at this point.

There is now considerable evidence that signal transduction and transmembrane signalling mechanisms lead to the activation of PKC, a lipid dependent enzyme (Nishizuka, 1984; Berridge and Irvine, 1984). PKC as well as other protein molecules involved in signal transduction have been identified and the primary structure of several of these proteins have been described. In the past five years evidence has accumulated to suggest that this signalling mechanism has a critical role in the information storage process in the central nervous system.

PKC is now known to exist as a family of subtypes (Coussens et al., 1986; see Nishizuka, 1988, for review) which is activated by two different mechanisms. As originally described (see Takai et al., 1982, for review) phospholipase C cleaves the diacylglycerol (DAG) moiety from the parent phospholipid producing a signal in the plane of the membrane to activate PKC. This is a calcium and phospholipid dependent activation mechanism, here dubbed the DAG

pathway, which had been considered the sole mechanism for PKC
activation.

A second mechanism was proposed by Murakami et al. (1986) based
on the discovery that PKC could be fully activated by cis
unsaturated fatty acids (CUFAs) in the absence of calcium or
phospholipid. It was suggested that calcium-activated
phospholipase A2 (PLA2) released CUFAs from the 2 position of
phospholipids and in their monomeric form activated PKC (Murakami
et al., 1986).

In the presynaptic side of the model it can be seen that calcium,
DAG and CUFA are proposed to translocate/activate PKC. It is
left unanswered, however, how these interact. Recently however,
Chen and Murakami (1991) and Shinomura et al. (1991) have studied
the potential interactions among these three PKC regulating
factors. They have shown that these act synergistically to
activate PKC. While subtype-specific effects have been noted by
Chen and Murakami, the evidence presented suggests that at a
given set of concentrations, synergism can be demonstrated for
each PKC subspecies.

IN SUMMARY, endogenous inhibitors, glial cell regulation by S100
and converging activations onto beta PKC presynaptically function
to provide major control points over the sequence of biochemical
and cellular events leading to synaptic plasticity and then

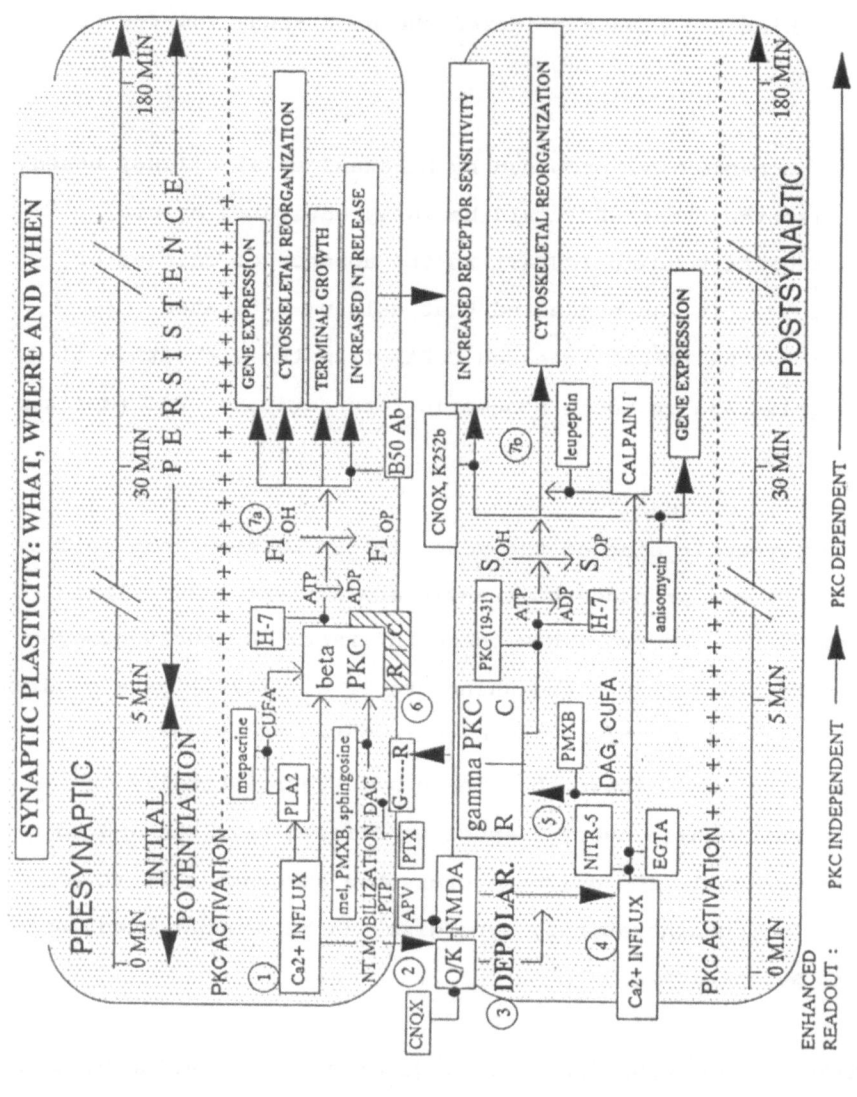

Figure 1 - A model of synaptic change (derived from the model proposed by Colley and Routtenberg, 1992).

information storage. Memory formation is not inevitable,

helterskelter or promiscuous. Rather, it is under multiple

regulatory controls to ensure that the only information that is

finally stored is that which is of utility to the organism.

REFERENCES

Benowitz, L I and Routtenberg, A (1987) A membrane
    phosphoprotein associated with neural development,
    axonal regeneration, phospholipid metabolism, and
    synaptic plasticity. Trends Neurosci 10: 527-532.
Berridge, M J and Irvine, R F (1984) Inositol trisphosphate,
    a novel second messenger in cellular signal
    transduction. Nature 312: 315-321.
Chan, S Y, Murakami, K, and Routtenberg, A (1986)
    Phosphoprotein F1: Purification and characterization
    of a brain kinase C substrate related to plasticity.
    J Neurosci 6: 3618-3627.
Chan, S Y, Murakami, K, and Routtenberg, A (1986b)
Characterization of purified protein F1, a specific
    protein kinase C substrate related to synaptic
    plasticity. Soc. Neurosci. 12: 1168.
Chen, S G and Murakami, K, Synergistic activation of (1991)
    type III protein kinase C by cis-fatty acid and
    diacylglyerol. Biochem J, in press.
Colley, P A, Sheu, F-S and Routtenberg, A (1990)
    Inhibition of protein kinase C blocks two components of
    LTP persistence leaving initial potentiation intact.
    J Neurosci 10: 3353-3360.
Colley, P A and Routtenberg, A (1991) Long-term potentiation
    as synaptic dialogue, submitted.
Collier, T J, Quirk, G S, and Routtenberg, A (1987)
    Separable roles of hippocampal granule cells in
    forgetting and pyramidal cells in remembering spatial
    information. Brain Res 409: 316-328.
Coussens, L, Parker, P J, Rhee, L, Yang-Feng, T L, Chen,
    E, Waterfield, M D, Francke, U and Ullrich, A (1986)
    Multiple, distinct forms of bovine and human protein
    kinase C suggest diversity in cellular signalling
    pathways. Science 233: 859-866.

162

McDonald, J R, Groschel-Stewart, U and Walsh, M P (1987) Properties and distribution of the protein inhibitor (M,17000) of protein kinase C. Biochem J 242: 695-705.

Murakami, K, Chan, S Y, and Routtenberg, (1986) A Protein kinase C activation by cis-fatty acid in the absence of Ca2+ andphospholipids. J. Biol. Chem 261: 15424-15429.

Nishizuka, Y, (1984) The role of protein kinase C in cell surface signal transduction and tumour promotion. Nature 308: 693-697.

Nishizuka, Y The molecular heterogenecity of protein (1988) kinase C and its implications for cellular regulation. Nature 334: 661-665.

Routtenberg, A (1978) The reward system of the brain. Sci Amer 239: 122-131.

Routtenberg, A, Strop, M, and Jerdan, J (1978) Response of infant rat to light prior to eyelid opening: Mediation by superior colliculus. Develop Psychobiol 11: 469-478.

Sheu, F -S, Marais R M, Parker, P J, Bazan, N G and Routtenberg, A (1990) Neuron-specific protein F1/GAP-43 shows substrate specificity for the beta subtype of protein kinase C. Biochem Biophys Res Commun 171: 1236-1243.

Sheu, F-S, Azmitia, E C, Marshak, D R, Parker, P J and Routtenberg, A (1991) Glial-derived S-100 protein selectively inhibits the neuron-specific protein F1/GAP-43 phosphorylation by beta 1 recombinant protein kinase C: Implications for a glial-neuronal interaction. Society for Neuroscience, New Orleans, Louisiana.

Shinomura, T, Asaoka, Y, Oka, M, Yoshida, K and Nishizuka, Y (1991) Synergistic action of diacylglycerol and unsaturated fatty acid for protein kinase C activation: Its possible implications. Proc Natl Acad Sci, USA, in press.

Staubli, U, Baudry, M and Lynch, G (1984) Leupeptin, a thiol proteinease inhibitor, causes a selective impairment of spatial maze performance in rats. Behav Neural Biol 40: 44-57.

Tarrant, S and Routtenberg, A (1977) The synaptic spinule in the dendritic spine: Electron microscopic study of the hippocampal dentate gyrus. Tissue and Cell 9: 461-473.

Takai, Y, Minakuchi, R, Kikkawa, U, Sano, K, Kaibuchi, K, Yu, B, Matsubara, T and Nishizuka, Y (1982) Membrane phospholipid turnover, receptor function and protein phosphorylation. Prog Brain Res 56: 287-301.

B-50 PHOSPHORYLATION IN RESPONSE TO DIFFERENT PATTERNS OF ELECTRICAL
STIMULATION IN RAT HIPPOCAMPAL SLICES

R. Corradetti, Maria Grazia Nunzi*, Willem Hendrick Gispen**, Cristina
Gianotti*
Department of Preclinical and Clinical Pharmacology,
University of Florence,
Viale G.B., Morgagni 65,
I-50134 Firenze, Italy.

During the past decade relevant research efforts have been directed to
identify some of the key steps within the multitude of biochemical and
electrophysiological events which constitute the cellular mechanisms
involved in learning and memory. A substantial advancement in this field
has been obtained with the discovery of the electrophysiological
phenomenon of long-term potentiation (Bliss and Lømo, 1973) which provided
the possibility to reproduce, either *in vivo* or *in vitro*, an enduring
enhancement of electrophysiological responses in several pathways of the
hippocampus, a brain structure which is involved in learning. Since then,
long-term potentiation (LTP) has been proposed as a model to study
cellular mechanisms of learning (Teyler and DiScenna, 1987), and has
provided the possibility of combining several approaches to the study of
biochemical and electrophysiological events involved in information
processing in the hippocampus. The more representative among them,
ranging from release of glutamate to mobilization of arachidonic acid or
activation of protein kinases and phosphorylation of protein substrates,
have been outlined in the preceding contributions by Dr. Lynch and by Dr.
Routtenberg. The LTP produced by a train of high-frequency stimulation in
the hippocampus appears at least partly due to increased glutamate release
from presynaptic terminals (Bliss et al., 1986; Bekkers and Stevens, 1990;
Malinow and Tsien, 1990). Among the mechanisms which may underlie this
increase in synaptic release, recent work has focussed attention on the
activation of protein kinase C (PKC), a calcium/phospholipid dependent
enzyme (Nishizuka, 1984) involved in several neuronal processes, from
differentiation to transmitter release (see Chiarugi et al., 1989). It is
relevant that the most specific substrate for PKC-induced phosphorylation
is the protein B-50, also called GAP-43 or F1 (Benowitz and Routtenberg,
1987). The phosphorylation of this protein which is present in

* Fidia Research Laboratories, via Ponte della Fabbrica 3/A, I- 35031
Abano Terme, Italy. **Rudolf Magnus Institute of Pharmacology, Padualaan
8, 3584 CH, Utrecht, The Netherlands.

NATO ASI Series, Vol. H 70
Phospholipids and Signal Transmission
Edited by R. Massarelli, L. A. Horrocks,
J. N. Kanfer, and K. Löffelholz
© Springer-Verlag Berlin Heidelberg 1993

presynaptic terminals, reversibly bound to calmodulin, has been functionally correlated to neurotransmitter release (Dekker et al., 1989a,b). The common problem with all these proposed mechanisms resides in the difficulty of developing experimental conditions in which direct quantitative evidence of the hypothesized biochemical changes can be provided.

Activation of PKC, following a train of high-frequency electrical stimulation able to induce a LTP of hippocampal electrophysiological responses in vivo or in vitro, has been demonstrated (for a review see Linden and Routtenberg, 1989). Accordingly, an increase in PKC activity in the hippocampus during learning has been recently shown (Olds et al., 1989). Consistently, much evidence of increased phosphorylation of the B-50 protein during LTP has been collected (Browning et al., 1979; Bar et al., 1980; Chan et al., 1986). The increase in phosphorylation of the PKC-specific substrate(s) has been only indirectly demonstrated. A critical review of the limitations of these approaches has been recently published (Linden and Routtenberg, 1989).

We therefore attempted to measure *directly* the changes in phosphorylation of one of the specific substrates of PKC, namely the protein B-50, following the induction of LTP. The phosphorylation of B-50 occurring during electrically induced LTP in the CA1 region of rat hippocampal slices was directly estimated using a quantitative method (DeGraan et al., 1989).

Materials and Methods

Adult male Wistar-Kyoto rats, 200-350 g of body weight, were purchased from Charles-River, Italy. Hippocampal slices (400 μm), obtained as previously described (Corradetti et al., 1989), were incubated for 1h in an oxygenated (95% O_2-5% CO_2) artificial cerebrospinal fluid (aCSF) comprised of (mM) NaCl 124, KCl 3.33, KH_2PO_4 1.25, $MgSO_4$ 2, $CaCl_2$ 2, $NaHCO_3$ 25, D- glucose 10, (pH 7.3) at room temperature. Three slices were then transferred at 35 ± 1 °C in an interface-type static bath (0.7 ml) containing a virtually phosphate-free aCSF (phosphate omitted). The slices were placed on a nylon net and oxygenated with a humidified O_2-CO_2 mixture. The three slices were tested for electrophysiological viability by delivering test stimuli in the stratum radiatum through bipolar nichrome electrodes with extracellular recording from the dendritic or somatic CA1 pyramidal cell region. Control experiments showed that under these conditions stable recording of evoked responses in the 3 slices could be followed for 4-6 h, in spite of the absence of phosphates. Then

100-120 μCi of [^{32}P]phosphate were added and allowed to equilibrate for 90 min before starting stimulation to collect baseline responses. Two slices were stimulated at 0.17 Hz (0.12 ms duration) and the third remained as the unstimulated control. Evoked responses were recorded on a digital tape recorder for later analysis. After obtaining 10 min of control responses, using test stimuli evoking 30-40% of maximal response in the preparation, one of the slices received either a train of high-frequency stimulation (100 Hz, 1 s, near maximal stimulus strength) or a single test pulse. Then the slices were stimulated at 0.17 Hz for 1h. At the end of the stimulation period the slices were rapidly transferred from the incubation bath to biochemical processing. The quantification of the *in situ* phosphorylation of B-50 was carried out as described by De Graan et al. (1989). ^{32}P labelling of B-50 was determined by quantitative immunoprecipitation of B-50 from the homogenate suspended in the stop-mix solution. Precipitates were collected by centrifugation, washed, resuspended in stop-mix solution and boiled for 10 min. The suspension was analyzed by 11% SDS-PAGE in triplicate, followed by autoradiography. Quantification of ^{32}P incorporation in B-50 was done using densitometric scanning of the autoradiogram and the ^{32}P incorporation of B-50 was corrected for the values of total ^{32}P incorporation in the slices obtained with TCA precipitation. Values of stimulated slices were expressed as percentage of the value of the unstimulated control in each experiment to allow comparison of results from different experiments. Statistical evaluation of data was carried out with ANOVA analysis and Student's t-test for unpaired or paired samples as appropriate.

Results

Figure 1 shows the marked improvement in measuring phosphorylated B-50 introduced by immunoprecipitation. It is evident that in the autoradiogram obtained from the total homogenate the changes in B-50 phosphorylation induced by PDB are more difficult to quantify than after immunoprecipitation.

Stimulation of the stratum radiatum with a train of high- frequency pulses was followed by a LTP in six of six preparations. The increase in synaptic efficacy was established immediately after the train, then slightly declined within 10 min to a stable potentiation which persisted for longer than 1 h (Figure 2). At 60 min after the induction of LTP the average increase in slope of dendritic evoked potentials was 64 ± 23% (P<0.05; n=3) and that of evoked population spikes was 118 ± 40% (P<0.05; n=3) indicating that under our conditions the essential characteristics of LTP were similar to those observed in submerged slices in the presence of

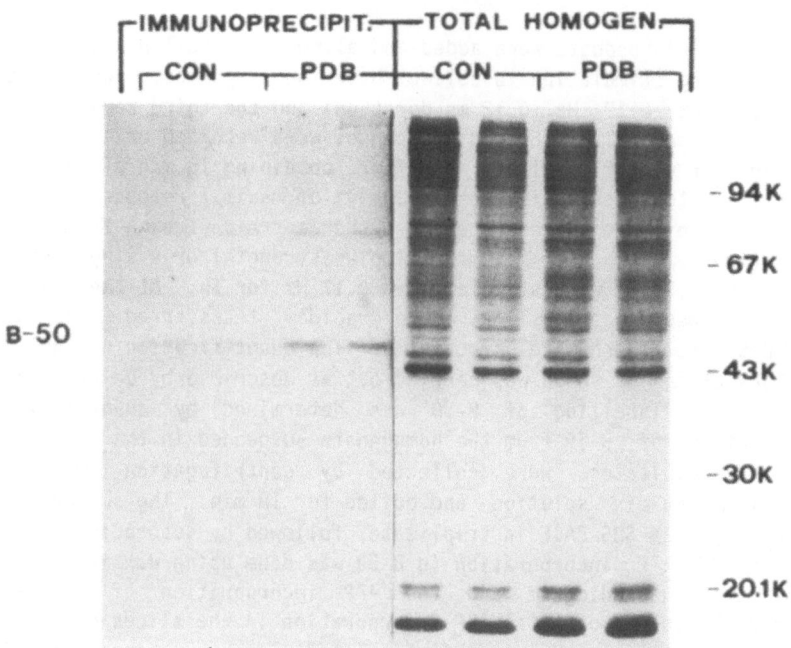

Fig. 1. Two autoradiograms were obtained from homogenates of hippocampal slices, labelled with [^{32}P]phosphate for 90 min and incubated for 30 min either in control conditions (CON) or in the presence of 1 μM 4-beta-phorbol 12,13-dibutyrate (PDB). The left autoradiogram obtained after immunoprecipitation shows the increase in phosphorylated B-50 induced by PDB. This increase is barely visible in the corresponding autoradiogram from total homogenate. Molecular weights are in Kdaltons (K).

phosphate (Schwartzkroin and Wester, 1975). Slices which did not receive a train of high-frequency stimulation showed no significant changes in evoked responses throughout the experimental period (Figure 2). In six experiments quantitative estimation of phosphorylated B-50, expressed as percent of that of unstimulated preparations, was carried out on potentiated slices matched with unpotentiated slices receiving only low-frequency test pulses (Figure 3).

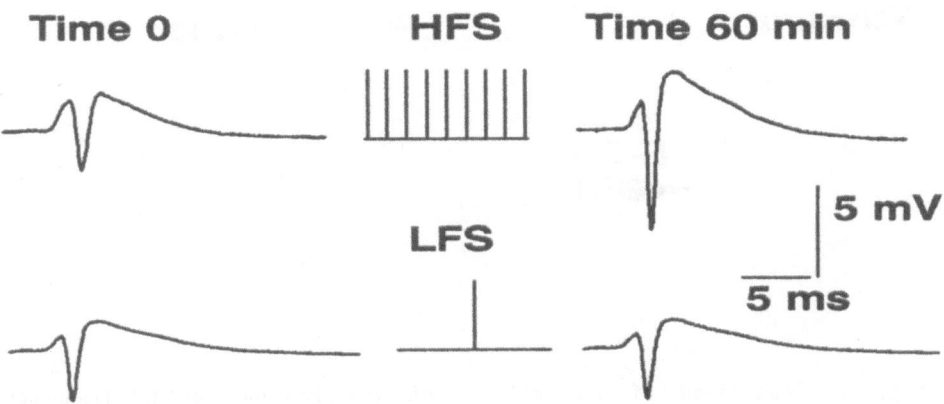

Fig. 2. A train of high-frequency stimulation produces a LTP in slices after incubation in [^{32}P]phosphate. Traces are representative somatic potentials evoked by test stimuli before (left traces) and after induction of LTP by a train of high-frequency stimulation (HFS; upper right trace) or in unpotentiated slices which received only low-frequency stimuli (LFS; lower right trace).

A significant increase in B-50 phosphorylation was found in the potentiated slices (47 ± 8 %; $P<0.02$) but not in control slices (16 ± 8 %; $P>0.1$). Moreover, B-50 phosphorylation in the potentiated slices was significantly ($P<0.05$) higher than that obtained in slices which did not receive the high-frequency train.

Discussion

Our results demonstrate that, after LTP induction by a train of high frequency stimulation in the stratum radiatum of rat hippocampal slices, the phosphorylation of B-50 is significantly increased in comparison with that of unpotentiated slices. In transverse slices most of B-50 appears to be localized in the CA1 region (Oestreicher and Gispen, 1986) and in lesser amounts in the CA3 and dentate regions. Therefore, since we measured the changes in phosphorylated B-50 in the whole slice, this could lead to an underestimation of B-50 phosphorylation related to LTP in the CA1 region. Unfortunately, due to the need of a given amount of proteins to run replicate samples and to estimate ^{32}P incorporation, it

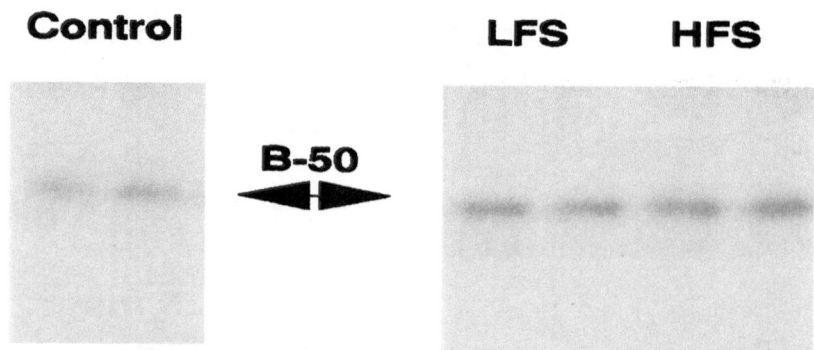

Fig. 3. Comparison of the effects of low-frequency or high-frequency stimulation on B-50 phosphorylation. Autoradiograph of a representative experiment showing the level of phosphorylation of B-50 in one unstimulated slice (Control) in comparison with that of the matched slices in which LTP was (HFS) or was not (LFS) induced.

was impossible to circumvent this limitation. On the other hand, it was unlikely that the other regions contribute to the increased B-50 phosphorylation since our stimulation was limited and did not excite dentate granule cells.

The putative excitatory neurotransmitter in the CA1 region of the hippocampus is glutamate (Corradetti et al., 1983) and an increase in glutamate release in the CA1 region has been demonstrated during LTP (Bliss et al., 1986; see also M. Lynch this volume), and during application of phorbol esters known to directly activate PKC (Malenka et al., 1987). Therefore, it has been repeatedly suggested (Bliss and Lynch, 1988; Chiarugi et al., 1989; Linden and Routtenberg, 1989) that, following LTP induction, activation of PKC leads to an increased release of the excitatory neurotransmitter from presynaptic nerve endings in the CA1 region of the hippocampus. Although the exact mechanism by which the phosphorylated B-50 might increase the electrophysiologically recorded synaptic efficacy is still a matter of investigation, it is worth noting that the phosphorylation of B-50 by PKC has been directly correlated with the release of some neurotransmitters in the hippocampus (Dekker et al., 1989a,b). It is therefore conceivable that the phosphorylation of B-50 is involved in the LTP-dependent increase in the release of excitatory neurotransmitters from Schaffer collaterals/ commissural nerve endings.

Our data provide direct evidence that, in presynaptic terminals of the pathway receiving a train of high-frequency stimulation able to induce LTP, the phosphorylation of B-50, a PKC specific substrate, was increased after the train for at least 1 h, possibly leading to a sustained higher release of neurotransmitters from the presynaptic terminals.

Acknowledgments

We thank Dr. P.N.E. De Graan and Dr. B. Oestreicher for supplying the antibodies against B-50 and for helpful advice. Partly supported by a C.N.R. grant (N°91.00595.04) to R.C.

References

Bär PR, Schotman P, Gispen WH, Tielen AM, Lopes da Silva FH (1980) Changes in synaptic membrane phosphorylation after tetanic stimulation in the dentate area of the rat hippocampal slice. Brain Res 198:478-484

Bekkers JM, Stevens CF (1990) Presynaptic mechanism for long-term potentiation in the hippocampus. Nature 346:724-729

Benowitz LI, Routtenberg A. (1987) A membrane phosphoprotein associated with neural development, axonal regeneration, phospholipid metabolism, and synaptic plasticity. Trends Neurosci 10:527-532

Bliss TVP, Lømo T. (1973) Long-lasting potentiation of synaptic transmission in the dentate area of anaesthetized rabbit following stimulation of the perforant pathway. J Physiol 232:331-356

Bliss TVP, Lynch MA (1988) Long-term potentiation of synaptic transmission in the hippocampus: properties and mechanisms. In: Landfield PW, Deadwyler SA (Eds) Long-term potentiation: from biophysics to behavior. Alan R. Liss, Inc., New York, pp 3-72

Bliss TVP, Douglas RM, Errington ML, Lynch MA (1986) Correlation between long-term potentiation and release of endogenous amino acids from dentate gyrus of anaesthetized rats. J Physiol 377:391-408

Browning M, Dunwiddie T, Nennet W, Gispen W, Lynch G (1979) Synaptic phosphoproteins: specific changes after repetitive stimulation of the hippocampal slice. Science 203:60-62

Chan SY, Murakami K, Routtenberg A (1986) Phosphoprotein F1: purification and characterization of a brain kinase C substrate related to plasticity. J Neurosci 8:3618-3627

Chiarugi VP, Ruggiero M, Corradetti R (1989) Oncogenes, protein kinase C, neuronal differentiation and memory. Neurochem Int 14:1-9

Corradetti R, Pugliese AM, Ropert N (1989) The protein kinase C inhibitor 1-(5-isoquinolinesulphonyl)-2-methylpiperazine (H7) disinhibits CA1 pyramidal cells in rat hippocampal slices. Brit J Pharmacol 98:1376-1382

Corradetti R, Moneti G, Moroni F, Pepeu G, Wieraszko A (1983) Electrical stimulation of the stratum radiatum increases the release and neosynthesis of aspartate, glutamate, and gamma-aminobutyric acid in rat hippocampal slices. J Neurochem 41:1518-1525

De Graan PNE, Dekker LV, Oestreicher AB, Van der Voorn L, Gispen WH (1989) Determination of changes in the phosphorylation state of neuron-specific protein kinase C substrate B-50 (GAP43) by quantitative immunoprecipitation. J Neurochem 52:17-23

Dekker LV, De Graan PNE, Oestreicher AB, Versteeg DHG, Gispen WH (1989a) Inhibition of noradrenaline release by antibodies to B-50 (GAP43). Nature 342:74-76

Dekker LV, De Graan PNE, Versteeg DHG, Oestreicher AB, Gispen WH (1989b) Phosphorylation of B-50 (GAP43) is correlated with neurotransmitter release in rat hippocampal slices. J Neurochem 52:24-30

Linden DJ, Routtenberg A (1989) The role of protein kinase C in long-term potentiation: a testable model. Brain Res Rev 14:279-296

Malenka RC, Ayoub GS, Nicoll RA (1987) Phorbol esters enhance transmitter release in rat hippocampal slices. Brain Res 403:198-203

Malinow R, Tsien RW (1990) Presynaptic enhancement shown by whole-cell recording of long-term potentiation in hippocampal slices. Nature 346:177-180

Nishizuka Y (1984) Studies and perspectives of protein kinase C. Science 233:305-312

Olds JL, Anderson ML, McPhie DL, Staten LD, Alkon DL (1989) Imaging of memory-specific changes in the distribution of protein kinase C in the hippocampus. Science 245:866-869

Oestreicher AB, Gispen WH (1986) Comparison of immunocytochemical distribution of the phosphoprotein B-50 in the cerebellum and hippocampus of immature and adult rat brain. Brain Res 375:267-279

Schwartzkroin PA, Wester K (1975) Long-lasting facilitation of a synaptic potential following tetanization in the *in vitro* hippocampal slice. Brain Res 89:107-119

Teyler TJ, DiScenna P. (1987) Long-term potentiation. Ann Rev Neurosci 10:131-161

PARTICIPATION OF INOSITOL PHOSPHATES, PROTEIN KINASE C, G-PROTEINS AND CYCLIC AMP IN SIGNAL TRANSDUCTION IN PRIMARY RAT ASTROGLIAL CULTURES

G. Hertting, P. Gebicke-Haerter and A. Seregi
Department of Pharmacology
University of Freiburg
Hermann-Herder Str. 5
D-7800 Freiburg
Germany

INTRODUCTION

Glial cells account for about ninety per cent of the cell population of brain tissue. Astrocytes represent about one half of the glial cells. There is increasing evidence that glia, especially astrocytes, substantially contribute to brain function. They regulate extracellular ionic milieu and volume, biophase concentration of various transmitters and control neuronal growth, guidance and development by producing various trophic factors (Hertz and Richardson, 1984; Manthorpe et al., 1986; Walz, 1987). Astrocytes are endowed with a number of binding sites for various neurotransmitters like amines, amino acids and peptides, some of which are known to be coupled to intracellular biological responses, thus representing true functional receptors (Murphy and Pearce, 1987). However, little is known about the biological events subsequent to receptor mediated astroglial responses. Astrocytes are the main source of prostanoids in the brain (Seregi et al., 1987; cf. Hertting and Seregi, 1989).

Prostanoids, which are biologically active enzymatic oxidation products of arachidonic acid (AA), have numerous effects on neuronal functions. They exhibit postsynaptic effects, like enhancements of intracellular concentrations of free calcium and of the second messengers cAMP and cGMP, as well as presynaptic effects, like inhibition of evoked noradrenaline and serotonin release, or stimulation of aspartate, GABA and taurine release from certain brain areas. They can also modulate postsynaptic actions of some excitatory and inhibitory amino acids (cf. Hertting and Seregi, 1990). In addition, PGD_2 and PGE_2 display anticonvulsant and CNS-depressant activities (cf. Hertting and Seregi, 1989).

Prostanoid synthesis in cells is initiated by the following sequence of events: an increase of free intracellular calcium eliciting the activation

NATO ASI Series, Vol. H 70
Phospholipids and Signal Transmission
Edited by R. Massarelli, L. A. Horrocks,
J. N. Kanfer, and K. Löffelholz
© Springer-Verlag Berlin Heidelberg 1993

of phospholipases, which liberate AA from membrane phospholipids. The rate-limiting step of prostanoid formation seems to be the availability of free AA. Arachidonic acid is further converted into cyclic endoperoxides by the action of cyclooxygenase. Depending on the cell type, endoperoxides are transformed into various prostanoids by the various synthetases.

In astrocytes, the major prostanoid is PGD_2 followed by thromboxane (TX). Only negligible amounts of PGF_2 and PGE_2 are synthesized (Keller et al., 1985). PG synthesis in astrocytes was stimulated by the calcium ionophore A_{23187}, by PLA_2 as well as by the PLA_2 activator mellitin. In the absence of extracellular calcium, in the presence of the intracellular calcium-immobilizing agent TMB-8 or of PLA_2-inhibitors, these compounds proved to be ineffective, suggesting the crucial role of Ca^{2+}-dependent PLA_2 in astroglial prostanoid synthesis.

In vivo, increased neuronal activity during spontaneous or experimental seizures elicits cerebral prostanoid synthesis. As it can be considered that the major source of prostanoids are the astroglial cells, the question arises how the neuronal elements communicate with the astrocytes? Neither the increase of extracellular K^+ (up to 60 mM), nor the application of putative neurotransmitters or receptor ligands, like dopamine, noradrenaline, serotonin, muscarine, carbachol, GABA, glutamate, aspartate, taurine, adenosine, angiotensin II, vasopressin, bradykinin, isoproterenol, kainic acid, and opiate receptor agonists, increased astroglial prostanoid synthesis (cf. Hertting and Seregi, 1990). Only ATP, which occurs as a co-transmitter in various neurone, induced prostanoid formation in astrocytes, presumably by acting on $P_{2\gamma}$-receptors (Gebicke-Haerter et al., 1988; Pearce et al., 1989). Astrocytes, being in close proximity to nerve terminals, can be expected as a target for the ATP released by nerve stimuli.

In addition, stimulation of astroglial P_2-receptors by ATP has been shown to mobilize intracellular Ca^{2+} (Pearce et al., 1989; McCarthy and Salm, 1991) and increase influx of extracellular Ca^{2+} (Neary et al., 1988), to initiate phosphoinositide hydrolysis giving inositol phosphates and 1,2-diacylglycerol (DAG) (Pearce et al., 1989), as well as to liberate AA (Bruner and Murphy, 1990). This points to a possible involvement of phospholipase C (PLC) and protein kinase C (PKC) in the receptor mediated prostanoid synthesis. It has been shown previously that phorbol esters, which mimic DAG in stimulating PKC, initiate PG formation and increase the effect of subthreshold concentrations of the ionophore A_{23178} on prostanoid synthesis in an overadditive manner (cf. Hertting and Seregi, 1990). This suggests a cooperation between PLA_2 and PLC in regulation of astroglial prostanoid synthesis. Puzzling is that several agonists (noradrenaline, acetylcholine, glutamate, histamine), which do stimulate receptor-mediated PI hydrolysis by phospholipase activation (Murphy and Pearce, 1987), did practically not induce AA liberation and prostaglandin synthesis (Keller et al., 1985; DeGeorge et al., 1986; Pearce et al., 1987). This seems to suggest that the supply of free AA via PI, and DAG hydrolysis is not sufficient for PG synthesis.

Astrocytes are endowed with at least three pertussis toxin (PTX)-sensitive G-proteins with molecular weights of 41, 40 and 39 kDa (Gebicke-Haerter et al., 1988). PTX-pretreatment of astrocytes reduced substantially both basal as well as ATP-induced PGD_2 synthesis. This suggests the involvement of PTX-sensitive G-proteins in some step(s) of the signal transduction of prostanoid synthesis. As shown in other cell types, cAMP, another second messenger, seems to exert an inhibitory effect on prostanoid synthesis (Malemud et al., 1986; Tan et al., 1987). We attempted to analyze the participation of the above mentioned pathways (AA liberation, PI breakdown, involvement of PKC and adenylate cyclase and possible participation of regulatory G-proteins) in the signal transduction of P_2 receptor-mediated PG synthesis in astrocytes.

METHODS

Cell cultures
Primary astroglial cultures were grown from neonatal rat brain hemispheres as described previously (Booher and Sensenbrenner, 1972, Keller et al., 1985). In brief, hemispheres were washed and cleaned of meninges, chopped and gently dissociated by trituration without enzymic treatment. Cells were separated from debris by sedimentation at low centrifugal speed (200 g, 10 min). They were seeded into 6-well plates (35 mm, Costar, Fernwald, F.R.G.) in Dulbecco's Minimum Essential Medium (DMEM) supplemented with 10% endotoxin-free fetal calf serum (Gibco, Heidelberg, F.R.G.) plus 10 ng lipopolysaccharide/ml and cultured for 21 days with two changes of media per week (Gebicke-Haerter et al., 1989).

Cell characterization
Cultures were fixed in ethanol-acetic acid (95:5 v/v) at -80°C for 12 min and incubated with GFAP (rabbit polyclonal) antisera (generous gift by L. Eng, Stanford) for type 1 astrocytes or with A2B5 (mouse monoclonal) antibodies against type 2 astrocytes for two h at 37°C. A2B5 antibodies were generously supplied by M.C. Raff, London. After removal of antibodies, cultures were incubated with fluorescein-conjugated anti-rabbit or rhodamine-conjugated anti-mouse antisera for another hour, washed, mounted in PBS-buffered glycerol and viewed with a Zeiss

fluorescence microscope under the appropriate wavelengths. Less than 0.1% of all cells proved to be type 2 and more than 90% type 1 astrocytes.

Prostaglandin synthesis
Cultures were washed 3 times in Hepes-buffered DMEM, pH 7.4, and incubated for 15 min in the same media at 37°C. Then they were incubated in Hepes/DMEM with substances as indicated in the respective experiments for another 15 min. Reactions were stopped by mixing supernatants into ice-cold 10 mM phosphate-buffered gelatin (0.1%) solution, pH 7.4, containing 1 μM indomethacin. 200-500 μl of supernatant-gelatin mixture were used for radioimmunoassays specific for PGD_2. Radioimmunoassays were done as previously described (Anhut et al., 1978). Cells were solubilized in 0.1 M NaOH and protein was determined according to Lowry et al. (1951).

cAMP formation
Cultures were washed in Dulbecco's phosphate buffered saline (DPBS), pH 7.4. One ml 0.3 M perchloric acid was then added and cultures were allowed to stand for 30 min at 4°C. Supernatants were transferred into plastic vials, neutralized with excess $CaCO_3$ and centrifuged (2,000 x g, 10 min, 4°C). 50 μl aliquots from the resultant supernatants were used for cAMP determinations. Cyclic AMP was measured by specific radioimmunoassay using an antibody developed and characterized in our laboratory (Ortmann, 1978). Protein was determined according to Lowry et al. (1951). Results are expressed in pmol cAMP per mg protein.

Arachidonic acid release
Cells were incubated overnight in presence of 0.25 μCi [^{3}H]arachidonic acid (spec.act. 207 Ci/mmol) and 10 ng/ ml unlabeled AA (to ensure an even distribution into all cellular compartments). After removal of media, they were incubated sequentially in DPBS (5 min), DPBS-1% BSA (fatty acid free) (5 min) and stimulated with ATP, A_{23178} or TPA in PBS/BSA (30 min). Supernatants were transferred into scintillation vials and radioactivity was determined. AA release was calculated as percentage of total radioactivity incorporated into cells. As verified in previous experiments by HPLC and TLC, radiolabel in those supernatants was exclusively associated with AA. Preferential binding of AA to albumin has been reported by others (Birkle and Bazan, 1985; Dieter et al., 1990).

Inositol phosphate formation
Cultures were incubated for 48 h in serum-free DMEM in the presence of 2 μCi myo-[^{3}H]inositol. They were washed 3 times in DPBS pH 7.4 and preincubated for 15 min in the same media at 37°C and incubated in 1 ml DPBS in presence of substances to be tested for another 15 min. Both preincubations and incubations were carried out in presence of 5 mM LiCl. Reactions were stopped by treating supernatant-free cultures with 1 ml 0.3

M perchloric acid. Cells were allowed to stand for 30 min at 4°C, extracts were neutralized by K_2CO_3 and analyzed for intracellular inositol phosphates (IPs). Radioactive IPs were separated by anion exchange chromatography as described by Berridge et al. (1982) and counted by liquid scintillation. Five to ten per cent of labeled *myo*-inositol was found intracellularly, and about five per cent of intracellular inositol was incorporated into lipids. Uptake of inositol varied between different cultivations, but the ratio of incorporated and free intracellular inositol remained unchanged. IP values were normalized relative to 10^6 cpm intracellular free inositol, and were expressed as percentage of unstimulated controls.

Pretreatment of cultures with toxins
Cultures were preincubated with 50 ng pertussis toxin (PTX) μml for 18 h or with 250 ng cholera toxin (CTX) μml for 6 or 18 h in serum-supplemented (PG and cAMP experiments) or serum free (IP experiments) DMEM prior to preincubations.

RESULTS AND DISCUSSION

Involvement of pertussis toxin and cholera toxin sensitive G-proteins on ATP-, ionophore-, or phorbol ester-evoked PGD_2 formation and arachidonic acid liberation.
Figure 1 shows that PTX pretreatment of astrocytes decreases both basal as well as the evoked PGD_2 synthesis of all three stimuli. Considering the decrease in basal PGD_2 synthesis following PTX pretreatment, the net formation of PGD_2 by the ionophore was not significantly affected. Basal AA release was not influenced by PTX, whereas similarly to PG synthesis, the effect of ATP or TPA on AA liberation was strongly inhibited. The ionophore-induced AA release was only negligibly reduced by PTX. Taken together, these results suggest the presence of a PTX-sensitive G-protein positively coupling the PKC and the receptor signal to a Ca^{2+}-channel. This would explain why the effect of the ionophore, which promotes the entry of Ca^{2+} independently of physiological channels, remains unaffected. ATP-induced Ca^{2+}-influx in astrocytes has already been reported (Neary et al., 1988). Measurements of free intracellular Ca^{2+} in astrocytes following the above and other stimuli, currently in progress in our laboratories, should provide direct evidence for this proposal. The additional involvement of a PTX-sensitive G-protein located with PLA_2 can also not be excluded. A G-protein coupled to PLA_2 has already been described in a variety of other cell types (cf. Burch, 1989).

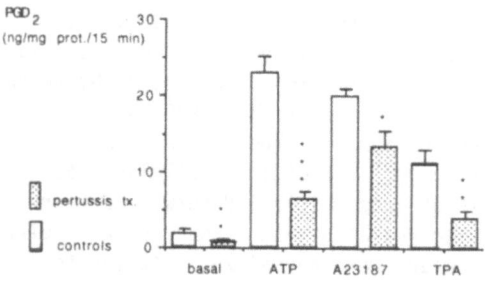

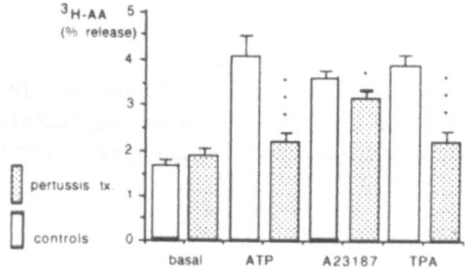

Figure 1. Effect of pertussis toxin pretreatment on ATP-, TPA- or A_{23187}-induced PGD_2 formation and arachidonic acid release in cultured astrocytes. 2 to 3 week-old astroglial cultures were pretreated with PTX (50 ng/ml) for 18 h. For arachidonic acid release cells were prelabeled with [^{3}H]arachidonic acid (0.25 μCi/ml plus 10 ng/ml unlabeled arachidonic acid) for 20 h. After removal of media, cultures were preincubated for 15 min and then stimulated for another 15 min with ATP (10^{-3} M), A23187 (10^{-6} M) or TPA (10^{-5} M). Results are means of three independent experiments performed in triplicate (n=9). Vertical bars: S.E.M.
*** p< 0.001; ** p< 0.01; * p< 0.05

In contrast to PTX, pretreatment of the cells with CTX strongly inhibited basal and evoked PGD_2 formation irrespective of the stimulus (Fig. 2). Surprisingly, basal and evoked AA liberation was not reduced, but slightly increased. This clearly indicates that a CTX-sensitive G-protein dependent step, distal to PLA_2, is involved in prostanoid synthesis. CTX-sensitive stimulatory G_s-proteins are known to be coupled to adenylate cyclase. In various cell types increased cAMP levels inhibit prostanoid synthesis (Malemud et al., 1986; Tan et al., 1987), whereas in some other cell types the effect of CTX on PG synthesis was independent of intracellular cAMP concentrations (Burch et al., 1988).

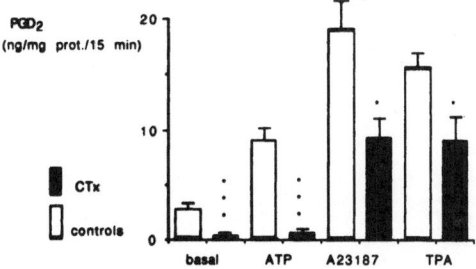

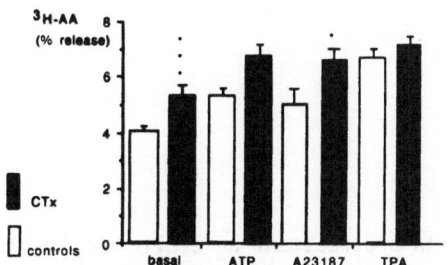

Figure 2. Effect of cholera toxin pretreatment on ATP-, TPA- or A23187-induced PGD_2 formation and arachidonic acid release in cultured astrocytes. 2 to 3 week-old astroglial cultures were pretreated with CTX (250 ng/ml) for 18 h. For arachidonic acid release cells were prelabeled with [^{3}H]arachidonic acid (0.25 $\mu Ci/ml$ plus 10 ng/ml unlabeled arachidonic acid) for 20 h. After removal of media, cultures were preincubated for 15 min and then stimulated for another 15 min with ATP (10^{-3} M), A23187 (10^{-6} M) or TPA (10^{-5} M). Results are means of three independent experiments performed in triplicate (n=9). Vertical bars: S.E.M.
*** p< 0.001; * p< 0.05

Therefore it was of interest to correlate cAMP formation and PG synthesis in astrocytes. CTX-pretreatment, indeed, markedly increased intracellular cAMP content (Fig. 3), suggesting that the second messenger cAMP may interfere with prostanoid synthesis, possibly at the level of cyclooxygenase or prostanoid synthetases. Adenylate cyclase activation in astrocytes can be achieved via β-adrenoceptor, adenosine-, PGE- (Murphy and Pearce, 1987) as well as PGI_2-receptor (Seregi et al., 1988) stimulation. Studies are in progress to clarify whether or not the decreased PG formation by CTX-pretreatment is due to increased levels of cAMP.

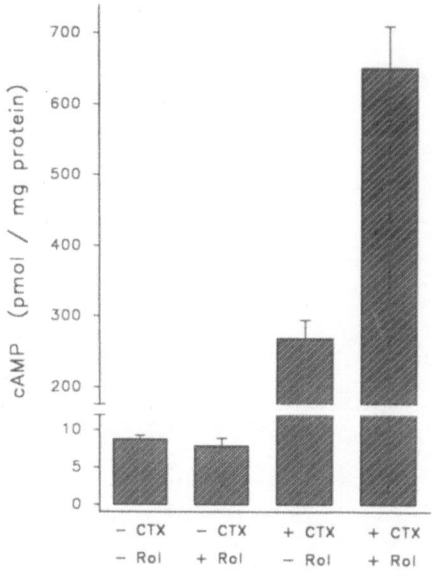

Figure 3. Effect of cholera toxin pretreatment on cyclic AMP levels in cultured astrocytes. 2 to 3 week-old astroglial cultures were pretreated with CTX (250 ng/ml) in the absence or presence of 10^{-5} M rolipam for 6 h. After removal of media intracellular cAMP levels were estimated. Results are means of three independent experiments performed in triplicate (n=9). Vertical bars: S.E.M.

Involvement of inositol phosphate pathway and G-proteins in PGD_2 synthesis

ATP, ATP[γS] and ADP markedly stimulated IP formation, adenosine was less potent and AMP was ineffective (Fig. 4). This order of potency closely resembles the effect of these compounds on PGD_2 formation, suggesting that both effects are initiated via the same or a similar $P_{2\gamma}$-receptor. Noradrenaline (NA), which is not able to induce prostanoid synthesis in astrocytes (Keller et al., 1987), also concentration-dependently increased total IP-formation. This effect of NA was inhibited by prazosine (IC_{50} ca. 5 x 10^{-8}M) and mimicked by similar concentrations of phenylephrine, suggesting that an α_1-adrenoceptor is involved. The effect of NA was more pronounced than that of ATP (Fig. 5). This finding seemed to contradict the hypothesis that PI-breakdown products participate in the stimulation of PG synthesis.

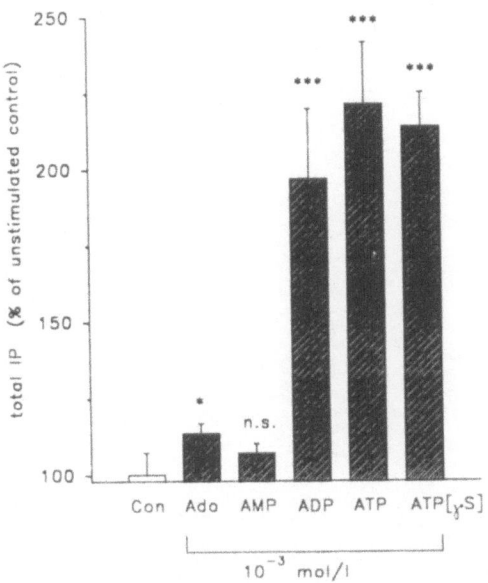

Figure 4. Effect of adenine nucleotides on inositol phosphate formation in cultured astrocytes. 2 to 3 week-old astroglial cultures were preloaded with myo-[³H]inositol (2 µCi/ml) for 48 h. After removal of media, cultures were preincubated for 15 min and then stimulated for another 15 min by adenosine (Ado), adenosine 5'-monophosphate (AMP), adenosine 5'-diphosphate (ADP), adenosine 5'-triphosphate (ATP), or adenosine 5'-O-(3-thiotriphosphate) (ATP[γS]) at 10^{-3} M concentrations each. Both preincubations and incubations were carried out in the presence of 5 mM LiCl. Results are means of three independent experiments performed in triplicate (n=9). Vertical bars: S.E.M.

Pearce et al. (1989) have shown that both ATP and NA induced accumulation of IPs and Ca^{2+}-mobilization in astrocytes, but only the ATP stimulus provoked prostanoid synthesis. The effects of ATP and NA were additive on IP formation. In contrast, there was non-additivity of the combined treatment by these agonists on Ca^{2+}-mobilization. ATP-evoked prostanoid synthesis was even decreased in the presence of NA. To explain this phenomenon, these authors suggest that although the activation of both receptors initiates the breakdown of phosphoinositides, possibly using different pools, the Ca^{2+} mobilized by NA may be not sufficient to activate PLA_2.

The comparison of the concentrations of IP_3, IP_2 and IP_1 during the time-course of their formation following the two different stimuli (i.e.

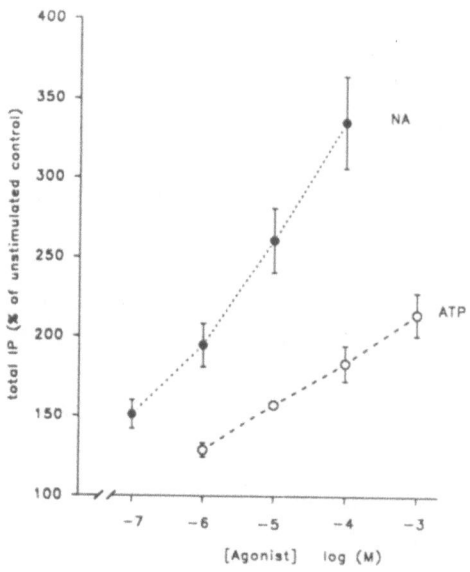

Figure 5. Concentration-response relationship of ATP- and noradrenaline-induced inositol phosphate formation in cultured astrocytes. 2 to 3 week-old astroglial cultures were preloaded with myo-[^{3}H]inositol (2 µCi/ml) for 48 h. After removal of media, cultures were preincubated for 15 min and then stimulated for another 15 min by ATP or by NA at concentrations as indicated. Both preincubations and incubations were carried out in the presence of 5 mM LiCl. Results are means of three independent experiments performed in triplicate (n=9). Vertical bars: S.E.M.

ATP or NA) exhibited completely different patterns (Fig. 6). ATP elicited a marked (2.1-fold over zero time value) but transient peak of IP_3 at 15 sec following application; from then on the concentration of IP_3 continuously declined. In contrast, IP_3 concentrations were only slightly (by 1.4-fold of zero time values) enhanced upon NA stimulation. The ability of ATP to promote AA release and PGD_2 synthesis seems to coincide with the more marked IP_3 formation soon after the initiation of the signal. This could serve as an explanation for the smaller increase of intracellular Ca^{2+}-concentrations after NA stimulation in comparison to ATP simulation, as observed by Pearce et al. (1989).

After ATP stimulation, IP_2 levels showed a profile with a delayed peak, similar to that of IP_3, whereas IP_1 levels rose steeply up to one min and continued to increase slowly later reaching twofold of initial values

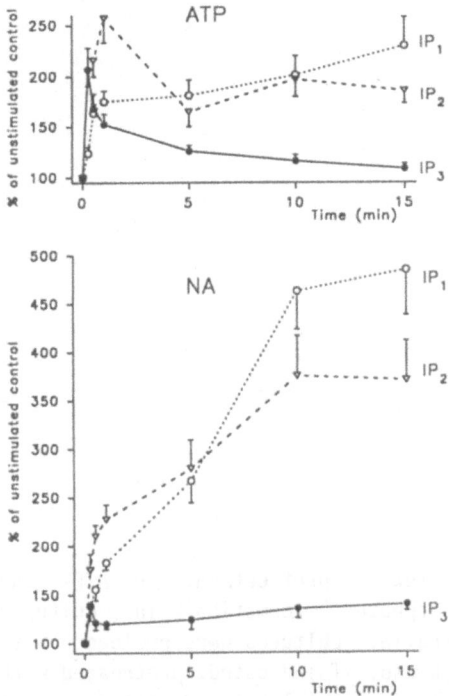

Figure 6. Time-course of ATP- and noradrenaline-induced formation of individual inositol phosphates in cultured astrocytes. 2 to 3 week-old astroglial cultures were preloaded with myo-[^{3}H]inositol (2 μCi/ml) for 48 h. After removal of media, cultures were preincubated for 15 min and then incubated with ATP (10^{-3} M) or NA (10^{-4} M) for different times, as indicated. Both preincubations and incubations were carried out in the presence of 5 mM LiCl. Control (zero time) IP radioactivity values (cpm/dish) were 4642 ± 392 for IP$_1$, 1016 ± 110 for IP$_2$ and 1509 ± 149 for IP$_3$ in the ATP experiments and 4628 ± 632 for IP$_1$, 902 ± 132 for IP$_2$ and 1405 ± 232 for IP$_3$ in the NA experiments, respectively. Results are means of six independent experiments performed in triplicate (n=18). Vertical bars: S.E.M.

at 15 min. Noradrenaline, on the other hand, increased IP$_1$ and IP$_2$ formation much more effectively in comparison to ATP. IP$_2$ and IP$_1$ rose rapidly, reaching maximal levels at 10 min (3.5-fold and 5.5-fold of initial values, respectively).

Calculating on the basis of absolute values and considering that the kinetics of both IP$_3$ and IP$_2$ hydrolysis are comparable and independent of the stimuli used, it seems unlikely that IP$_3$ can be the essential

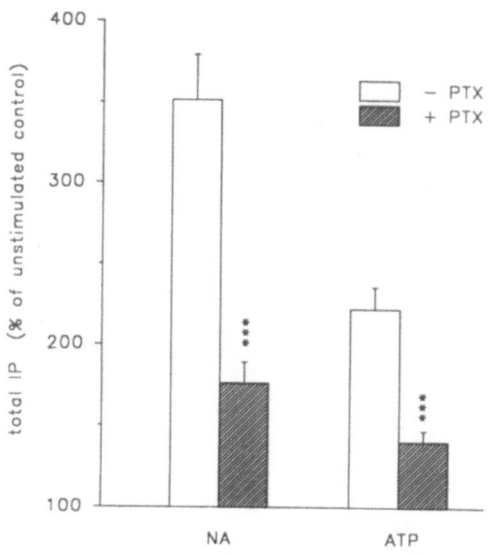

Figure 7. Effect of pertussis toxin pretreatment on ATP- and noradrenaline-induced inositol phosphate formation in cultured astrocytes. 2 to 3 week-old astroglial cultures were preloaded with myo-[^{3}H]inositol (2 μCi/ml) for 48 h and, if indicated, pretreated with PTX (50 ng/ml) during the last 18 h of inositol labeling. After removal of media, cultures were preincubated for 15 min and then stimulated for another 15 min by ATP (10^{-3} M) or NA (10^{-4} M). Preincubations and incubations were carried out in the presence of 5 mM LiCl. Results are means of three independent experiments performed in triplicate (n=9). Vertical bars: S.E.M.   *** p< 0.001;

source of NA-induced IP$_1$ formation. One possible explanation for this phenomenon could be that, whereas the ATP-receptor is linked to a PIP$_2$-specific PLC, the α_1-adrenoceptor is coupled to a PLD which hydrolyses PIP and/or PIP$_2$, yielding IP$_2$/IP$_1$ and phosphatidic acid (PA). The possibility that PA is transformed in astrocytes into the second messenger DAG seems unlikely, since in contrast to the effect of the DAG-mimetic TPA, there is no PG formation following α_1-adrenoceptor activation. Certainly, sequestrations of the various intracellular messengers as IPs, DAG and Ca^{2+} into different compartments could also explain the differences of the PG response following different stimuli. Functional heterogeneity of the cultured astroglia type 1 population (Hansson et al., 1987; El-Etr et al., 1989; McCarthy and Salm, 1991) may also account for the qualitative and quantitative difference in the ability to respond to the different stimuli.

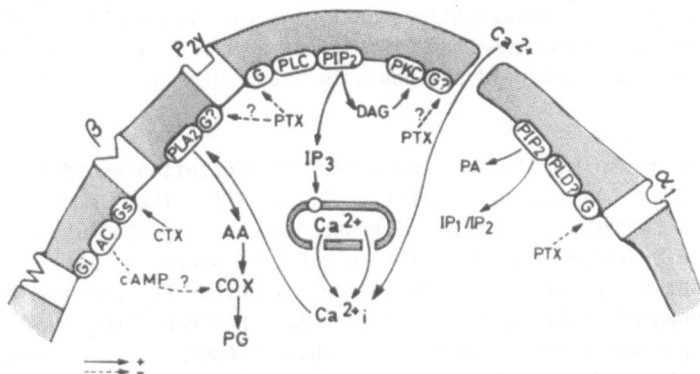

Figure 8. Receptor-mediated signal transduction pathways and prostanoid synthesis in astroglia.

Pretreatment with PTX markedly inhibited the formation of total IPs (Fig. 7) in response to both stimuli. This was the case for all three individual IP fractions (data not shown). This indicates that both receptors operate via a PTX-sensitive propagatory G-protein which couple the receptor to PLC (ATP) or PLD (NA). Inhibition of NA-induced PI breakdown by PTX-pretreatment was recently reported by others as well (Wilson and Minneman, 1990). In contrast to the effect of PTX, pretreatment of astrocytes by CTX did not attenuate ATP- or NA-induced PI breakdown (data not shown).

SUMMARY AND CONCLUSIONS

Increased neuronal activity initiates cerebral prostanoid synthesis. Astrocytes seem to be the main source of prostanoid formation. PG synthesis in astrocytes can be elicited by the Ca^{2+}-ionophore A_{23187} or by DAG-mimetic phorbol esters, which activate PKC. Free Ca^{2+} stimulates PLA_2 liberating AA, the precursor of prostanoids. DAG originates from phospholipid breakdown by PLC or PLD. PIP_2 hydrolysis by PLC yields, along with DAG, an additional second messenger, IP_3. Among several neuroligands which stimulate receptor-mediated PI hydrolysis, ATP exclusively initiates prostanoid synthesis in astrocytes.

The results of the present investigations analyzing the various steps (AA release, IP formation, involvement of G-proteins and cAMP) of the signal transduction pathway of prostanoid synthesis in astrocytes are summarized in Fig. 7.

Pertussis toxin treatment diminished markedly AA liberation as well as PGD_2 formation induced by ATP and the PKC-activator TPA. The PTX effect on the ionophore, however, was much less pronounced. These findings can be explained by G-proteins coupling PKC to a Ca^{2+}-channel and possibly also the $P_{2\gamma}$-receptor to PLA_2. Stimulation of both $P_{2\gamma}$-purino- or α_1-adrenoceptor enhanced intracellular IP levels, however, IP_3 levels rose only after purinoceptor stimulation. This strongly suggests that this receptor is coupled to PLC, releasing the Ca^{2+}-mobilizing IP_3 and simultaneously the PKC-activator DAG. Stimulation of the α_1-adrenoceptor, on the other hand, predominantly increases IP_1 (and possibly also PA), which suggests the coupling of this receptor to PLD. These intracellular signals do not appear to increase free intracellular Ca^{2+} sufficiently, which would activate PLA_2 to supply free AA for PG synthesis. This would explain why this latter stimulus is incapable of initiating PG synthesis in astrocytes. In addition, PTX also inhibited effectively IP formation induced by both $P_{2\gamma}$-purino- or α_1-adrenoceptor activation. This latter observation suggests that PTX-sensitive propagatory G-proteins are coupling these two receptors to the corresponding phospholipases. A coupling of the purinoceptor via a PTX-sensitive G-protein to PLA_2 cannot be excluded. CTX pretreatment solely inhibited PG formation, whereas AA release and IP formation were not attenuated. CTX treatment markedly increases intracellular cAMP concentration which has been shown to inhibit PG formation in other cell types.

REFERENCES

Anhut H, Peskar BA, Wächter W, Gräbling B, Peskar BM (1978) Radioimmunological determination of prostaglandin D_2 synthesis in human thrombocytes. Experientia 34:1494-1496

Berridge MJ, Downes CP, Hanley MR (1982) Lithium amplifies agonist-dependent phosphatidylinositol responses in brain and salivary glands. Biochem J 206:587-595

Birkle DL, Bazan NG (1985) Metabolism of arachidonic acid in the central nervous system. The enzymatic cyclooxygenation and lipoxygenation of arachidonic acid in the mammalian retina. In: Horrocks LA, Kanfer JN, Porcellati G, (eds) Phospholipids in the Nervous System, Vol. 2, (eds.) Raven Press, New York, pp 193-208

Booher J, Sensenbrenner M (1972) Growth and cultivation of dissociated neurons and glial cells from embryonic chick, rat and human brain in flask cultures. Neurobiology 2:97-105

Bruner G, Murphy S (1990) ATP-evoked arachidonic acid mobilization in astrocytes is via a $P_{2\gamma}$-purinergic receptor. J Neurochem 55:1569-1575

Burch RM, Jelsema C, Axelrod J (1988) Cholera toxin and pertussis toxin stimulate prostaglandin E_2 synthesis in a murine macrophage cell line. J Pharm Exptl Ther 224:765-773

Burch RM (1989) G protein regulation of phospholipase A_2. Molec Neurobiol 3:155-171

DeGeorge JJ, Morell P, McCarthy KD, Lapetina EG (1986) Adrenergic and cholinergic stimulation of arachidonate and phosphatidate metabolism in cultured astroglial cells. Neurochem Res 11:1061-1071

Dieter P, Krause H, Schulze-Specking A (1990) Arachidonate metabolism in macrophages is affected by albumin. Eicosanoids 3:45-51

El-Etr M, Cordier J, Torrens Y, Glowinski J, Prémont J (1989) Pharmacological and functional heterogeneity of astrocytes: regional differences in phospholipase C stimulation by neuromediators. J Neurochem 52:981-984

Gebicke-Haerter P.J., Bauer J., Schobert A. and Northoff H. (1989) Lipopolysaccharide-free conditions in primary astrocyte cultures allow growth and isolation of microglial cells. J Neurosci 9, 183-194

Gebicke-Haerter PJ, Wurster S, Schobert A, Hertting G (1988) P_2-purinoceptor induced prostaglandin synthesis in primary rat astrocyte cultures. Naunyn-Schmiedeberg's Arch Pharmacol 338:704-707

Gebicke-Haerter PJ, Seregi A, Wurster S, Schobert A, Allgaier C, Hertting G (1988) Multiple pertussis toxin substrates as candidates for regulatory G-proteins of adenylate cyclase coupled to the somatostatin receptor in primary rat astrocytes. Neurochem Res 13:997-1001

Hansson E, Simonsson P, Alling C (1987) 5-Hydroxytryptamine stimulates the formation of inositol phosphate in astrocytes from different regions of the brain. Neuropharmacology 26:1377-1382

Hertting G, Seregi A (1989) Formation and function of eicosanoids in the central nervous system. Ann NY Acad Sci 559:84-99

Hertting G, Seregi A (1990) Synthesis, origin and involvement of eicosanoids in the convulsing brain. In: Bazan NG (ed) Lipid Mediators in Ischemic Brain Damage and Experimental Epilepsy, New Trends in Lipid Mediators Res., vol.4, Karger, Basel, pp 162-189

Hertz L, Richardson JS (1984) Is neuropharmacology merely the pharmacology of neurons or are astrocytes important too? Trends Pharmacol Sci 5:272-276

Keller M, Jackisch R, Seregi A, Hertting G (1985) Comparison of prostanoid forming capacity of neuronal and astroglial cells in primary cultures. Neurochem Int 7:655-665

Keller M, Seregi A, Jackisch R, Hertting G (1985) Prostanoid formation in primary astroglial cell cultures: Ca^{2+}-dependency and stimulation by A_{23187}, melittin and phospholipases A_2 and C. Neurochem Int 4:433-443

Lowry OH, Rosebrough AL, Farr AL, Randall RJ (1951) Protein measurement with the Folin phenol reagent. J Biol Chem 193:265-275

Malemud CJ, Mills TM, Papay RS (1986) Supression of prostaglandin synthesis by analogues of cyclic AMP and forskolin in chondrocyte monolayer cultures. Prostaglandins 32:495-502

Manthorpe M, Rudge JS, Varon S (1986) Astroglial cell contributions to neuronal survival and neuritic growth. In: Fedoroff S and Vernadakis A (eds) Astrocytes, vol.2, Academic Press, New York, pp 315-376

McCarthy KD and Salm AK (1991) Pharmacologically-distinct subsets of astroglia can be identified by their calcium response to neuroligand. Neuroscience 41:325-333

Murphy S, Pearce B (1987) Functional receptors for neurotransmitters on astroglial cells. Neurosci 22:381-394

Neary JT, van Bremen C, Forster E, Norenberg LO, Norenberg MD (1988) ATP stimulates calcium influx in primary astrocyte cultures. Biochem Biophys Res Commun 157:1410-1416

Ortmann R (1978) Effect of PGI_2 and stable endoperoxide analogues on cyclic nucleotide levels in clonal cell lines of CNS origin. FEBS Lett 90:348-352

Pearce B, Jeremy J, Morrow C, Murphy S, Dandona P (1989) Inositol phospholipids are probably not the source of arachidonic acid for eicosanoid synthesis in astrocytes. FEBS Lett 211:73-77

Pearce B, Murphy S, Jeremy J, Morrow C, Dandona P (1989) ATP-evoked Ca^{2+} mobilization and prostanoid release from astrocytes: P_2-purinergic receptors linked to phosphoinositide hydrolysis. J Neurochem 52:971-977

Seregi A, Keller M, Hertting G (1987) Are cerebral prostanoids of astroglial origin? Studies on the prostasnoid forming system in developing rat brain and primary cultures of rat astrocytes. Brain Res 404:113-120

Seregi A, Schobert A, Hertting G (1988) The stable prostacyclin-analogue iloprost, unlike prostanoids and leukotrienes, potently stimulates cyclic adenosine monophosphate synthesis in primary astroglial cell cultures. J Pharm Pharmacol 40:437-438

Tan CH, Lam TJ, Wong LY, Pang MK (1987) Prostaglandin synthesis and its inhibition by cyclic AMP and forskolin in postpartum follicles of the guppy (Poecilia reticulata). Prostaglandins 34:697-715

Walz W (1978) Potassium channels and carriers in glial cell membranes. In: Grisar T, Frank G, Hertz L, Norton WT, Sensenbrenner M, Woodbury D (eds), Dynamic properties of glial cells II: cellular and molecular aspects, Pergamon Press, Oxford, pp. 145-154

Wilson KM, Minneman KP (1990) Pertussis toxin inhibits norepinephrine-stimulated inositol phosphate formation in primary brain cell cultures. Molec Pharmacol 38:274-281

SPHINGOSINE, A BREAKDOWN PRODUCT OF CELLULAR SPHINGOLIPIDS, IN CELLULAR PROLIFERATION AND PHOSPHOLIPID METABOLISM

Sarah Spiegel
Department of Biochemistry and Molecular Biology
Georgetown University Medical Center
357 Basic Science Building
3900 Reservoir Road NW
Washington, DC 20007

Sphingolipids, important lipid constituents of plasma membranes, consist of several long chain bases of which sphingosine (4-*trans*-sphingenine) is the most prominent. Sphingolipids have long been implicated in diverse cellular functions, including cellular communication, transformation, proliferation, differentiation, and modulation of receptor function (reviewed in Hakomori,1985;1986; Sweeley,1985). However, the mechanism of action and signal transduction pathways modulated by these complex lipids have largely remained ill defined (Spiegel and Panagiotopoulos,1988; Spiegel,1989). A tantalizing link between sphingolipids and signal transduction surfaced recently when sphingosine was found to inhibit protein kinase C *in vitro* and to be an inhibitor of a variety of protein kinase C-dependent processes *in vivo* (Hannun and Bell,1987;1989; Merrill and Stevens,1989). Thus, it has been suggested that sphingosine is an endogenous inhibitor of protein kinase C, opposing the action of DAG (Hannun and Bell,1987;1989; Merrill and Stevens,1989).

Since protein kinase C appears to play a prominent role in cellular proliferation (Nishizuka,1986;1988), we investigated the effect of sphingosine on the growth of Swiss 3T3 fibroblasts, a convenient model system for the study of cell activation and growth. Confluent cultures of Swiss 3T3 fibroblasts become quiescent in the G_1 to G_0 phase of the cell cycle when deprived of serum (Rozengurt,1986). However, these quiescent cultures can be stimulated to initiate DNA synthesis not only by addition of serum but also by various mitogenic compounds.

NATO ASI Series, Vol. H 70
Phospholipids and Signal Transmission
Edited by R. Massarelli, L. A. Horrocks,
J. N. Kanfer, and K. Löffelholz
© Springer-Verlag Berlin Heidelberg 1993

Sphingosine Stimulates Proliferation of Quiescent 3T3 Fibroblasts

Surprisingly, in contrast to previous reports that sphingosine is highly cytotoxic for a variety of cell types (Hannun and Bell,1987;1989; Merrill and Stevens,1989), we have found that sphingosine at low concentrations stimulates DNA synthesis and acts synergistically with known growth factors to induce proliferation of quiescent Swiss 3T3 fibroblasts. Although sphingosine is only a weak mitogen by itself, other known mitogens, such as insulin or EGF, likewise are not potent mitogenic agents unless they are added in combination (Rozengurt,1986). Sphingosine also potentiated the growth promoting effects of optimal stimulatory concentrations of insulin, EGF and even unfractionated FCS. This synergistic interaction between sphingosine and growth factors was observed even in combinations with two growth factors, such as EGF and insulin or the B subunit and insulin. While any two of the growth factors synergized with each other, addition of sphingosine caused a further potentiation of [^{3}H]thymidine incorporation (Table I). A mitogenic effect was observed at a concentration of sphingosine as low as 0.2 µM and maximum stimulation was achieved at 10-20 µM. Up to this concentration, there was no loss of cell viability as measured by trypan blue exclusion. However, concentrations of sphingosine above 50 µM inhibited DNA synthesis and were strongly cytotoxic (Zhang et al,1990a).

TABLE I. SPECIFICITY OF THE EFFECT OF SPHINGOSINE ON DNA SYNTHESIS IN QUIESCENT CULTURES OF SWISS 3T3 FIBROBLASTS

Stimulants	[^{3}H]Thymidine Incorporation (percent of control)		
	Sphingosine	N-Stearoyl-sphingosine	Stearylamine
None	167±11	101±11	67±14
Insulin	395±15	98±12	90±5
EGF	230±23	97±16	96±6
EGF plus Insulin	225±21	96±11	92±5
B plus Insulin	483±26	111±19	68±4
TPA	514±14	118±18	43±5
TPA plus Insulin	339±21	105±16	62±6

Quiescent cultures of Swiss 3T3 cells were exposed to the indicated mitogens in the presence of sphingosine, N-stearoylsphingosine or stearylamine (10 µM) and [^{3}H]thymidine incorporation was measured (Zhang,1990a). The data is expressed as percent of the incorporation in the control cells not treated with a long-chain aliphatic compound. The concentrations of the mitogenic agents were as follows: Insulin, 2 µg/ml; EGF, 5 ng/ml; B subunit of cholera toxin (B), 1 µg/ml; TPA, 100 nM.

Specificity of the Sphingosine Effect

Since sphingosine has a common potentiation effect on mitogenesis with various growth factors, it could induce non-specific membrane perturbations which trigger subsequent changes in transmembrane signaling systems. However, this is an unlikely explanation due to the lack of effect of structurally related compounds. N-stearoylsphingosine, which lacks a free amino group, and other long chain aliphatic amines, including stearylamine and oleylamine, which have alkyl chain lengths similar to sphingosine but lack both hydroxyl groups, did not mimic the mitogenic effects of sphingosine nor did they potentiate the effect of any combination of other growth factors (Zhang et al,1990a).

Is the Mitogenic Effect of Sphingosine Mediated Through Protein Kinase C?

In contrast to previous studies, we have observed that the effect of sphingosine on cellular proliferation is clearly independent of protein kinase C (Zhang et al,1990a). Several lines of evidence suport this conclusion. Firstly, sphingosine which has been proposed to be a negative modulator of protein kinase C, also potentiates, rather than inhibits, the mitogenic effect of the tumor promoter, 12-O-tetradecanoylphorbol 13-acetate (TPA), whose action is known to be mediated through activation of protein kinase C. The synergistic effect of optimal concentrations of sphingosine and TPA suggests that they do not share a common pathway of mitogenic action. Secondly, in contrast to sphingosine, H-7, which is known to inhibit protein kinase C in intact cells, inhibits the mitogenic response to TPA. It is important to note that sphingosine, unlike H-7, did not inhibit the stimulation of TPA-induced phosphorylation of an acidic cellular 80 kDa protein, which is a specific marker for the activation of protein kinase C in Swiss 3T3 fibroblasts. Interestingly, stearylamine, which has been reported to be as potent an inhibitor of protein kinase C *in vitro* and *in vivo* as is sphingosine, did not mimic the mitogenic effect of sphingosine nor did it potentiate the effect of other growth factors. Rather, similar to the effect of H-7, it inhibited TPA-induced mitogenesis. Thirdly, mitogenic concentrations of sphingosine do not compete with phorbol dibutyrate for specific receptors (protein kinase C) on intact cells (Fig. 1). Finally, down regulation of protein kinase C, by prolonged treatment with TPA, has no effect on sphingosine-mediated mitogenesis.

In summary, our discovery that exogenously added sphingosine stimulates the proliferation of quiescent cultures of Swiss 3T3 fibroblasts raises the intriguing possibility that these breakdown products of cellular sphingolipids may play an important role as **positive** modulators of cell growth acting in a fundamentally different, protein kinase C-independent pathway. Recently, several other reports have also raised the possibility that sphingosine has complex biological effects which are at least in part protein kinase C-independent (Faucher et al,1988; Davis et al,1988; Igarashi et al,1989; Jefferson and Schulman,1988; Winicov and Gershengorn,1988). Thus, caution should be exercised in the use of sphingosine as a tool to unravel the involvement of protein kinase C in a variety of cellular processes. Only high, cytotoxic concentrations of sphingosine have any effect on protein kinase C in our system (Fig. 1), indicating that sphingosine may not be a physiological modulator of protein kinase C at all but may modulate some other novel signal transduction pathway and that the target of sphingosine action still remains to be uncovered.

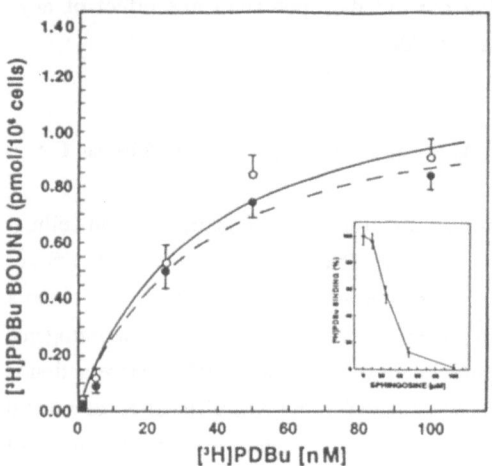

Fig. 1. Effect of sphingosine on specific binding of [3H]PDBu to Swiss 3T3 cells. Confluent and quiescent cultures were incubated in the absence (o) or presence of sphingosine (10 μM) for 1 h, and [3H]PDBu binding was measured (Zhang et al, 1990a). Where indicated (o), the cells were treated with sphingosine (10 μM) for 1 h before and during the binding assay. The curves were generated by computer analysis of the data solid line for control, dashed line for cells treated with sphingosine. Insert: Cells were pretreated with different concentrations of sphingosine for 1 h and specific [3H]PDBu binding (at 25 nM) was measured. The data are expressed as percent of control without sphingosine.

What Are the Cellular Targets of Sphingosine?

Although there is an abundance of evidence demonstrating that many growth factors mediate their action through increased phosphoinositide metabolism (Berridge,1984;1985; Nishizuka,1988), it is also clear that in many cases there is an increased breakdown of other phospholipids as a result of activation of phospholipases C and D (Exton,1990). It has been suggested that, in addition to the two lipid intermediates, diacylglycerol and inositol trisphosphate, phosphatidic acid also plays a crucial role in signal transduction and cellular proliferation (Exton,1990; Moolenaar et al,1986). In any event, there is no doubt that cellular proliferation is closely correlated with active metabolism of phospholipids (Hasegawa-Sasaki, 1985). Therefore, we initiated a study to determine whether sphingosine has any effects on phospholipid metabolism in quiescent cultures of Swiss 3T3 fibroblasts by determining ^{32}Pi incorporation into various phospholipids.

Increases in Phosphatidic Acid Levels Accompany Sphingosine Stimulated Proliferation of Quiescent Swiss 3T3 Cells

Addition of optimal mitogenic concentrations of sphingosine to cells prelabeled with ^{32}Pi for 24 h to isotopic equilibrium significantly stimulated ^{32}P incorporation into phosphatidic acid in (Table II). Sphingosine induced a 2-fold increase in the labeling of phosphatidic acid and decreased phosphatidylcholine labeling to 70% of the control without any accompanying changes in the other phospholipids (Table II). Since the cells were metabolically labeled to isotopic equilibrium, it is most likely that these changes reflect changes in phosphatidic acid mass rather than enhanced radioactive labeling. However, to further substantiate this point, the cells were double-labeled with both [^{3}H]-glycerol and ^{32}P$_i$ to isotopic equilibrium. Mitogenic concentrations of sphingosine induced identical stimulations of incorporation of [^{3}H]-glycerol (2.0 ± 0.1 fold increase) and ^{32}P (1.9 ± 0.1 fold increase) into phosphatidic acid, indicating that the measured changes likely reflect increases in phosphatidic acid mass. Structurally related analogs of sphingosine, such as N-stearoylsphingosine and other long chain aliphatic amines, which did not mimic the mitogenic effect of sphingosine did not elevate phosphatidic acid levels.

TABLE II. EFFECT OF SPHINGOSINE ON INCORPORATION OF ^{32}Pi AND [^{3}H]GLYCEROL INTO PHOSPHOLIPIDS IN QUIESCENT SWISS 3T3 FIBROBLASTS

	Incorporation of Radioactivity into Phospholipids (cpm)			
	CONTROL		SPHINGOSINE - TREATED	
	[^{32}P]Pi	[^{3}H]glycerol	[^{32}P]Pi	[^{3}H]glycerol
SM	681±17	ND	667±106	ND
PC	26776±457	10600±375	21022±866	9699±559
PE	4176±905	2734±466	3736±764	3031±627
PI	4038±312	1485±105	4168±377	1845±266
PA	564±89	220±58	1336±209	496±101
SP1P	355±66	ND	4928±312	ND

Quiescent cultures of Swiss 3T3 cells prelabeled with [2-^{3}H]glycerol (2 µCi/ml) and ^{32}P$_i$ (40 µCi/ml) were treated with sphingosine (20 µM) or vehicle, harvested and the lipids extracted and chromatographed. The indicated phospholipid bands were excised from the plates and the radioactivity determined (Zhang et al,1990b). PC, phosphatidylcholine; PE, phosphatidylethanolamine; PI, phosphatidylinositol; SM, sphingomyelin; PA, phosphatidic acid; SP1P, sphingosine-1-phosphate. ND, none detected (radioactivity not significantly different than background).

A significant increase in [^{32}P]-labeled phosphatidic acid was detected within 10 min after exposure of the cells to sphingosine. Thus, the response to sphingosine was rapid, reaching nearly maximal levels within 60 min, when 2-fold stimulation over the basal value was observed. After this period, the level decreases and returns to the basal value by 4 h. The rapid response suggests that the formation of phosphatidic acid should be considered to be an early event that precedes the entry into the S phase of the cell cycle.

The observation that sphingosine, at mitogenic concentrations, increases the level of phosphatidic acid has important implications for the mechanism of action of sphingosine since it is well known that phosphatidic acid is a potent mitogen for various cell types, including Swiss 3T3 cells (Moolenaar et al,1986; Yu et al,1988).

There is a great deal of similarity in the mitogenic properties of phosphatidic acid and sphingosine. First, both sphingosine and phosphatidic acid stimulated DNA synthesis in cells made protein kinase C-deficient by prolonged treatment with phorbol ester and sphingosine still stimulated similar increases in phosphatidic acid in these cells. Secondly, the time course of the increase in DNA synthesis in quiescent cultures of Swiss 3T3 cells induced by sphingosine treatment was essentially the same as that induced in response to phosphatidic acid. For both mitogens, the lag period preceding the onset of DNA synthesis was identical (15h) and the rate of entry into S phase was exactly the same. Furthermore, not only did sphingosine stimulate [^{3}H]thymidine incorporation with similar efficiency and kinetics as phosphatidic acid, it also induced similar morphological alterations.

The striking similarity in the mitogenic properties of phosphatidic acid and sphingosine together with the observation that sphingosine induces an early rise in the levels of phosphatidic acid raises the possibility that the mitogenic effect of sphingosine is mediated through phosphatidic acid generation. If these two lipid mitogens stimulate DNA synthesis by an identical pathway, they should not interact, either additively or synergistically, in stimulating DNA synthesis when tested at optimal concentrations. Both sphingosine and phosphatidic acid acted synergistically with a variety of growth factors, such as insulin, EGF, TPA, fibroblast growth factor, and unfractionated fetal calf serum. In sharp contrast, sphingosine and phosphatidic acid did not have additive or synergistic effects in either the presence or absence of other growth factors. Taken together, these findings indicate that sphingosine and phosphatidic acid control cellular responses in Swiss 3T3 cells by a common mechanism and that at least part of the mitogenicity induced by sphingosine may be mediated by a rapid rise in phosphatidic acid. If indeed the mitogenic effect of sphingosine is related to phosphatidic acid accumulation, then it should also modify the same effector systems already established for phosphatidic acid in these cells. Previously, it has been shown that incubation of Swiss 3T3 fibroblasts with phosphatidic acid decreases cellular cAMP levels due to inhibition of adenylate cyclase activity and also causes rapid generation of inositol phosphates, most probably due to activation of phospholipase C (Murayama and Ui,1987). Similar to the actions of phosphatidic acid on signal transduction in Swiss 3T3 cells, sphingosine also caused a drastic decrease in the cellular cAMP levels and provoked a rapid generation of inositol phosphates.

Phosphatidic acid is a potent mitogen for a variety of cell types that evokes growth factor-like effects: it raises cytoplasmic pH, induces expression of *c-fos* and *c-myc* protooncongenes and also triggers the hydrolysis of phosphoinositides (Moolenaar et al,1986; Yu et al,1988). It has been shown that phosphatidic acid is produced in quiescent 3T3 fibroblasts by activation of phospholipase D in response to certain mitogens and it has been suggested that elevation of endogenous phosphatidic acid is an essential step in mitogenic signal transduction cascades (Exton,1990; Ben-Av and Liscovitch,1989). Furthermore, it has been suggested that phosphatidic acid is related to the biological action of cellular *ras* activity (Yu et al,1988; Tsai et al,1989). Cellular *ras* protein participates as a molecular switch in the early steps of the signal transduction pathway associated with cell growth (Barbacid,1987). When the protein is in the GTP-complexed form it is active in the signal transduction pathway, whereas, it is inactive in its GDP-complexed form. A cytoplasmic GTPase activating protein (GAP) stimulates the conversion to the inactive form. Recently it has been shown that GAP activity is inhibited by phosphatidic acid (Tsai et al,1989). This finding suggests that phosphatidic acid might be involved in positively modulating *ras* activity during mitogenic stimulation through its effect on GAP (Tsai et al,1989). In view of the prominent role of phosphatidic acid in signal transduction and cellular proliferation, our observations that sphingosine, at mitogenic concentrations, increases the level of phosphatidic acid and also mimics the effects of phosphatidic acid on signal transduction, have important implications for the mechanism of action of sphingosine.

By what mechanism does sphingosine stimulate phosphatidic acid levels? One possibility is that sphingosine stimulates DAG kinase-catalyzed phosphorylation of DAG. *In vitro* studies demonstrated that sphingosine activates the 80 kDa isoenzyme form of DAG kinase while inhibiting another form of 150 kDa (Sakane et al,1990). Thus, the effect of sphingosine on DAG kinase could be different depending on the type of DAG kinase isozyme present in a particular cell type. In this regard, it has been observed that sphingosine causes accumulation of phosphatidic acid in the human leukemic Jurkat cell line which has been reported to be enriched in the 80 kDa isozyme form of DAG kinase (Sakane et al,1990). However, it should be mentioned that Swiss 3T3 fibroblasts do not contain immunologically detectable levels of this isozyme and in a preliminary study, we did not find evidence for the involvement of this pathway. A second possibility for which there is some indirect support is that phosphatidic acid is formed by the action of phospholipase D on phospholipids. From our results it seems clear that mitogenic concentrations of sphingosine cause decreases in the level of phosphatidylcholine. Higher concentrations of sphingosine drastically decrease phosphatidylcholine levels accompanying the large increases in phosphatidic acid. These results suggest that sphingosine activates phospholipase D. This presumption is supported by another study showing that sphingosine stimulates the release of ethanolamine from phosphatidylethanolamine in NIH 3T3 fibroblasts (Kiss and Anderson,1990). In addition, in a recent report it has been demonstrated that high concentrations of sphingosine activate phospholipase D in various cell types (Lavie and Liscovitch,1990). However, it must be stressed that the high concentrations of sphingosine which strongly activate phospholipase D also cause severe cell damage and the physiological significance of these observations in relation to cell growth regulation still remains to be elucidated. In addition, recent evidence indicates that sphingosine also inhibits the activity of phosphatidic acid phosphohydrolase, an enzyme that catalyzes the hydrolysis of phosphatidic acid to produce diacylglycerol (Lavie et al,1990; Mullmann et al,1990).

Although the precise mechanism by which sphingosine modulates the level of phosphatidic acid is still unclear, our results raise the important question of whether cellular sphingosine, formed by degradation of sphingolipids, acts in a similar manner as exogenous sphingosine to regulate phosphatidic acid levels and cell proliferation (Fig. 2). If so, sphingosine could function in a positive feedback loop to amplify the cascade of events following receptor stimulation. In addition, through this mechanism, sphingosine can affect a key constituent of signal transduction, *ras* GAP activity.

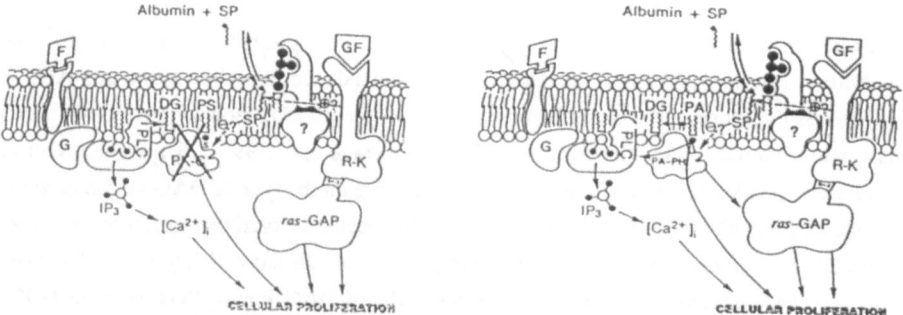

Figure 2. Proposed Models for Involvement of Sphingosine in Cellular Proliferation and Signal Transduction. Left, a depiction of sphingosine as a negative modulator of protein kinase C. Right, modulation of phosphatidic acid levels by sphingosine which in turn modulates signal transduction and cellular proliferation.

Conclusive evidence is still lacking that sphingolipids, similar to phospholipids, can generate active metabolites in response to cell agonists (Wilson et al,1988). However, a recent study demonstrated the existence of a "sphingolipid cycle" during differentiation of human promyelocytic leukemia HL60 cells (Okazaki et al,1989;Kim et al,1991). There are also some observations suggesting an important role for endogenous sphingosine in the mediation of corticosteroid action (Ramachandran et al,1990). Furthermore, recently it was found that DAG stimulates sphingomyelin hydrolysis in GH3 pituitary cells leading to an increase in sphingosine levels (Kolesnick,1987). In addition, it now appears that most, if not all, sphingomyelin is synthesized in mammalian cells by transfer of phosphocholine from phosphatidylcholine to ceramide, leading also to the formation of DAG (Pagano,1988; Merrill and Jones,1990). Thus, it has been suggested that the DAG cycle could influence the sphingolipid cycle (Kolesnick,1987). An important additon to these pathways is the stimulation of the formation of phosphatidic acid by sphingosine, which indicates that sphingolipid turnover could regulate the DAG cycle. The crosstalk between these lipid metabolites which act as second messengers and the function of these interlocking pathways remains unclear. Our results suggest that sphingolipid turnover could regulate the diacylglycerol cycle and that crosstalk between these lipid metabolites, which have been proposed to serve as intracellular second messengers, may be important.

Sphingosine-1-Phosphate, a Novel Lipid, Involved in Cellular Proliferation Induced by Sphingosine

Phosphatidic acid is not the only mitogenic lipid that is produced in response to a mitogenic concentration of sphingosine. Two-dimensional TLC analysis revealed that in addition to phosphatidic acid, mitogenic concentration of sphingosine also significantly stimulated incorporation of ^{32}P into another phospholipid spot which did not comigrate with any of the major known phospholipids of Swiss 3T3 fibroblasts (phosphatidylcholine, phosphatidylethanolamine, phosphatidylinositol, or sphingomyelin). To further identify this unknown phospholipid, the cells were double-labeled with both [^{3}H]-glycerol and ^{32}Pi to isotopic equilibrium (Table I). In untreated cells, there was some incorporation of ^{32}P but no labeling with ^{3}H in this lipid. As mentioned above, identical stimulations of incorporation of [^{3}H]-glycerol and ^{32}Pi into phosphatidic acid were found in response to sphingosine. In contrast, sphingosine induced significant increases in incorporation of ^{32}P into this unidentified phospholipid without corresponding increases in labeling with ^{3}H. Thus, these results indicate that the new phospholipid, which was resistant to mild alkali treatment, does not contain the glycerol moiety and that sphingosine enhances ^{32}P incorporation into this phospholipid. This unknown phospholipid could be a phosphorylated derivative of sphingosine. It is known that cells contain sphingosine kinase, an enzyme which catalyzes the phosphorylation of free sphingosine at the 1-OH position (Stoffel et al,1970,1973; Hirschberg et al,1970; Louis et al,1976). To further substantiate the identity of this new compound, we prepared sphingosine-1-phosphate by a facile enzymatic synthesis from sphingosinephosphocholine using phospholipase D. This elegant procedure, described recently by Bell's group (Van Veldhoven et al,1989), enabled us to prepare milligram quantities of pure sphingosine-1-phosphate. The material obtained from Swiss 3T3 cells co-migrated with sphingosine-1-phosphate standard in a variety of TLC systems and in addition, was converted to ethyleneglycol monophosphate by periodate/borohydride treatment (Hirschberg et al,1970). These results established unequivocally the identification of this unique phospholipid as sphingosine-1-phosphate.

We demonstrated that mitogenic concentrations of sphingosine induce a rapid rise in the levels of ^{32}P-labeled sphingosine-1-phosphate. A significant increase in ^{32}P-sphingosine-1-phosphate was detected within 5 min after exposure of the cells to sphingosine and within 60 min, 4 to 6-fold stimulation over the basal value was observed. After 1-2 h, sphingosine-1-phosphate levels decrease but are still elevated up to 8 h. The accumulation of ^{32}P-labeled-sphingosine-1-phosphate in response to sphingosine was also concentration dependent, with a maximal accumulation at 20 μM. This dose response correlated closely with the dose response for sphingosine-induced stimulation of DNA synthesis (Zhang et al,1990b). Similar to sphingosine, low concentrations of sphingosine-1-phosphate also stimulated DNA synthesis and cell division in quiescent cultures of Swiss 3T3 cells (Fig. 3). It should be noted that sphingosine-1-phosphate alone at optimal concentrations is more mitogenic than insulin and it is as effective as EGF, a potent growth stimulator for Swiss 3T3 cells. Although both sphingosine and sphingosine-1-phosphate acted synergistically with a wide variety of growth factors, such as insulin, EGF, TPA, fibroblast growth factor, and unfractionated FCS, there was no additive or synergistic effect in response to a combination of sphingosine. Both sphingosine and sphingosine-1-phosphate stimulate DNA synthesis in cells made protein kinase C-deficient by prolonged treatment with phorbol ester and sphingosine still elicited similar increases in sphingosine-1-phosphate levels in these cells. Furthermore, not only did sphingosine-1-phosphate stimulate [^{3}H]thymidine incorporation with similar efficiency and kinetics as sphingosine, it also induced similar morphological transformation. It should be noted that sphingosine-1-phosphate stimulated DNA synthesis to a greater extent than sphingosine and required a lower concentration for the maximum response. Therefore, sphingosine-1-phosphate is more potent in stimulating [^{3}H]thymidine incorporation than sphingosine, suggesting that it may mediate the mitogenic activity of sphingosine.

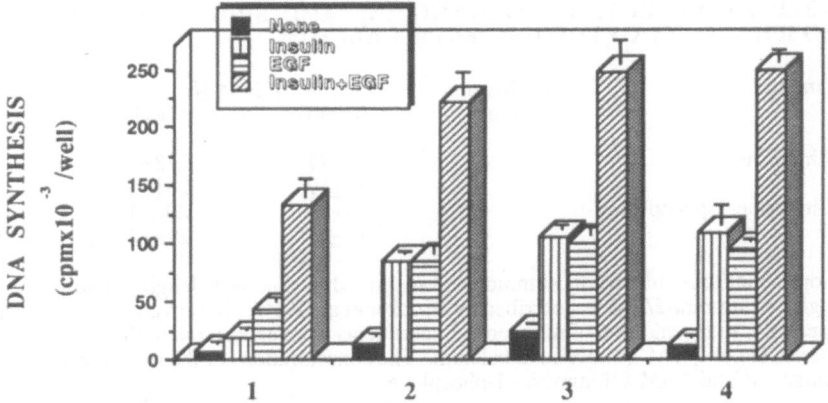

Figure 3. Effects of sphingosine and sphingosine-1-phosphate on DNA synthesis. Quiescent cultures of Swiss 3T3 cells were exposed to the indicated mitogens in the absence (1) or in the presence of sphingosine (2) or sphingosine-1-phosphate (3) or both (4) and [^{3}H]thymidine incorporation was measured (Zhang et al,1990b). The concentrations of the mitogenic agents were as follows: insulin, 4 µg/ml; EGF, 10 ng/ml; sphingosine, 20 µM; sphingosine-1-phosphate, 5 µM.

Recently a new action of sphingosine in mediating rapid and profound translocation of Ca^{2+} from intracellular stores has been described in permeabilized smooth muscle cells (Ghosh et al, 1990). Ghosh et al (1990) have proposed that sphingosine must be converted intracellularly to sphingosine-1-phosphate which affects calcium release from an intracellular calcium pool that includes the inositol trisphosphate-sensitive and insensitive pools. However, direct experimental evidence that sphingosine-1-phosphate itself caused the release of calcium from internal sources was not presented. Using a digital imaging system for measurement of calcium changes, we observed that at 37^0 both sphingosine and sphingosine-1-phosphate are potent calcium-mobilizing agonists in viable 3T3 fibroblasts. Upon addition of mitogenic concentrations of sphingosine-1-phosphate, there was an immediate and transitory rise in cytosolic free calcium concentration even at room temperature. The onset in the rise in calcium was almost instantaneous, reaching a peak response within a short time (<60 sec) after stimulation with sphingosine-1-phosphate and decaying over the next 2 min. In contrast to the action of sphingosine-1-phosphate, sphingosine itself had almost no detectable effects on cytosolic free calcium at room temperature (Zhang et al,1991). However, at 37^0 sphingosine increased cytosolic free calcium by 3-6-fold over the basal value. In most of the single cells that were examined, this peak was followed by another wave of increasing cytosolic free calcium (Zhang et al,1991). The rapid rise in cytosolic free calcium was independent of the presence of calcium in the external medium, indicating that the response is due to the mobilization of calcium from internal stores (Table III).

TABLE III. EFFECTS OF SPHINGOSINE AND SPHINGOSINE-1-PHOSPHATE ON CYTOSOLIC FREE CALCIUM

Stimulus	Medium Calcium	[Ca^{2+}]i Basal	Maximum
Sphingosine	-	61±27	278±65
	+	60±19	261±88
Sphingosine-1-phosphate	-	68±11	206±17
	+	75±21	298±49

Cytoplasmic free calcium concentration was measured by dual wavelength fluorescence imaging using fura-2/AM as described by Brooker et al (1990). Basal values were determined 10 sec prior to the addition of mitogens. Experiments were carried out in the absence (-) or presence (+) of 2 mM extracellular calcium. The concentrations of the mitogens were 50 µM sphingosine and 2 µM sphingosine-1-phosphate.

The advantage of the digital imaging system for measurement of free Ca^{2+} changes is that local concentrations of cytoplasmic free Ca^{2+} can be continuously imaged with great accuracy and resolution in single living cells within a large population of cells. This system makes it possible to detect small fluctuations in the [Ca^{2+}]i concentration within individual cells that would have been averaged out in the population of cells since the changes do not occur in exact synchrony. With this method, it is also possible to follow short term changes in the spatial distribution of Ca^{2+} with great accuracy, sensitivity, and resolution. Both sphingosine and sphingosine-1-phosphate induced a wave of increased free Ca^{2+} which spread to involve almost the entire fibroblast. The rate of this spread varied slightly among cells. However, there was a population of cells unresponsive to sphingosine or sphingosine-1-phosphate which did not have any obvious characteristic morphological features. These results are similar to recent observations on heterogeneous responses to other growth factors, such as bombesin, EGF, and vasopressin (Hesketh et al,1988), where 40% of cells were unresponsive to any mitogens.

We found that only sphingosine-1-phosphate releases calcium from internal sources at room temperature and the response is almost instantaneous. Furthermore, the concentration of sphingosine required to release calcium at 37° is much higher than the effective concentration of sphingosine-1-phosphate. The temperature dependency of the effect of sphingosine on intracellular calcium, together with the increased lag in its action, support the suggestion that sphingosine must require a temperature-dependent enzymatic conversion to sphingosine-1-phosphate for its function (Ghosh et al,1990). Sphingosine-1-phosphate has long been known to be produced from sphingosine in a variety of cell types (Stoffel et al, 1970,1973) and it is known that cells contain sphingosine kinase, an enzyme which catalyzes the phosphorylation of free sphingosine at the 1-OH position (Stoffel et al,1973). Sphingosine-1-phosphate is thought to be rapidly degraded by the action of a microsomal lyase which cleaves sphingosine-1-phosphate to trans-2-hexadecanal and phosphoethanolamine (Stoffel et al,1970). Thus, sphingosine-1-phosphate could be a suitable candidate for an intracellular second messenger as it can be rapidly induced and degraded. Our results are consistent with such an hypothesis. Sphingosine-1-phosphate is rapidly produced in response to mitogenic concentrations of sphingosine, is mitogenic by itself for 3T3 fibroblasts, and mediates calcium release in viable cells. Thus, sphingosine-1-phosphate may be an important component of the intracellular second messenger system that is involved in calcium release and the regulation of cell growth. Whether this is a plausible speculation for the cellular function of sphingosine and its metabolite sphingosine-1-phosphate awaits further study.

Acknowledgements - This work was supported by Research Grants 1RO1 GM43880 from the National Institutes of Health.

REFERENCES

Barbacid M (1987) Ras genes. Ann Rev Biochem 56:779-827
Ben-Av P, Liscovitch M (1989) Phospholipase D activation by the mitogens platelet-derived growth factor and 12-O-tetradecanoylphorbol 13-acetate in NIH 3T3 cells. FEBS Lett 259:64-66
Berridge MJ (1984) Inositol phosphate and diacylglycerol as second messengers. Biochem J 220:345-360
Berridge MJ (1985) The molecular basis of communication within the cell. Sci Amer 253:95-106
Brooker G, Seki T, Croll D, Wahlestedt C (1990) Calcium wave evoked by activation of endogenous or exogenously expressed receptors in Xenopus oocytes. Proc Natl Acad Sci USA 87:2813-2817.
Davis RJ, Girones N, Faucher M (1988) Two alternative mechanisms control the interconversion of functional states of the epidermal growth factor receptor. J Biol Chem 263:5373-5379
Exton JH (1990) Signaling through phosphatidylcholine breakdown. J Biol Chem 265:1-4
Faucher M, Girones N, Hannun YA, Bell RM, Davis RJ (1988) Regulation of epidermal growth factor receptor phosphorylation state by sphingosine in A431 human epidermoid carcinoma cells. J Biol Chem 263:5319-5327
Ghosh TK, Bian J, Gill DL (1990) Intracellular calcium release mediated by sphingosine derivatives generated in cells. Science 248:1653-1656
Hakomori S (1985) Aberrant glycosylation in cancer cell membranes as focused on glycolipids: overview and perspectives. Cancer Res 45:2405-2415
Hakomori S (1986) Glycosphingolipids. Sci Amer 254:44-53
Hannun YA, Bell RM (1987) Lysosphingolipids inhibit protein kinase C: Implications for the sphingolipidoses. Science 235:670-674
Hannun YA, Bell RM (1989) Functions of sphingolipids and sphingolipid breakdown products in cellular regulation. Science 243:500-506
Hasegawa-Sasaki H (1985) Early changes in inositol lipids and their metabolites induced by platelet-derived growth factor in quiescent Swiss mouse 3T3 cells. Biochem J 232:99-109
Hesketh TR, Morrison DH, Moore JP, Metcalfe JC (1988) Ca^{2+} and pH responses to sequential additions of mitogens in single 3T3 fibroblasts: correlations with DNA synthesis. J Biol Chem 263:11879-11886
Hirschberg CB, Kisic A, Schroepfer GJ (1970) Enzymatic formation of dihydrosphingosine

1-phosphate. J Biol Chem 245:3084-3090

Igarashi Y, Hakomori S, Toyokuni T, Dean B, Fujita S, Sugimoto M, Ogawa T, Ghendy KE, Racker E (1989) Effect of chemically well-defined sphingosine and its N-methyl derivatives on protein kinase C and src kinase activities. Biochemistry 28:3138-3145

Jefferson AB, Schulman H (1988) Sphingosine inhibits calmodulin-dependent enzymes. J Biol Chem 263:15241-15244

Kim MY, Linardie C, Obeid L, Hannum Y (1991) Identification of sphingomyelin turnover as an effector mechanism for the action of tumor necrosis factor alpha and gamma-interferon. Specific role in cell differentiation. J Biol Chem 266:484-489

Kiss A, Anderson WB (1990) ATP stimulates the hydrolysis of phosphatidylethanolamine in NIH 3T3 cells.Potentiating effects of guanosine triphosphate and sphingosine. J Biol Chem 265:7345-7350

Kolesnick RN (1987) 1,2-Diacylglycerols but not phorbol esters stimulate sphingomyelin hydrolysis in GH3 pituitary cells. J Biol Chem 262:16759-16762

Lavie Y, Liscovitch M (1990) Activation of phospholipase D by sphingoid bases in NG108-15 neural-derived cells. J Biol Chem 265:3868-3872

Lavie Y, Piterman O, Liscovitch M (1990) Inhibition of phosphatidic acid phosphohydrolase activity by sphingosine. Dual action of sphingosine in diacylglycerol signal termination. FEBS Lett 277:7-10

Louis DD, Kisic AK, Schroepfer GJ (1976) Sphingolipid base metabolism. Partial purification and properties of sphinganine kinase of brain. J Biol Chem 251:4557-4564

Merrill AH, Jones JJ (1990) An update of the enzymology and regulation of sphingomyelin metabolism. Biochim Biophys Acta 1044:1-12

Merrill AH, Stevens VL (1989) Modulation of protein kinase C and diverse cell functions by sphingosine - a pharmacologically interesting compound linking sphingolipids and signal transduction. Biochim Biophys Acta 1010:131-139

Moolenaar WH, Kruijer W, Tilly BC, Verlaan I, Bierman AJ, deLaat SW (1986) Growth factor-like action of phosphatidic acid. Nature 323:171-173

Mullmann TJ, Siegel ML, Egan RW, Billah MM (1991) Sphingosine inhibits phosphatidate phosphohydrolase in human neutrophils by a protein kinase C-independent mechanism. J Biol Chem 266:2013-2016

Murayama T, Ui M (1987) Phosphatidic acid may stimulate membrane receptors mediating adenylate cyclase inhitibion and phospholipid breakdown in 3T3 fibroblasts. J Biol Chem 262:12463-12467

Needleman P, Turk J, Jakschik BA, Morrison AR, Lefkowith JB (1986) Arachidonic acid metabolism. Annu Rev Biochem 55:69-102

Nishizuka Y (1988) The molecular heterogeneity of protein kinase C and its implications for cellular regulation. Nature 334:661-665

Okazaki T, Bell RM, Hannun YA (1989) Sphingomyelin turnover induced by vitamin D3 in HL-60 cells. J Biol Chem 264:19076-19080

Pagano RE (1988) What is the fate of diacylglycerol produced at the Golgi apparatus? Trends Biochem Sci 13:202-205

Pai JK, Siegel MI, Egan RW, Billah MM (1988) Phospholipase D catalyzes phospholipid metabolism in chemotactic peptide-stimulated HL-60 granulocytes. J Biol Chem 263:12472-12477

Ramachandran CK, Murray DK, Nelson DH (1990) Dexamethasone increases neutral sphingomyelinase activity and sphingosine levels in 3T3-L1 fibroblasts. Biochem Biophys Res Commun 167:607-613

Rozengurt E (1986) Early signals in the mitogenic response. Science 234:161-166

Sakane F, Yamada K, Kanoh H (1990) Different effects of sphingosine, R59022 and anionic amphiphiles on two diacylglycerol kinase isozymes purified from porcine thymus cytosol. TIBS 15:47-50

Spiegel S, Panagiotopoulos C (1988) Mitogenesis of 3T3 fibroblasts induced by endogenous ganglioside GM1 is not mediated by cAMP, protein kinase C, or phosphoinositides turnover. Exp Cell Res 177:414-427

Spiegel S (1989) Possible involvement of a GTP-binding protein in a late event during endogenous ganglioside GM1-modulated cellular proliferation. J Biol Chem 264:6766-6772

Stoffel W, Assmann G, Binczek E (1970) Metabolism of sphingosine bases. XIII. Enzymatic synthesis of 1-phosphate esters of 4t-sphingenine (sphingosine), sphinganine (dihydrosphingosine), 4-hydroxysphinganine (phytosphingosine) and 3-dehydrosphingosine by erythrocytes. Hoppe-Seyler's Z Physiol Chem 351:635-642

Stoffel W, Heimann G, Hellenbroich B (1973) Sphingosine kinase in blood platelets. Hoppe-Seyler's Z Physiol Chem 354:562-566

Sweeley CC (1985) Sphingolipids. In:Biochemistry of lipids and membranes, Vance DE, Vance JE, The Benjamin/Cummings Publishing Co, Menlo Park, CA, pp 361-403

Tsai MH, Yu CL, Wei FS, Stacey DW (1989) The effect of GTPase activating protein upon ras is inhibited by mitogenically responsive lipids. Science 243:522-526

Van Veldhoven PP, Foglesong RJ, Bell RM (1989) A facile enzymatic synthesis of sphingosine-1-phosphate and dihydrosphingosine-1-phosphate. J Lipids Res 30:611-616

Wilson E, Wang E, Mullins RE, Liotta DC, Merrill AH, Lambeth JD (1988) Modulation of the free sphingosine levels in human neutrophils by phorbol esters and other factors. J Biol Chem 263:9304-9309

Winicov I, Gershengorn MC (1988) Sphingosine inhibits thyrotropin-releasing hormone binding to pituitary cells by a mechanism independent of protein kinase C. J Biol Chem 263:12179-12282

Yu CL, Tsai MH,Stacey DW (1988) Cellular ras activity and phospholipid metabolism. Cell 52:63-71

Zhang H, Buckley NE, Gibson K, Spiegel S (1990a) Sphingosine stimulates cellular proliferation via a protein kinase C-independent pathway. J Biol Chem 265:76-81

Zhang H, Naishadh ND, Murphey JM, Spiegel S (1990b) Increases in phosphatidic acid levels accompany sphingosine stimulated proliferation of quiescent Swiss 3T3 cells. J Biol Chem 265:21309-21316

Zhang H, Naishadh ND, Olivera A, Seki T, Brooker G, Spiegel S (1991) Sphingosine-1-phosphate, a novel lipid, involved in cellular proliferation. J Cell Biol 114:155-167

REGULATION OF PHOSPHATIDYLSERINE SYNTHESIS DURING T CELL ACTIVATION.

Claude AUSSEL
Immunologie Cellulaire et Moleculaire
INSERM U210
Faculté de Médecine
06034 Nice cédex, France.

The importance of phospholipids in biological systems has become increasingly recognized in recent years. In T cells as in other cells, a majority of investigators have focused their research on the role of the phosphoinositide (PtdIns) cycle as a system that generates the lipidic second messengers. Our studies on T lymphocytes have shown that other phospholipids and particularly phosphatidylserine (PtdSer) could play an essential role in T cell activation monitored by the measurement of interleukin-2 synthesis. The present paper takes stock of our current knowledge on the regulation of PtdSer synthesis during T cell activation and on its possible role in the regulation of interleukin-2 synthesis in Jurkat cells, a leukemic T cell line widely used as a model for the studies on lymphocyte activation.

T lymphocyte activation.

T lymphocytes can be activated to produce lymphokines, to express cell surface proteins and to proliferate. In vivo, T cells are activated by antigen presented by macrophages in the context of the major histocompatibility complex. This interaction is mediated by the T cell receptor complex (TCR). The TCR is composed of a disulfide-linked heterodimer glycoprotein (Ti). Associated with Ti are several peptides named T3 that probably play a role in the transduction of signals delivered to the Ti. In fact, the full activation of T

NATO ASI Series, Vol. H 70
Phospholipids and Signal Transmission
Edited by R. Massarelli, L. A. Horrocks,
J. N. Kanfer, and K. Löffelholz
© Springer-Verlag Berlin Heidelberg 1993

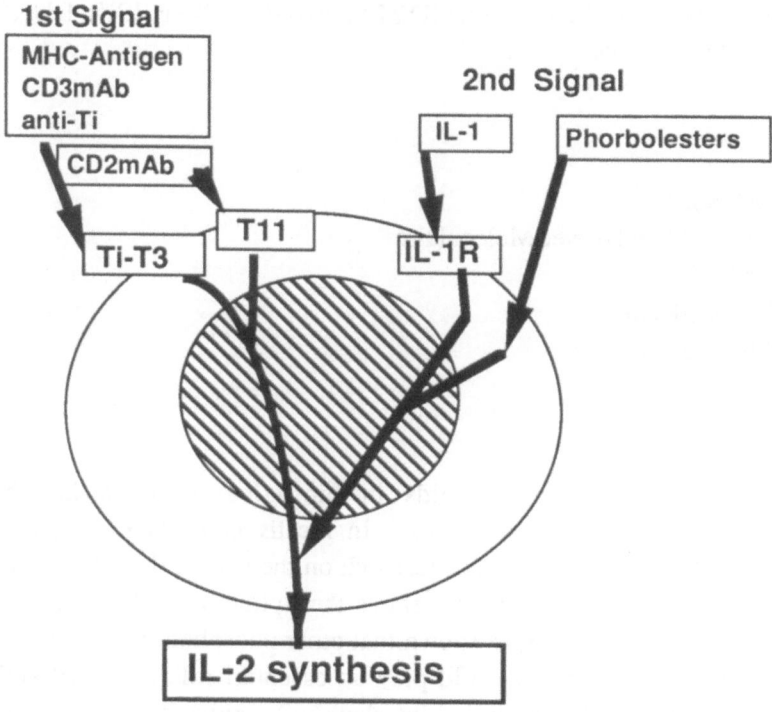

Fig. 1 Schematic representation of the different pathways leading to T cell activation and interleukin-2 synthesis

cells that culminates in interleukin-2 (IL-2) synthesis and proliferation necessitates a second signal delivered through the interaction of interleukin-1 (IL-1) with its specific receptor on the surface of T cells (Davis & Lipsky, 1985; Roska & Lipsky, 1985; Larsson et al, 1980; Palacios, 1982; Chiller et al, 1982). Activation of the T cell line Jurkat to produce IL-2 also necessitates two signals. In vitro, these two signals are delivered by lectins such as phytohaemagglutinin (PHA) or monoclonal antibodies (mAb) directed against either Ti, T3 (CD3mAb) or T11 (CD2mAb) that represents an alternative pathway for T cell activation. The second signal is generated either with recombinant IL-1 or a phorbolester known to activate the protein kinase C (PKC) such as tetradecanoyl phorbol acetate (TPA) or phorbol dibutyrate (PDBu) (Wiskocil et al, 1985) (Fig.1).

Transduction through the TCR

The transduction pathway of the first signal involves the activation of phospholipase(s) C that generate the inositol phosphate (IP3) from triphosphoinositides (Imboden et al, 1985; Weiss et al, 1984) and diacylglycerol (DAG) from both phosphoinositides and phosphatidylcholine (Aussel et al, 1990a). The mechanism of transduction from the TCR to the phospholipase C is not entirely established since it has been proposed as either a coupling of the TCR to the phospholipase C via a GTP-binding protein or alternatively an activation of the phospholipase C through phosphorylation of a tyrosine residue mediated by a kinase associated with the TCR. Whatever the route used, the DAG produced by phospholipase(s) C is then rapidly phosphorylated, by a DAG-kinase, into phosphatidic acid (PtdOH). Both IP3 and PtdOH release Ca^{2+} from intracellular stores (Breittmayer et al, 1991). The mobilization of Ca^{2+} from intracellular stores, is rapidly followed by a sustained Ca^{2+} influx possibly through a yet poorly characterized membrane Ca^{2+} channel. The interaction of Ca^{2+} with calmodulin induces the activation of a Ca^{2+} and calmodulin-dependent protein kinase that phosphorylates serine residues on two cytoplasmic proteins pp21 and pp23 (Peyron et al, 1989; Mary et al, 1989). Beside this pathway, DAG activates protein kinase C resulting in the phosphorylation of other protein substrates (Nishizuka, 1984; Kaibuchi et al, 1985). Our current knowledge of transmembrane signaling in T cells is schematized in Fig.4.

Phosphatidylserine biosynthesis

The biosynthesis of PtdSer in mammalian cells appears to occur solely by an exchange of L-serine for either the ethanolamine moiety of phosphatidylethanolamine (PtdEtn) or the choline moiety of phosphatidylcholine (PtdCho) (Miura & Kanfer, 1976; Kanfer, 1972) . In cell- free sytems, PtdSer synthesis occurs predominantly by a Ca^{2+} stimulated, energy-independent exchange reaction. The use of rat brain and liver microsomes has demonstrated that PtdSer synthesis is a calcium- but also a calmodulin-dependent process (Bjerve, 1973a; Bjerve, 1973b; Gaiti et al, 1974). In addition, PtdSer synthesis could be regulated, at least in vitro, by a phosphorylation / dephosphorylation process (Kanfer et al, 1988).

Inhibition of PtdSer synthesis during T cell activation.

The full activation of Jurkat cells that leads to IL-2 synthesis, i.e. by using PHA in combination with TPA, is accompanied by changes in the incorporation of the polar head groups into their respective phospholipids. While the incorporation of [3H]ethanolamine and [3H]choline into PtdEtn and PtdCho respectively remained unchanged, the incorporation of [3H]myo-inositol into PtdIns was raised probably as a consequence of the activation of the PtdIns cycle during the activation process (Imboden et al, 1985; Weiss et al, 1984). Conversely, the incorporation of [3H]serine into PtdSer was markedly diminished (Fig. 2). The decreased labeling of PtdSer occured without noticeable change in the total uptake of [3H]serine by the cells and without modification of the distribution of [3H]serine into the different intracellular pools of serine (solvent and water soluble materials, TCA-precipitated materials and serine metabolites). Since in [3H]serine prelabeled cells, the activation procedure did not result in a decrease of the PtdSer pool (Fig.2), nor a change in the metabolism of PtdSer into PtdEtn by decarboxylation, it was concluded that the biosynthesis of PtdSer was specifically inhibited during T cell activation (Pelassy et al, 1989a). Since this unexpected inhibition was obtained in the presence of the two activating signals (PHA and TPA) leading to the induction of IL-2 synthesis, we have searched whether one or both signals were required. It was found that the first signal (i.e. when using anti-Ti, CD3, CD2 (Pelassy et al, 1989a), mAbs or the lectin, PHA) inhibited PtdSer synthesis while the second signal, either TPA or PDBu, did not change PtdSer biosynthesis. IL-1, by contrast, partially reversed the inhibition induced by the first activation signals (Mary et al, 1988).

Regulation of PtdSer synthesis by Ca^{2+}

Signals generated upon receptor triggering are known to induce changes in the cytosolic free calcium concentration. These Ca^{2+} changes are due both to a release of Ca^{2+} ions from intracellular stores and to a Ca^{2+} influx (Fig.3). These two Ca^{2+} movements induced by T cell activators can be

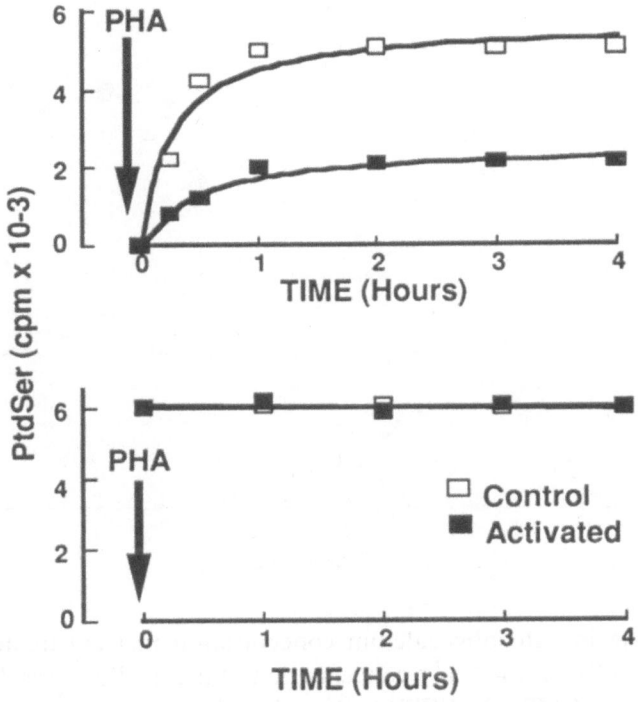

Fig. 2 Effect of phytoheamagglutinin (PHA) on phosphatidylserine synthesis (upper panel) and phosphatidylserine degradation (lower panel). For synthesis experiments, the cells were incubated as a function of time in the presence of both PHA and tritiated serine. For degradation studies, tritiated serine-prelabeled cells were incubated with PHA at time 0. The amount of tritiated PtdSer was measured after thin layer chromatography.

mimicked by calcium-ionophores. The use of A23187 on Jurkat cells inhibited PtdSer synthesis, suggesting that Ca^{2+} participates to the inhibition process (Pelassy et al, 1989a). The role of calcium as regulator of PtdSer synthesis in other systems has been pointed out by several authors (Bjerve, 1973a; Bjerve, 1973b; Gaiti et al, 1974; Buchanan & Kanfer, 1980). In Jurkat cells, PtdSer synthesis is rapidly augmented by increasing the Ca^{2+} concentration in the culture medium (in the absence of detectable changes in intracellular Ca^{2+} concentration) and strongly inhibited when the extracellular Ca^{2+} is chelated by EGTA (Aussel et al, 1991a). This EGTA -induced

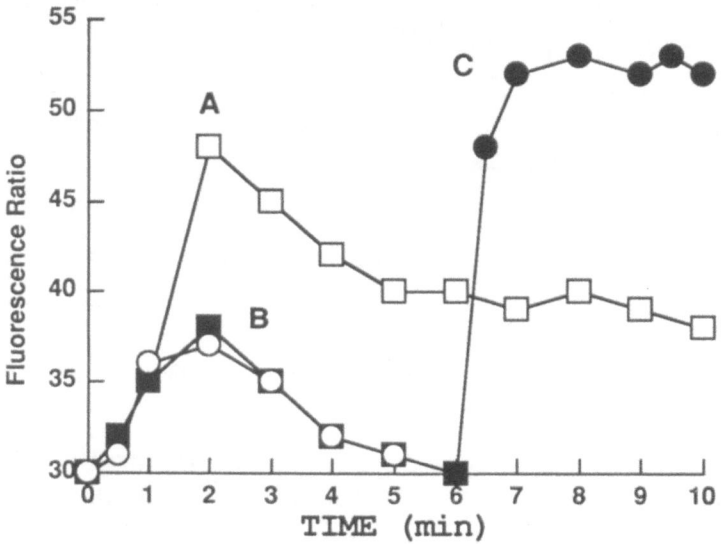

Fig. 4 Increase in cytosolic calcium concentration in CD3 treated, Indo-1-loaded Jurkat cells. A: the Ca^{2+} concentration of the medium was 1 mM. B: in the absence of Ca^{2+} (2 mM EGTA). C : as in B but 2 mM Ca^{2+} was added in the medium at time 6 min.

inhibition was totally reversed, within a few minutes, after the addition of Ca^{2+} in the experimental medium. Since the activation of Jurkat cells in the presence of EGTA also resulted in a net inhibition of PtdSer synthesis that was not reversed by the addition of Ca^{2+} in the medium, we looked at the possible influence of intracellular Ca^{2+} movements on PtdSer synthesis. In T cells, and especially in Jurkat cells, the free Ca^{2+} rise due to a release from intracellular stores can be abolished by N-ethylmaleimide (NEM) (Ng et al, 1987) or the intracellular Ca^{2+} chelator, BAPTA. On the other hand, the Ca^{2+} influx is specifically inhibited by either TPA, TPCK or NDGA (Ng et al, 1987; Dornand et al, 1987; Rickard et al, 1985; Auberger et al, 1989). Evaluation of the effects of these drugs on PtdSer synthesis have shown that blocking the release of Ca^{2+} ions from intracellular stores with either NEM or BAPTA reversed the inhibition induced by CD3 mAb or PHA, while blocking the Ca^{2+} influx with either TPA, NDGA or TPCK did not change the percentage of inhibition of PtdSer synthesis induced by the T cell activators (Aussel et al, 1991a). It was thus concluded that the inhibition of PtdSer synthesis was probably due to the release of Ca^{2+} from intracellular compartments. To confirm this surprising result, we have tested whether other mAbs or other Ca^{2+} mobilizing agents also induced inhibition of PtdSer synthesis.

CD2 mAb and Phosphatidic acid induced inhibition of PtdSer synthesis.

Among CD2 mAbs, able to activate T cells, D66 has been shown to be able to mobilize intracellular Ca^{2+} but unable to induce a Ca^{2+} influx. D66 mAb also inhibited PtdSer synthesis confirming that the release of Ca^{2+} from endoplasmic reticulum is sufficient to inhibit the synthesis of this phospholipid. Recent works (Moolenaar et al, 1986; Jalink et al, 1990) have pointed out that exogenously added phosphatidic acid (PtdOH) induced Ca^{2+} mobilization in A431 cells. In Jurkat cells, this phospholipid also induced a release of Ca^{2+} from intracellular stores but was unable to open the Ca^{2+} channel that is responsible for the Ca^{2+} influx generated by the classical T cell activators (Breittmayer et al, 1991). As found with the CD2 mAb D66, PtdOH induced in Jurkat cells a net decrease of PtdSer synthesis, pointing out that elevation of cytosolic Ca^{2+} due to the release of Ca^{2+} from intracelluar stores allows the inhibition of the synthesis of this phospholipid (Pelassy et al, 1991b). Comparisons of the ability of synthetic PtdOH differing in their fatty acid content to affect PtdSer synthesis demonstrated that PtdOH containing long chain unsaturated fatty acids were more active than PtdOH with saturated long or saturated short chains. In contrast to what was found on A431 cells (Moolenaar et al, 1986; Jalink et al, 1990), PtdOH did not activate the phosphoinositol cycle in Jurkat cells (Breittmayer et al, 1991).

Effects of DAG-kinase inhibitors.

Since exogenously added PtdOH induced both the release of Ca^{2+} from intracellular stores and the inhibition of PtdSer synthesis and since T cell activators generated PtdOH by phosphorylation of DAG, we have tested whether DAG-kinase plays a role in the inhibition of PtdSer synthesis. Using three DAG-kinase inhibitors, monoacylglycerol (MOG), dioctylethyleneglycol (DOEG) and R59022 (Bishop et al, 1986), we found that PtdOH generated by the DAG-kinase is for a large part responsible both for the release of Ca^{2+} from intracellular stores (Breittmayer et al, 1991) and the inhibition of PtdSer syntheis (Aussel et al, 1991b).

Effect of K+ channel blockers.

Jurkat T cells, as well as normal human peripheral blood T lymphocytes, express voltage-gated K^+ channels. These channels are blocked by quinine, 4-aminopyridine and tetraethylammonium and are suspected to play a role in T cell activation since the three channel blockers cited above are able to inhibit IL-2 synthesis in Jurkat cells and both IL-2 synthesis and proliferation in peripheral blood lymphocytes (Gallin, 1986; Gelfand, 1989). It was recently shown (Aussel et al, 1990b) in Jurkat cells that quinine, 4-aminopyridine and tetraethylammonium induced a rise in PtdSer synthesis suggesting that K+ channels also participate in the regulation of membrane phospholipid synthesis. More recently, we have shown that verapamil, a drug known as a blocker of voltage-gated Ca^{2+} channels, but quite capable, in the mM range, to inhibit the K^+ channel in T lymphocyte also enhanced PtdSer synthesis.

Effect of exogenously added PtdSer on T cell activation.

Ponzin et al (1990) have shown that in mouse spleen cells, PtdSer-containing liposomes decreased both IL-2 production and the expression of IL-2-receptors and subsequently T cell proliferation induced by mitogens. This prompted us to test whether the addition of PtdSer to the culture medium of Jurkat cells resulted in similar effects. It was found that the addition of PtdSer markedly inhibited the incorporation of [3H]serine, [3H]choline and [3H] ethanolamine into their respective phospholipids namely, PtdSer, PtdCho and PtdEtn respectively. In addition it was shown that exogenously added PtdSer resulted in a strong decrease of [3H]oleoyl-labeled DAG production in cells activated with either CD3 mAb or phytohaemagglutinin. Consequently IL-2 synthesis also was markedly inhibited (Pelassy et al, 1991c).

Effect of the serine analog, serinol.

Results obtained by Kanfer (1972) on rat brain microsomes have shown that some serine analogs such as iso-serine and serinol markedly increased and decreased respectively, the incorporation of serine into PtdSer. In Jurkat cells, it was found that serinol was able to decrease by 75% the amount of PtdSer synthesized and also to markedly decrease the synthesis of PtdCho and PtdEtn. On the contrary, PtdIns synthesis was not modified. Concomitantly, in serinol treated cells, IL-2 production was strongly inhibited. Monitoring the production of the second messengers generated by T cell activators showed that in serinol treated cells the production of DAG was impaired while Ca^{2+} mobilization remained unaffected. Serinol thus appeared to be a potential immunoregulatory molecule active at the level of protein kinase C regulation either through its interaction with PtdSer or through DAG production (Pelassy et al, 1991a).

Mechanisms regulating PtdSer synthesis in Jurkat cells.

Fig. 4 summarizes the proposed pathway leading to PtdSer inhibition in activated Jurkat T cells. T cell effectors first induce the activation of phospholipase(s) generating the second messenger, DAG. DAG is rapidly metabolized into PtdOH by a DAG-kinase. This DAG-kinase is probably

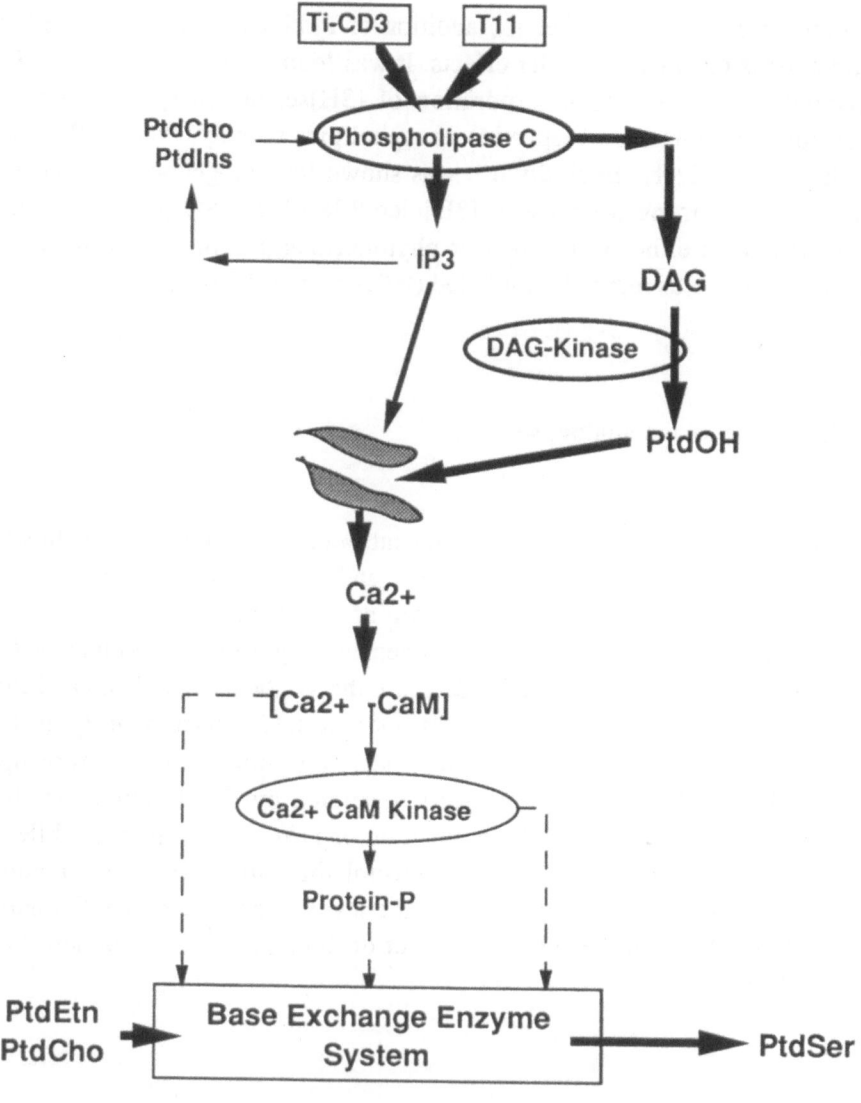

Fig. 4 Schematic representation of the pathways leading to the formation of second messengers after triggering either the T cell receptor complex or the T11 molecules and resulting in the regulation of the phosphatidylserine synthesis through the base exchange enzyme system. Dotted lines represent hypothetic routes. Large arrows summarize our current knowledge on the regulation of PtdSer synthesis and regulation in Jurkat T cells.

the 80 kDa enzyme described in T lymphocytes as well as in Jurkat cells by Sakane et al (1989) and Kanoh & Sakane (1990). PtdOH, beside IP3, releases Ca^{2+} from the endoplasmic reticulum. Then Ca^{2+} binds to calmodulin (CaM) and the complex $[Ca^{2+}/CaM]$ interferes with the base exchange enzyme system responsible for PtdSer synthesis. What happen from the interaction of Ca^{2+} with CaM and the inhibition of PtdSer, at that time, remains in a black box. Nevertheless, several hypotheses can be proposed. Either the Ca^{2+}/CaM complex activates a protein kinase leading to the phosphorylation of protein substrates interfering with the base exchange enzyme system or the Ca^{2+}/CaM-dependent kinase directly phosphorylates the base exchange enzyme system. Alternatively, it can be also proposed that the Ca^{2+}/CaM complex directly regulates the activity of the PtdSer synthase. Further works are necessary to elucidate the exact mechanism of the regulation of PtdSer synthesis in T cells.

Conclusions.

Our results show that a good correlation appears between the ability of Jurkat cells to synthesize IL-2 and modifications of PtdSer synthesis. Several drugs such as DAG-kinase inhibitors and K^+ channel inhibitors that were found to increase PtdSer synthesis were also good inhibitors of IL-2 synthesis. As stated above the addition of exogenous PtdSer reduced IL-2 synthesis. In addition, changes in PtdSer synthesis induced with the serine analog, serinol, also inhibited IL-2 synthesis. Since T cell activation is accompanied by a decrease of PtdSer synthesis and since all means used to enhance PtdSer are able to inhibit IL-2 production, our experiments strongly suggest that PtdSer plays a role in the regulation of T cell activation.

References

Auberger, P., Mary, D., Breittmayer, J.P., Aussel, C., and Fehlmann, M. (1989) Chymotryptic type protease inhibitors blocked Ca++ response and IL-2 production in Jurkat T cells. J. Immunol. 142: 1253-1259.
Aussel, C., Pelassy C., and Rossi, B. (1990a) Breakdown of a phosphatidylcholine pool arising from the metabolic conversion of

phosphatidylethanolamine as a novel source of diacylglycerol in
activated T cells. J. Lipid Mediators 2: 103-116.

Aussel, C., Pelassy, C., Mary, D., Choquet, D., and Rossi, B. (1990b)
Regulation of interleukin-2 production and phosphatidylserine synthesis
in Jurkat T lymphocytes by K+ channel antagonists.
Immunopharmacology 20: 97-103.

Aussel, C., Pelassy, C., Mary, D., Breittmayer, J.P., Cousin, J.L., and Rossi,
B. (1991a) Calcium-dependent regulation of phosphatidylserine synthesis
in control and activated Jurkat cells. J. Lipid Mediators 3: 267-282.

Aussel, C., Pelassy, C., and Breittmayer, J.P. (1991b) A diacylglycerol
kinase is involved in the regulation of interleukin 2 synthesis in Jurkat T
cells. Cellular Immunol.139: 333-341.

Bishop, W.R., Ganong, B.R., and Bell, R.M. (1986) Attenuation of sn-1,2-
diacylglycerol second messengers by diacylglycerol kinase. Inhibition by
diacylglycerol analogs in vitro and in human platelets. J. Biol. Chem.
261: 6993-6998.

Bjerve, K.S. (1973a) The phospholipid substrates in the calcium stimulated
incorporation of nitrogen bases into microsomal phospholipids.
Biochim. Biophys. Acta 306: 396-403.

Bjerve, K.S. (1973b) The Ca2+ dependent biosynthesis of lecithin,
phosphatidylethanolamine and phosphatidylserine in rat liver subcellular
particles. Biochim. Biophys. Acta 296: 549-562.

Breittmayer, J.P., Aussel, C., Farahifar, D., Cousin, J.L., and Fehlmann, M.
(1991) A phosphatidic acid sensitive intracellular pool of calcium is
released by anti-CD3 in Jurkat T cells. Immunology 73: 134-139.

Buchanan, A.G. and Kanfer, J. (1980) The effect of various incubation
temperature, particulate isolation and possible role of calmodulin on the
activity of the base exchange enzyme of rat brain. J. Neurochem. 35:
814-822.

Chiller, J.M., DeFreitas, E., Chesnut, R., Grey, H., and Skidmore, B. (1982)
Signal requirements for T lymphocyte activation. In Isolation
Characterization and Utilization of T lymphocytes clones. C.Fathman
and F. Fitch, Eds, Academic Press, N.Y. p.83-94.

Davis, L., and Lipsky, P.E. (1985) Signals involved in T-cell activation. 1)
Phorbolesters enhance responsiveness but cannot replace intact accessory
cells in the induction of mitogen-stimulated T cell proliferation.
J.Immunol. 135: 2946-2952.

Dornand, J., Sekkat, C., Mani, J.C., and Gerber, M. (1987) Lipoxygenase
inhibitors suppress IL-2 synthesis. Relationships with rise in Ca++ and
the events dependent on protein kinase C activation. Immunol. Lett.
16:101-105.

Gaiti, A., De Medio, E., Brunetti, M., Amaducci, L., and Porcellati, G. (1974)
Properties and function of the calcium dependent incorporation of
choline, ethanolamine and serine into the phospholipids of isolated rat
brain microsomes. J. Neurochem. 23: 1153-1159.

Gallin, E.K. (1986) Ionic channels in leukocytes. J.Leuko. Biol. 39: 241-250.

Gelfand, E.W. (1989) Transmembrane ion fluxes in T lymphocytes: Prerequisites or consequences of ligand-induced activation. In Advances in regulation of cell growth. Vol.1 Mond, J.J., Cambier, J.C. and Weiss, A. Eds Raven Press, N.Y. 6: 119-123.

Imboden, J.B., Weiss, A., and Stobo, J. (1985) The antigen receptor on a human T cell line initiates activation by increasing cytoplasmic free calcium. J. Immunol. 134: 633-638.

Jalink, K., Van Corven, E.J., and Moolenaar, W.H. (1990) Lysophosphatidic acid but not phosphatidic acid is a potent Ca2+ mobilizing stimulus for fibroblasts. Evidence for an extracellular site of action. J. Biol. Chem. 265, 12232-12237.

Kaibuchi, K., Takai, Y., and Nishizuka, Y. (1985) Protein kinase C and calcium ions in mitogenic response of macrophage depleted human peripheral lymphocytes. J. Biol. Chem. 260: 1366-1372.

Kanfer, J. (1972) Base exchange reactions of the phospholipids in rat brain particles. J. Lipid Res. 13, 468-476.

Kanfer, J., McCartney, D., and Hattori, H. (1988) Regulation of the choline, ethanolamine and serine base exchange activities of rat brain microsomes by phosphorylation and dephosphorylation. FEBS Lett. 240: 101-104.

Kanoh, H. Yamada, K. and Sakane, F. (1990) Diacylglycerolkinase. A key modulator of signal transduction? Trends Biol. Sci. 15: 47-50.

Larsson, E.A., Coutinho, A., and Martinez, C. (1980) A suggested mechanism for T lymphocyte activation. Implication on the acquisition of functional reactivities Immunol. Rev. 51: 61-64.

Mary, D., Aussel, C., Pelassy, C., and Fehlmann, M. (1988) IL-1 signaling for IL-2 production in T cells involves a rise in phosphatidylserine synthesis. J. Immunol. 141: 3078-3080.

Mary, D., Peyron, J.F., Auberger, P., Aussel, C., and Fehlmann, M. (1989) Modulation of T cell activation by differential regulation of the phosphorylation of two cytosolic proteins. Implication of both Ca2+ and cyclic AMP-dependent protein kinases. J. Biol. Chem. 264: 14498-14502.

Miura, T. and Kanfer, J. (1976) Studies on base-exchange reactions of phospholipids in rat brain. Heterogeneity of base exchange enzymes. Arch. Biochem. Biophys. 175: 654-660.

Moolenaar, W.H., Kruijer, W., Tilly, B.C., Verlaan, I., Bierman, A.J., and DeLaat, S.W. (1986) Growth factor like action of phosphatidic acid. Nature 323, 171-173.

Ng, J., Fredholm, B., Jondal, M., and Anderson, T. (1987) Characterization of PHA and anti-T3 induced transduction mechanisms in a human T cell leukemia line. Int. J. Immunopharmacol. 9: 17-21.

Nishizuka, Y. (1984) The role of protein kinase C in cell surface signal transduction and tumor promotion. Nature 308:693-696.

Palacios, R. (1982) Mechanism of T cell activation. Role and functional relationships of HLA-DR antigens and interleukins. J. Immunol. 63: 76-82.

Pelassy, C., Aussel, C., and Fehlmann, M. (1989a) Phospholipid metabolism and T cell activation. Receptor triggering is associated with the inhibition of phosphatidylserine synthesis. Cellular Signaling 1: 99-105.

Pelassy, C., Dallanegra, A., Aussel, C., and Fehlmann, M. (1989b) Inhibition of phosphatidylserine synthesis induced by triggering CD2 or the CD3-TCR complex in a human T cell line. Relationships with G proteins and receptors modulation. Mol. Immunol. 26: 1081-1086.

Pelassy, C., Mary, D., and Aussel, C. (1991a) Effects of the serine analogues isoserine and serinol on interleukin-2 synthesis and phospholipid metabolism in a human T cell line Jurkat. J. Lipid Mediators 3:79-89.

Pelassy, C., Breittmayer, J.P., Mary, D., and Aussel, C. (1991b) Inhibition of phosphatidylserine synthesis by phosphatidic acid in the Jurkat cell line: Role of calcium ions released from intracellular stores. J. Lipid Mediators 4: 199-210.

Pelassy, C., Mary, D. and Aussel, C. (1991c) Mechanism of phosphatidylserine induced inhibition of interleukin-2 synthesis in the human T cell line Jurkat. Immuno. Pharmacol. Toxicol. 12: 633-645.

Peyron, J.F., Aussel, C., Ferrua, B., Haring, H., and Fehlmann, M. (1989) Phosphorylation of two cytosolic proteins. An early event of T cell activation. Biochem. J. 258: 505-510.

Ponzin, D., Mancini, C., Toffano, G., Bruni, A., and Doria, G. (1989) Phosphatidylserine induced modulation of the immune response in mice. Effect of intravenous administration. Immunopharmacology 18: 167-173.

Rickard, J.E., and Sheterline, P. (1985) Evidence that phorbolesters interfere with stimulated Ca++ redistribution by activating Ca++ efflux in neutrophil lymphocytes. Biochem. J. 231: 623-628.

Roska, A.K. and Lipsky, P.E. (1985) Dissection of the functions of antigen-presenting cells in the induction of T cell activation. J.Immunol. 135: 2953-2959.

Sakane, F., Yamada, K., and Kanoh, H. (1989) Different effects of sphingosine, R59022 and anionic amphiphiles on two diacylglycerol kinase isozymes purified from porcine thymus. FEBS Lett. 255:409-413.

Weiss, A., Imboden, J., Shoback, D., and Stobo, J. (1984) Role of T3 surface molecules in human T cell activation. T3 dependent activation results in an increase in cytoplasmic free calcium. Proc. Natl. Acad. Sci. USA 81: 4169-4174.

Wiskocil, R., Weiss, A., Imboden, J., Kamin-Lewis, R., and Stobo, J. (1985) Activation of a human T cell line. A two stimulus requirement in the pretranslational events involved in the coordinate expression of interleukin-2 and interferon-g genes J.Immunol. 134: 633-639.

ANTI-PHOSPHATIDYLSERINE MONOCLONAL ANTIBODY: STRUCTURAL TEMPLATE FOR STUDYING LIPID-PROTEIN INTERACTIONS AND FOR IDENTIFICATION OF PHOSPHATIDYLSERINE BINDING PROTEINS

Masato Umeda, Koji Igarashi, Shigeru Tokita, Farooq Reza, and Keizo Inoue

Department of Health Chemistry, Faculty of Pharmaceutical Sciences, The University of Tokyo, 7-3-1 Hongo, Bunkyo-ku, Tokyo 113, Japan.

INTRODUCTION

Although phosphatidylserine (PS) is an essential component for the formation of membrane bilayers, it has been shown to contribute to many regulatory processes in biological responses (Kaibuchi et al., 1981, Zwaal et al., 1986, Madsen et al., 1989, Inoue et al., 1989). Pharmacological effects of PS have also been reported (Bruni et al., 1976, 1989). Some PS may exhibit biological activity through interacting with specific binding proteins. Some of the proteins were shown to interact with PS in a highly specific manner and precise configurations of PS molecule are required for the interaction. The typical examples are the receptor for lyso-PS on rat mast cells (Horigome et al., 1986, Chang et al., 1988), aminophospholipid translocase (Morrot et al., 1989) and protein kinase C (Lee and Bell, 1989). The PS-binding proteins so far reported are listed in Table 1. Although much effort has been focused on understanding the molecular mechanism involved in the interactions between PS and PS-binding proteins, the difficulties of handling these proteins because of their limited aqueous solubility and scarcity in the biological systems have hampered progress in this area.

Our approach to this problem has involved the production of monoclonal antibodies (mAb) which specifically recognize PS with binding profiles similar to those of other PS-binding proteins. The mAb with definite specificity to the phospholipids may provide a valuable information about the lipid-protein interactions, and may also represent a structural template for the production of anti-idiotypic antibody (anti-Id) which can cross-react with the actual receptor molecules. Our previous analyses have shown that the anti-Id against a phosphatidylcholine-specific mAb cross-reacted extensively with the phosphatidylcholine-specific lipid transfer protein from bovine liver, supporting the idea that the anti-phosphatidylcholine mAb and phosphatidylcholine-specific transfer protein share a common structure, and that some of the anti-Id may carry the internal image of phosphatidylcholine molecule (Nam et al., 1990a, 1990b). In this chapter, we will describe the

NATO ASI Series, Vol. H 70
Phospholipids and Signal Transmission
Edited by R. Massarelli, L. A. Horrocks,
J. N. Kanfer, and K. Löffelholz
© Springer-Verlag Berlin Heidelberg 1993

Table 1.
PHOSPHATISYLSERINE-BINDING PROTEINS

Membrane proteins

Aminophospholipid translocase	Morret et al. (1989)
Synaptic vesicle protein, p65	Perin et al. (1990)
Synapsin I	Benfenati et al. (1989)
PS-receptors (lysoPS-receptor) on	
macrophage	Schroit et al. (1985)
mast cell	Horigome et al. (1986)
fibroblast	Martin et al. (1987)
reticuloendothelial cell	Allen et al. (1988)

Cytosolic proteins

Adducin (brain)	Wolf et al. (1986)
Annexins	Kaetzel et al. (1989)
Phospholipase A1 (platelet)	Horigome et al. (1987)
Phospholipase A2 (mast cell)	Murakami et al. (1991)
Platelet	Burgener et al . (1990)
Protein 4.1 (erythrocytes)	Rybicki et al. (1988)
Protein kinase C	Lee et al. (1989)

Serum proteins

Blood coagulation factors	Mann et al. (1988)
Factor V, VIII, X	
Apolipoprotein H	Wurm H. (1984)

production and structural analyses of the anti-PS mAb, and the reactivities of anti-Id mAb raised against the PS-specific mAb will be shown.

ANTI-PS MONOCLONAL ANTIBODIES

PRODUCTION

Although anti-phospholipid antibodies have been frequently detected in patients with autoimmune, infectious, and other disorders, the majority of the anti-phospholipid antibodies so far reported cross-reacted extensively with other phospholipids (reviewed in Alving C.R., 1986; Janoff et al., 1986;Ghoson et al, 1986 Harris et al., 1988). These anti-phospholipid antibodies form a family of poorly characterized antibodies and no information has been available regarding the precise nature of the interaction between the antibodies and the phospholipid. We found that the immune response to phospholipids could be effectively induced when the lipids coated on the acid treated *Salmonella* were administered through the intrasplenic route (Miyazawa et al, 1986; Umeda et al, 1989; Nam et al

Table 2
Phospholipid specificities and isotypes of anti-PS mAb

Group	Immunization	Clones	Ig class	Specificity
I	Long-term	PS4A7	IgM	PS
	Long-term	PS1G3	IgG3	PS
	Short-term	PS3A	IgM	PS,PE
	Short-term	PSF6	IgG3	PS,PE
II	Short-term	PSF7	IgM	PS,PE
	Short-term	PSB4	IgG2b	PS,PE
	Long-term	PS3H1	IgG2a	PS,PE
	Long-term	PS3E10	IgM	PS,PE
	Short-term	PSC8	IgM	PS,PE,CL,PA
	Short-term	PSF11	IgM	PS,PE,CL,PA
	Short-term	PSG3	IgG2b	PS,PE,CL,PA
III	Short-term	PSD11	IgM	PS,PE,PA
	Short-term	PSF10	IgM	PS,CL
	Long-term	PS3D12	IgM	PS,PE,CL,PA
	Long-term	PS2C11	IgG3	PS,PA

1990). In the short term immunization protocol, BALB/c mice were immunized with PS coated on the acid-treated *Salmonella* by intrasplenic injection. They received the second intrasplenic injection of the antigen 1 week after the first immunization, and the fusion was performed 3 days after the second immunization. In the long-term immunization protocol, the splenic immunization was followed by five injections of the antigen via lateral tail vein (3-wk intervals between each immunization) and the fusion was performed 3 days after the final immunization. The reactivity of the mAb to phospholipid was examined by ELISA and mAb that bind to PS but not to PC were selected. Among 61 clones obtained by five fusions, 15 mAb were established (Umeda et al., 1989).

REACTIVITY PROFILES

The mAb exhibited three distinct reactivity profiles ranging from highly specific to broadly cross-reactive (Table 2). Typical reactivity profiles are shown in Figure 1. Group I mAb (PS4A7, PS1G3) are highly specific to PS and bind only weakly to other acidic phospholipids. PS3A represents the Group II mAb that bind to PS and phosphatidylethanolamine (PE) almost equally, and this family of mAb does not cross-react with other acidic phospholipids. PSC8 represents Group III mAb that cross-react extensively with other acidic phospholipids. In Figure 1, relative reactivities of the cellular PS-recognizing proteins such as scavenger receptor of mouse macrophages (Nishikawa et al., 1990), phospholipase A1 of rat platelets (Higashi et al., 1988) and phospholipase A2 of rat mast cells (Murakami et al., 1991), are also indicated. The scavenger receptor of mouse peritoneal

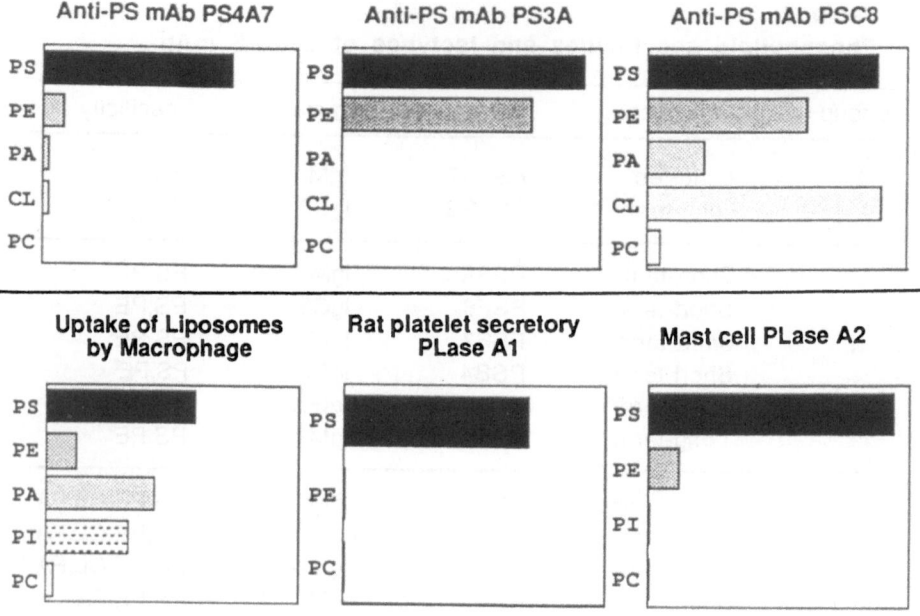

Figure 1. Relative phospholipid specificities of PS-binding proteins

Reactivities of anti-PS mAb were measured by ELISA (Umeda et al, 1989). Microtiter plates were coated with phospholipids and the mAb bound were detected with biotinylated anti-mouse Ig and streptavidin-conjugated alkaline phosphatase.

Data about "Uptake of liposome by macrophage", "Rat platelet secretory PLase A1" and "Mast cell PLase A2" were quoted from Nishikawa et al., 1990, Higashi et al., 1988, and Murakami et al., 1991, respectively.

macrophage showed a broad specificity to acidic phospholipids, showing a similar reactivity profile with those of group III anti-PS mAb. It is generally accepted that scavenger receptors may recognize the anionic surface of a ligand. Although it could bind to various ligands with negatively charged surface such as acetylated low density lipoprotein (acetyl-LDL), oxidized LDL and dextran sulfate, it could not bind to other polyanionic substances such as heparin, polyadenylic acid and polycytidylic acid, suggesting that it could recognize a particular array of acidic substances on the surface. Acidic phospholipids on the surface of liposome may provide a proper environment for the recognition. Among Group III anti-PS mAb, the mAb showed slightly different reactivity profiles. Some anti-PS mAb, such as PSC8, PSF11, PSG3 and PS3D12 cross-reacted with all of the acidic phospholipids, while PS2C11 reacts preferentially with PS and PA, and PSF10 has an even higher affinity to CL than to PS. The cross-reactivities of the mAb may not be simply a result of a high density of negative charge in the reactive antigen, and presumably, each antibody may

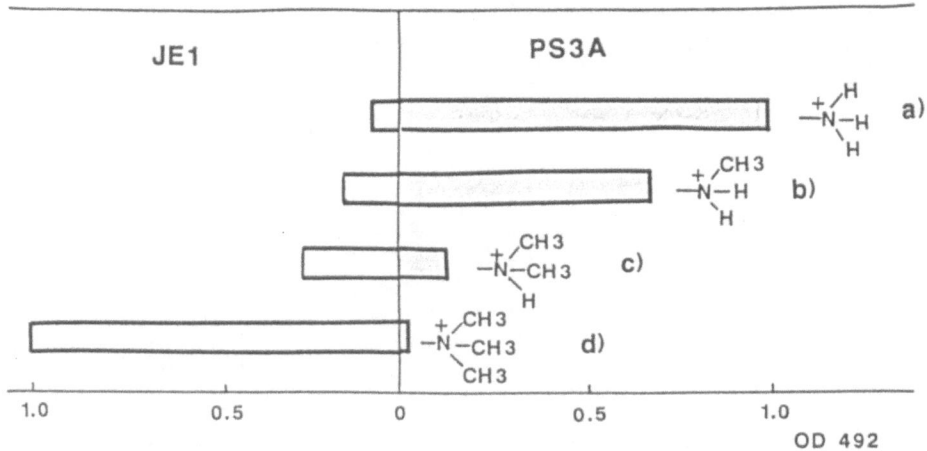

Figure 2. Reactivities of PS3A and JE-1 with synthetic phosphatidylethanolamine and its analogs.

Microtiter plates were coated with phospholipid antigens and the mAb bound were detected with biotinylated anti-mouse Ig and streptavidin-conjugated alkaline phosphatase. JE-1 is a monoclonal antibody that specifically reacts with phosphatidylcholine. Phospholipid antigens used were a)1,2-dipalmitoylphosphatidylethanolamine (PE), b)monomethyl-PE, c)dimethyl-PWE, d)PC.

prefer a particular geometry of the epitopes that will be presented by each phospholipids. Both phospholipase A1 of rat platelets and phospholipase A2 of rat mast cells are highly specific to PS, showing a similar reactivity profile with that of Group I mAb.

The epitopes recognized by the Group I and Group II mAb were further studied by using synthetic phospholipid analogs. Reactivities of PS3A (Group II) with synthetic PE analogs with modified polar head group are shown in Figure 2. As a control, the reactivity of anti-PC mAb, JE-1, are also indicated. The order of the reactivity of PS3A to the analogs is PE>monomethyl-PE>>dimethyl-PE>trimethyl-PE (PC), while JE-1 showed a reverse order of specificities to the analogs. PS3A recognizes both PS and PE through interacting with the amino group of the polar head group of the phospholipids, while the methyl groups on the quarternary nitrogen of the choline moiety of PC is responsible for the binding of JE-1. The reactivity profile of PS3A to the analogs shows the close similarity with those of aminophospholipid translocase, which were determined by the uptake and transbilayer movement of the spin-labeled PE analogs (Morrot et al., 1989).

Reactivities of PS4A7 and PS3A with synthetic PS analogs are shown in Figure 3. PS4A7 binds to PS (phosphatidyl-L-serine) but not to either phosphatidyl-D-serine or phosphatidyl-L-homoserine. PS3A bind not only to PS, but also to phosphatidyl-D-serine and phosphatidyl-L-homoserine. PS4A7 shows some affinity to phosphatidyl-serineamide, while PS3A shows an even higher affinity to phosphatidyl-serineamide than to PS. Another Group I mAb, PS1G3 has a higher reactivity with phosphatidyl-L-serine than with phosphatidyl-D-serine

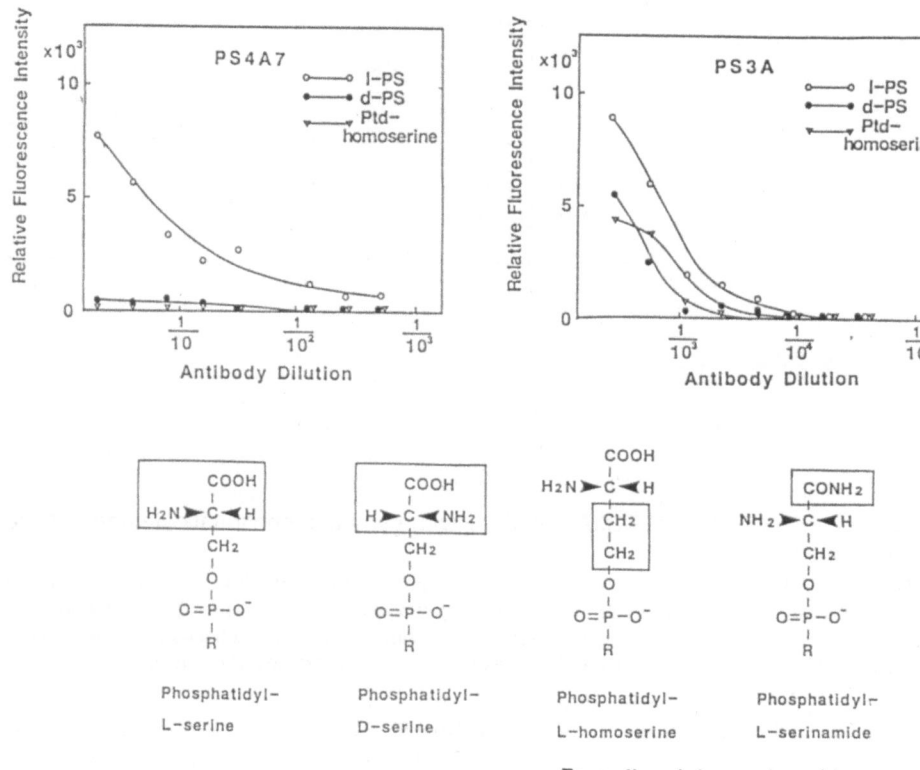

Figure 3. Reactivities of PS3A and PS4A7 with PS analogs.

or phosphatidyl-L-homoserine, though it cross-reacts extensively with phosphatidyl-serineamide. These observations suggest that Group I mAb can recognize the stereo-specific configuration of the serine residue of PS and that a precise distance between the phosphodiester bond and the amino and carboxyl group of serine residue is required for the interaction. Group II mAb could not distinguish the conformational structure of the serine residue and they cross-react extensively with phosphatidyl-serineamide, indicating that the amino group of the phospholipids may play an important role in the interactions. Reactivity of the anti-PS mAb was not inhibited by the water-soluble PS analogs such as serine, phosphoserine, or glycerophosphoserine. The mAb may interact not only with the head group of PS but also with hydrophobic portion of the phospholipid. Further analysis of the reactivities of the mAb (Igarashi et al., 1991), and application of the mAb to the histochemical analysis of the distribution of PS (Ito et al., 1991) were described elsewhere.

STRUCTURES

In order to obtain a further insight into the structural basis of the antigen binding properties of these antibodies, we determined the amino acid sequences of the variable regions in heavy and light chains of the mAb, and attempted to make computer-aided models of the combining sites of the mAb. A few aspects of the structure of immunoglobulin protein will be reviewed before presenting data on the amino acid sequences and the predicted structures of anti-PS mAb. Comparative studies performed on the crystal structures of the immunoglobulins with different specificities have revealed that the fundamental structure of the variable fragment is highly conserved, especially in the framework regions where they are not directly involved in the actual binding interactions. The framework regions of H and L chains consist of nine antiparallel beta-strands, forming a nearly perfect cylindrical barrel structure (Novotny et al., 1983)(Figure 4). The antigen-combining site is formed by the juxtaposition of six complementarity determining regions (CDRs), three deriving from the light chain (designated CDR1-, 2- and 3-L) and three from the heavy chain (CDR1-, 2- and 3-H). The CDRs form loops protruding from the backbone barrel structure and this hypervariable loops construct the non-conserved antigen-binding sites. Although the prediction of the three dimensional structure of immunoglobulins in the past could not always correctly predict the structure (Chothia et al., 1986), recent models have provided reasonably successful predictions of the structures of the binding sites of various antibodies (Chothia et al., 1989).

Variable regions of the heavy and light chains of five anti-PS mAb, one (PS4A7) from Group I, two (PS3A, PSB4) from Group II and two (PS3D12, PSG3) from Group III, were sequenced by a combination of amino acid and cDNA sequencing (Kimura et al. 1988). The variable regions of heavy chain are encoded by three gene segments, VH, DH and JH, and light chain variable regions are encoded by VL and JL segments. The gene segments used by the anti-PS mAb are summarized in Table 3. PS4A7 and PS3A used the same VH gene, showing

Table 3.

Gene compositions of anti-PS monoclonal antibodies

Clone	Group	VH	DH	JH	VK	JK
PS4A7 (IgM, K)	I	J558	S.P.2.2	JH4	VK9	JK4
PS3A (IgM, K)	II	J558	FL16.1	JH3	VK4	JK4
PSB4 (IgG2b,K)	II	J558	FL16	JH2	VK1A	JK2
PS3D12(IgM,K)	III	Q52	FL16.1	JH3	VK3	JK4
PSG3 (IgG2b,K)	III	J606	SP.2(DQ52)	JH3	VK5	JK4

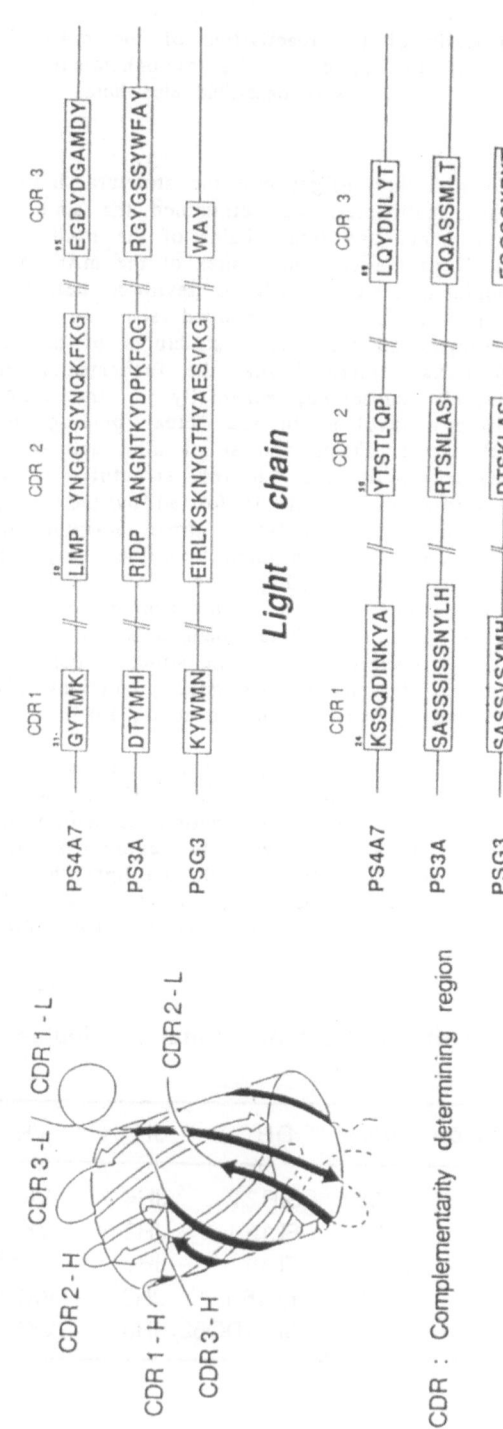

Figure 4. Amino acid sequences of anti-PS monoclonal antibodies

In the left panel, topology of the binding site barrel (Novotny et al, 1983) is indicated.
In the right panel, CDRs of H and L chains of anti-PS mAb are indicated.

homologous amino acid sequences, while Group III mAb used an entirely different VH gene. When the amino acid sequences of CDRs of anti-PS mAb were compared, they show a remarkable heterogeneity in both heavy and light chain sequences (Figure 4). No mAb, so far reported, has the same combinations of VH and VL gene segments with those of the anti-PS mAb. The results show a clear contrast to those obtained with anti-glycolipid mAb where they used very restricted heavy and light chain segments (Kimura et al., 1989, 1990; Zenita et al., 1990), suggesting that the structure encoded by the gene segments is particularly well adapted for forming a binding site to carbohydrate determinants.

The computer-aided modeling of variable fragments of the typical two mAb, PS4A7 (Group I) and PSG3 (group III), was performed by the program entitled FRGMNT written by Dr.Haruki Nakamura (Institute for Protein Research, Osaka University). The structures were subject to a restrained energy minimization using a modified version of AMBER (Singh et al., 1986). The proposed combining sites for PS4A7 and PSG3 are very different in the length of the CDRs and in the nature of their amino acids (Figure 5). The combining site surface proposed for PS4A7 has a deep pocket-like (cavity) structure, while that proposed for PSG3 is a relatively shallow groove-like structure (Padlan and Kabat, 1988). The cavity of PS4A7 is formed by the CDR3 of heavy chain (CDR3-H) and CDR1 of light chain (CDR1-L) that protrudes from the backbone barrel structure, whereas the protrusion of CDR3-H is absent from PSG3 because of the extremely shorter CDR3-H of PSG3. PS4A7 is highly specific to PS and could also recognize the stereo-specific configuration of serine residue of PS. The basic (Lys and Arg) and acidic amino acid (Glu and Asp) residues in the cavity of the combining site of PS4A7 face each other, and these residues may be responsible for the specific interaction. In contrast, PSG3 shows an extensive cross-reaction with other acidic phospholipids as well as double-stranded DNA (Igarashi et al., 1991). Since PSG3 has a cluster of the positively charged amino acids from CDR1-H and CDR2-H

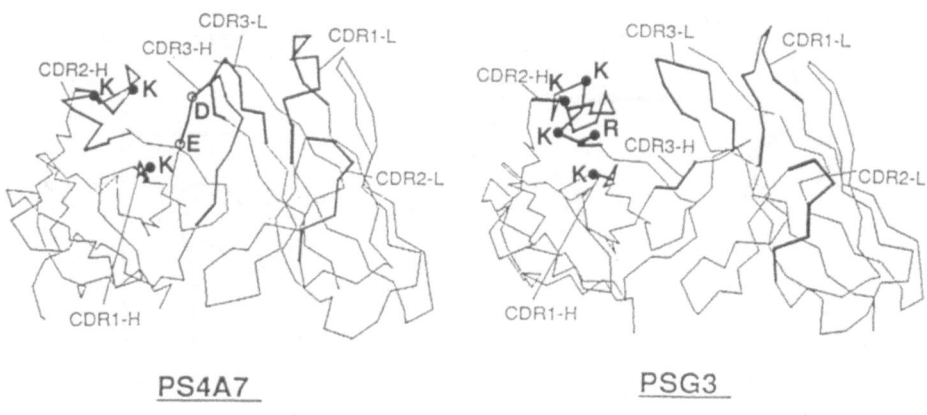

PS4A7   PSG3

Figure 5. Predicted structure of combining sites of anti-PS monoclonal antibodies
The α-carbon backbone of the Fv of PS4A7 (left panel) and PSG3 (right panel) are shown.
Bold lines indicate the amino acid residues in CDRs.

on the surface of the binding site, the particular distribution of the positively charged amino acids is responsible for the cross-reaction, possibly through the formation of salt bridges with complementary charged groups on the antigen. In fact, a modification of Lys residues of PSG3 by biotinylation completely abolished the reactivity of the mAb, whereas biotinylation had no effect on the reactivity of PS4A7. Further insight into the structural basis of the interaction will be obtained by x-ray crystallographic and NMR studies of the antigen-antibody complexes.

ANTI-IDIOTYPIC MONOCLONAL ANTIBODIES

RATIONALE

Many anti-idiotypic antibodies have been shown to recognize cross-reactive structures shared by the antibody and the receptor molecules, indicating the presence of structural similarity between the antibody and the ligand-binding receptors (Gaulton and Greene, 1986). Our previous analysis showed that an anti-idiotypic antibody against a monoclonal antibody that specifically binds to phosphatidylcholine (PC) showed an extensive cross-reaction with the PC-specific lipid transfer protein isolated from bovine liver (Nam et al., 1990b). The anti-idiotypic antibody also inhibited the lipid transfer activity of the protein. The peptide conformations of the binding sites, which are shared by the antibody and the transfer protein, might be recognized by the anti-idiotypic antibodies, and some of the anti-idiotypic antibody may carry the internal image property of PC molecule.

Since PS4A7 recognizes the stereo-specific configuration of the serine moiety of PS, the mAb will represent a structural template for the production of anti-idiotypic antibodies that can cross-react with the actual PS-binding proteins that also interact specifically with PS (Figure 6).

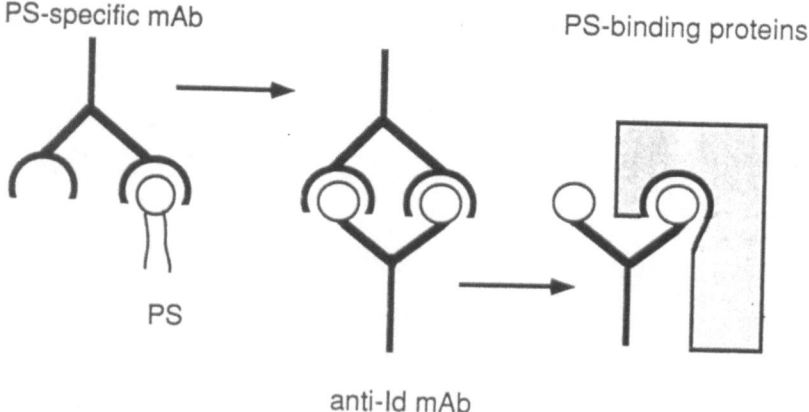

Figure 6. Schematic model of the interaction between anti-Id mAb against anti-PS mAb and PS-binding proteins

REACTIVITY PROFILES

BALB/c mice were immunized with PS4A7 conjugated to keyhole limpet homcyanin (Umeda et al. 1986) and 35 syngeneic anti-idiotypic mAb were established from 3 fusions. The anti-idiotypic mAb showed two distinct reactivity profiles; one group of anti-idiotypic mAb react only with PS4A7 and the other group cross-react with other anti-PS mAb. None of the anti-idiotypic mAb bound to anti-PC mAb, JE-1. Typical reactivity profiles of two clones, named Id11A2 and Id8F7, to the anti-PS mAb, PS4A7 (Group I), PS3A (Group II) and PS3D12 (Group III) are shown in Figure 7. Id8F7 reacts only with PS4A7, while Id11A2 reacts with both PS4A7 and PS3A but not with 3D12, suggesting that Id11A2 recognizes the common structure shared by PS4A7 and PS3A. Since the preincubation of PS4A7 with anti-idiotypic mAb totally inhibited the binding of PS4A7 to PS, the anti-idiotypic mAb recognize the idiotopes that are identical or spacially in proximity to the combining site of the PS4A7.

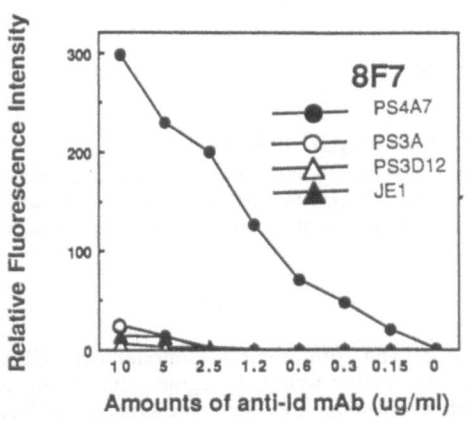 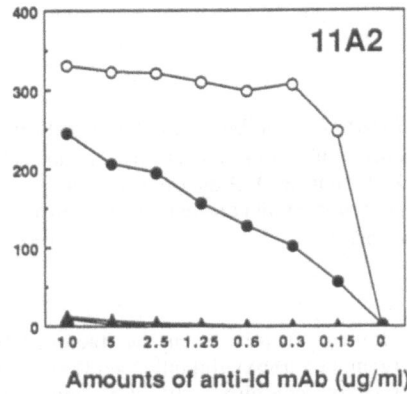

Figure 7. Binding of anti-idiotypic mAb Id8F7 and Id11A2 to anti-phospholipid mAb.

Microtiter plates were coated with PS4A7 (anti-PS, Group I), PS3A (anti-PS, Group II), PS3D12 (anti-PS, Group III) and JE-1 (anti-PC). The anti-idiotypic mAb bound were detected with biotinylated anti-mouse IgG (Fc) and alkalinephosphatase-conjugated streptavidin.

CROSS-REACTIONS WITH PLASMA OR CELLULAR PROTEINS

In the next series of experiments, the effect of anti-idiotypic mAb on the various biological reactions was investigated. In the course of this study, we found that one of the anti-idiotypic mAb, Id8F7, showed a strongly inhibitory effect on the activity of blood clotting factor VIII (Figure 8). This inhibitory effect was observed only with Id8F7. It is well known that the initiation of blood clotting requires the exposure of PS on the outer layer of the platelet membrane and blood clotting factors form an active clotting complexes on the surface. Blood

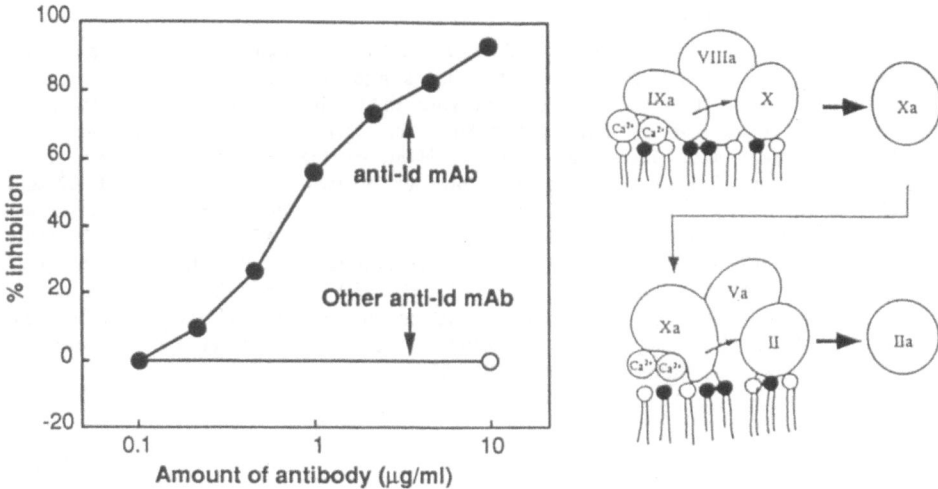

Figure 8. Inhibition of factor VIII activity by anti-idiotypic mAb 8F7

Factor VIII activity of human plasma was determined by the factor VIII-dependent generation of factor Xa from factor X (Kabi Coatest method). In the inhibition assay, 1/500 diluted human plasma was preincubated with anti-Id mAb for 30 min at room temperature, and then the activity of factor VIII was measured.

clotting factors such as factor VII, IX, X and prothrombin contain the 'GLA (gamma-carboxyglutamic acid) domain' at the amino-terminal region, and form a tertiary complex with Ca^{++} and acidic phospholipids on the membranes. Factor VIII, which functions as a cofactor during the activation of factor X by factor IXa, interacts with lipids through a different mechanism, since it does not contain any particular domain such as GLA domain for the interaction with acidic phospholipids. Although the phospholipid specificity of the binding of factor VIII was not thoroughly studied, factor VIII showed a preferential binding to PS (Lajimanovich et al., 1981; Gilbert et al., 1990). Computer-aided homology comparison of amino acid sequences between factor VIII and anti-PS mAb showed that a part of the sequence of factor VIII exhibited a striking similarity to that of CDR3-H of PS4A7 (Vehar et al., 1984)(Figure 9), while no significant homology was observed with other anti-PS mAb. Since the predicted structure of PS4A7 suggest that the CDR3-H protrudes from the immunoglobulin surface, the CDR3-H could be an antigenic determinant which might be recognized by Id8F7. In order to further define the epitope recognized by Id8F7, we synthesized the peptide corresponding either to the CDR3-H of PS4A7 (AREGDYDAMDY, amino acid residues 93-101), or to the consensus sequence in the B-domain of FVIII (TRQNVEGSYDGAY, amino acid residues 1251-1263). The peptide were coupled to BSA via functional -SH group of the additional cystein residue at the carboxy terminus and the resulting peptide-BSA complexes were coated onto the wells of

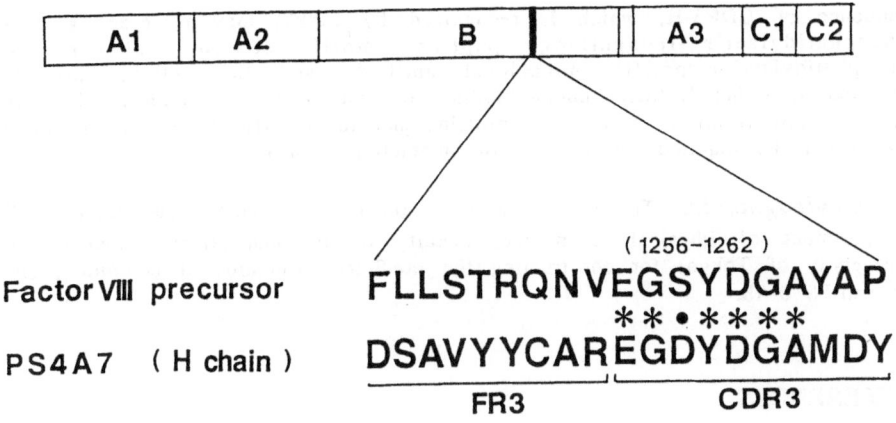

Figure 9. Comparison of amino acid sequences between anti-PS mAb PS4A7 and factor VIII.

microtiter plates. The binding of Id8F7 to plate coated peptides was examined by ELISA. Id8F7 strongly bound to both CDR3-H peptide and FVIII-B peptide. Id8F7 did not react with other peptides derived from the sequences corresponding to either CDR1 or CDR2 region of heavy chain of PS4A7 (data not shown). Although the direct interaction between Id8F7 and factor VIII remained to be studied, Id8F7 might bind to factor VIII, resulting in inhibition of the factor VIII activity. These observations raised the possibility that the FVIII-B peptide sequence may have functional significance for interaction with phosphatidylserine. Previous studies using recombinant factor VIII fragments and antibodies against factor VIII suggested that the major phospholipid-binding site resides within the C2 domain (Bloom,1987; Arai et al.,1989). Although the functional role of B domain is still unclear and further studies will be required to understand the mechanism of the inhibition, the present findings might provide a clue to understand the function of B-domain of factor VIII in the process of the blood clotting.

It is noteworthy that protein kinase C required precise configuration of PS for the activation, though it bound less specifically to the mixed micelles containing various acidic phospholipids (Lee and Bell, 1989). This observation implies that some proteins interact with acidic phospholipid in a less specific manner, whereas it may recognize a precise structure of PS by its specific binding site. This interaction may cause a conformational change of the protein, leading to the activation. Some particular domain structure of the protein may be responsible for the specific interaction. In our recent analyses, we found that the CDR3-H peptide itself could specifically bind to phosphatidylserine with similar specificity to that of PS4A7, albeit with markedly lower affinity. Since Id8F7 also showed an extensive cross-reaction with other cellular proteins, the sequence of CDR3-H, which is recognized by Id8F7, may represent a novel phosphatidylserine-recognizing peptide motif shared between the phosphatidylserine-specific monoclonal antibody and the cellular proteins. detected by either ELISA, immuno-blotting, or histochemical staining of the cells. We are now trying to isolate the proteins and to identify the structural module which will be required for the specific interaction with PS.

Acknowledgments. The authors wish to thank Dr. Arata Y and Mizutani R. (Department of Physical Chemistry, Faculty of Pharmaceutical Sciences, The University of Tokyo) for constructing the predicted structures of combining sites of anti-PS mAb.

REFERENCES

Allen TM, Willianson P, Schlegel RA (1988) Phosphatidylserine as a determinant of reticuloendothelial recognition of liposome models of the erythrocyte surface. *Proc.Natl.Acad.Sci.USA* 85:8067-8071

Alving C, (1986) Antibodies to liposomes, phospholipids and phosphate esters. *Chem.Phys.Lipids* 40:303-314

Arai M, Scandella D, Hoyer LW (1989) Molecular basis of factor VIII inhibition by human antibodies. *J.Clin.Invest.* 83:1978-1984

Benfenati F, Greengard P, Burunner J, Bahler M (1989) Electrostatic and hydrophobic interaction of synapsin I and synapsin I fragments with phospholipid bilayers. *J.Cell Biol.* 108:1851-1862

Bloom JW (1987) The interaction of rDNA factor VIII des-797-1562 and factor VIII des-797-1562-derived peptides with phospholipid. *Thromb.Res.* 48:439-448

Bruni A, Toffano G, Leon A, Boarato E. (1976) Pharmacological effects of phosphatidylserine liposome. Nature (London) 260:331-333

Bruni A, Mietto L, Bellini F, Boarato E, Toffano G. (1989) Pharmacological and autopharmacological action of phosphatidylserine. *Phospholipids in the Nervous System*: Bazan NG, Horrocks LA, Toffano G (eds) Fidia Research Series. Vol 17:217-224

Burgener R, Wolf M, Ganz T, Gaggiolini M (1990) Purification and characterization of major phosphatidylserine-binding phosphoprotein from human platelets. *Biochem.J.* 269:729-734

Chang HW, Inoue K, Bruni A, Boarato E, Toffano G (1988) Stereoselective effects of lysophosphatidylserine in rodents. *Br.J.Pharmacol.* 93:647-653

Chothia C, Lesk AM, Levitt M, Amit AG, Mariuzza RA, Phillips SEV, Poljak RJ (1986) The predicted structure of immunoglobulin D1.3 and its comparison with the crystal structure. *Science* 233:755-758

Chothia C, Lesk AM, Tramontano A, Levitt M, Smith-Gill SJ, Air G, Sheriff S, Padlan EA, Davies D, Tulip WR, Colman PM, Spinelli S, Alzari PM, Poljak RJ (1989) Comformations of immunoglobulin hypervariable regions. *Nature* 342:877-883

Gaulton GN, Greene MI (1986) Idiotypic mimicry of biological receptors. *Annu.Rev.Immunol.* 4:253-280

Ghoson S, Campbell AM (1986) Multispecific monoclonal antibodies. *Immonol.Today* 7:217-222

Harris EN, Asherson RA, Hughes GRV (1988) Antiphospholipid antibodies-autoantibodies with a difference. *Annu.Rev.Med.* 39:261-271

Higashi S, Kobayashi T, Kudo I, Inoue K (1988) Purification and characterization of lysophospholipase released from rat platelets. *J.Biochem.* 103:442-447

Horigome H, Tamori-Natori Y, Inoue K, Nojima S (1986) Effect of serine phospholipid structure on the enhancement of concanavalin A-induced degranulation in rat mast cells. *J.Biochem.* 100:571-579

Horigome K, Hayakawa M, Inoue K, Nojima S (1987) Selective release of phospholipase A2 and lysophosphatidylserine-specific lysophospholipase from rat platelets. *J.Biochem.* 101:53-61

Igarashi K, Umeda M, Tokita S, Nam KS, Inoue K (1991) Effective induction of anti-phospholipid and anticoagulant antibodies in normal mouse. *Thromb.Res.* 61:135-148

Inoue K, Kobayashi T, Kudo I (1989) Function and metabolism of lysophosphatidylserine in rat mast cell activation. *Phospholipids in the Nervous System*: Bazan NG, Horrocks LA, Toffano G (eds) Fidia Research Series. Vol 17:225-231

Ito E, Miyazawa A, Takagi H, Yoshioka T, Horikoshi T, Yanagisawa K, Nakamura T, Kudo Y, Umeda M, Inoue K, Mikoshiba K (1991) Developmental assembly of calcium-mobilizing systems for excitatory amino acids in rat cerebellum. *Neurosci. Res.* 11:179-188.

Janoff A, Rauch J (1986) The structural specificity of antiphospholipid antobodies in autoimmune disease. *Chem.Phys.Lipids* 40:315-332

Kaetzel MA, Hazarika P, Dedman JR (1989) Differential tissue exprossion of three 35-KDa annexin calcium-dependent phospholipid-binding proteins. *J.Biol.Chem.* 264:14463-14470

Kaibuchi K, Takai Y, Nishizuka Y (1981) Cooperative roles of various membrane phospholipids in the activation of calcium-activated, phospholipid-dependent protein kinase. *J.Biol.Chem.* 256:7146-7149

Kimura H, Cook R, Meek K, Umeda M, Ball E, Capra JD, Marcus DM. (1988) Sequences of the Vh and Vl regions of murine monoclonal antibodies against 3-fucosyllactosamine. *J.Immunol.* 140:1212-1217

Kimura H, Buescher ES, Ball ED, Marcus DM (1989) Restricted usage of Vh and Vk genes by murine monoclonal antibodies against 3-fucosyllactosamine. *Eur. J.Immunol.* 19:1741-1746

Lajmanovich A, Clergeon GH, Freyssinet JM, Marguerie G (1981) Human factor VIII procoagulant activity and phospholipid interaction. *Biochim.Biophys.Acta*, 678:132-136

Lee MH, Bell RM (1989) Phospholipid functional groups involved in protein kinase C activation, phorbol ester binding, and binding to mixed micelles. *J.Biol.Chem.* 264:14797-14805

Madsen J, Connor J, Schroit AAJ (1989) Recognition of phosphatidylserine by the reticuloendothelial system. *Phospholipids in the Nervous System*: Bazan NG, Horrocks LA, Toffano G (eds) Fidia Research Series. Vol 17:3-10

Mann KG, Jenny RJ, Krishnaswamy S (1988) Cofactor proteins in the assembly and expression of blood clotting enzyme complexes. *Ann.Rev.Biochem.* 57:915-956

Martin OC, and Pagano RE (1987) Transbilayer movement of fluorescent analogs of phosphatidylserine and phosphatidylethanolamine at the plasma membrane of cultured cells. J.Biol.Chem. 262:5890-5898

Miyazawa A, Umeda M, Horikoshi T, Yanagisawa K, Yoshioka T, Inoue K. (1988)

Production and characterization of monoclonal antibodies that bind to phosphatidylinositol 4,5-bisphosphate. *Molecular Immunol.* 25:1025-1031

Morrot G, Herve P, Zachowski A, Fellmann P, Devaux PF, (1989) Aminophospholipid translocase of human erythrocytes: phospholipid substrate specificity and effect of cholesterol. *Biochemistry* 28:3456-3462

Murakami M, Kudo I, Umeda M, Komada M, Fujimori Y, Takahashi K, Inoue K (1991) Cultured mast cells express three distinct phospholipase A2. submitted.

Nam KS, Igarashi K, Umeda M, Inoue K (1990a) Production and characterization of monoclonal antibodies that specifically bind to phosphatidylcholine. *Biochim.Biophys.Acta* 1046:89-96

Nam KS, Umeda M, Igarashi K, Inoue K (1990b) Anti-idiotypic antibody identifies the structural similarity between the phosphatidylcholine-specific monoclonal antibody and phosphatidylcholine-specific transfer protein. *FEBS Lett.* 269:394-397

Nishikawa K, Arai H, Inoue K (1990) Scavenger receptor-mediated uptake and metabolism of lipid vesicles containing acidic phospholipids by mouse peritoneal macrophages. *J.Biol.Chem.* 265:5226-5231

Novotny J, Bruccoleri R, Newell J, Murphy D, Haber E, Karplus M (1983) Molecular anatomy of the antibody binding site. *J.Biol.Chem.* 258:14433-14437

Padlan EA, Kabat EA (1988) Model-building study of the combining sites of two antibodies to α(1→6)dextran. *Pro.Nat.Acad.Sci.USA* 85:6885-6889

Perin MS, Fried VA, Mignery GA, Jahn R, Sudhof TC (1990) Phospholipid binding by a synaptic vesicle protein homologous to the regulatory region of protein kinase C. *Nature* 345:260-263

Rybicki AC, Heath R, Lubin B, Schwartz RS (1988) Human erythrocyte protein 4.1 is a phosphatiddylserine binding protein. J.Clin.Invest. 81:255-260

Schroit AJ, Madsen J, Tanaka Y (1985) In vivo recognition and clearance of red blood cells containing phosphatidylserine in their plasma membranes. *J.Biol.Chem.* 260:5131-5138

Umeda M, Diego I, Ball ED, Marcus DM (1986) Idiotypic determinations of monoclonal antibodies that bind to 3-fucosyllactosamine. *J.Immunol.* 136:2562-2567

Umeda M, Igarashi K, Nam K.S, Inoue K. (1989) Effective production of monoclonal antibodies against phosphatidylserine: Stereo-specific recognition of phosphatidylserine by monoclonal antibody. *J.Immunol.* 143;2273-2279

Vehar GA, Keyt B, Eaton D, Rodriguez H, O'Brien DP, Rotblat F, Oppermann H, Keck R, Wood WI, Harkins RN, Tuddenham EGD, Lawn RM, Capon DJ (1984) Structure of human factor VIII. *Nature* 312:337-341

Wolf M, Sahyoun N (1986) Protein kinase C and phosphatidylserine bind to Mr 110,000/115,000 polypeptides enriched in cytoskeletal and postsynaptic density. *J.Biol.Chem.* 261:13327-13332

Wurm H (1984) Beta 2-glycoprotein-I (apolipoprotein H) interactions with phospholipid vesicles. Int.J.Biochem. 16:511-515

Zenita K, Hirashima K, Shigeta K, Hiraiwa N, Takada A, Hashimoto K, Fujimoto, E, Yago, K, Kannagi R. (1990) Northern hybridization analysis of Vh gene expression in murine monoclonal antibodies directed to cancer-associated ganglioside antigens having various sialic acid linkages. J.Immunol. 144:4442-4451

Zwaal RFA, Hemker HC (eds) (1986) Blood coagulation.Elsevier Science Publishers B.V. 141-169

IMMUNOMODULATION BY SERINE PHOSPHOLIPIDS

A. Bruni, F. Bellini*, E. Caselli*, G. Monastra* and D. Ponzin*
Department of Pharmacology
University of Padova
Largo Meneghetti 2
35131 Padova
Italy

INTRODUCTION

When the body is under threat from invading pathogens or other type of injury, local cell populations initiate a coordinated defense program, known as the inflammatory response. The first wave of signalling molecules causes vasodilation and the expression of leukocyte-endothelial cell adhesion molecules, directing the migration of circulating cells into the injured area. Signalling compounds are then generated by the newly recruited cells while they are engaged in removing the pathogen. The tumultuous occurrence of cell arrival, the secretion of hydrolytic enzymes, together with a change in the equilibrium of the extracellular environment makes the inflamed area a difficult place for cell survival. The degradation products from injured cells form an additional wave of signalling molecules, further modulating the progress and the extent of inflammatory reactions. Under particular circumstances the inflammatory response is made more selective and efficient by a specific immunological component. However, in this case a powerful reaction can be also initi-

*Fidia Research Laboratories
Via Ponte della Fabbrica 3/A
35031 Abano Terme, Italy

NATO ASI Series, Vol. H 70
Phospholipids and Signal Transmission
Edited by R. Massarelli, L. A. Horrocks,
J. N. Kanfer, and K. Löffelholz
© Springer-Verlag Berlin Heidelberg 1993

ated against innocuous exogenous substances such as pollen or against the tissues of the body itself. Negative modulation of the inflammatory program is therefore of considerable clinical interest. The engulfment of microorganisms or other cells by phagocytes is accompanied by activation of phospho-lipid-degrading enzymes of both the ingesting and the ingest-ed cell. Furthermore, the membrane changes that accompany cell trauma or death may trigger net degradation of phospho-lipids, in which endogenous and exogenous phospholipases may participate (Elsbach and Weiss, 1988). The phospholipids be-come therefore a potential source of compounds modulating the progress of inflammatory disease. Our studies show that phos-phatidylserine (PS) and its hydrolytic product lysoPS, are indeed able to influence the activity of inflammatory cells, chiefly when immunological mechanisms are involved. The ac-tion of these phospholipids will be briefly summarized in this chapter.

IMMUNOMODULATION IN VIVO

Since the experiments on cultured cells are of limited value in the assessment of immunomodulatory properties of a given compound in vivo, our effort was directed at obtaining evidence of serine phospholipid-induced effects in the intact animal. As expected from compounds acting as transient modu-lators of inflammatory reactions, the circulation time of both PS and lysoPS is short. In rats, the i.v. injection of PS vesicles is followed by their transfer into liver and lymphoid organs (spleen) with a half-time of 0.85 min (Palatini et al., 1991). In mice, the disappearance of lysoPS from plasma is even faster (plasma half-life of 0.6 min, Bruni et al., 1990). In spite of brief circulation time, lysoPS reaches perivascular areas where mast cells are locat-ed causing their degranulation (Fig. 1). This action is specific and stereoselective (Bruni et al., 1984; Chang et

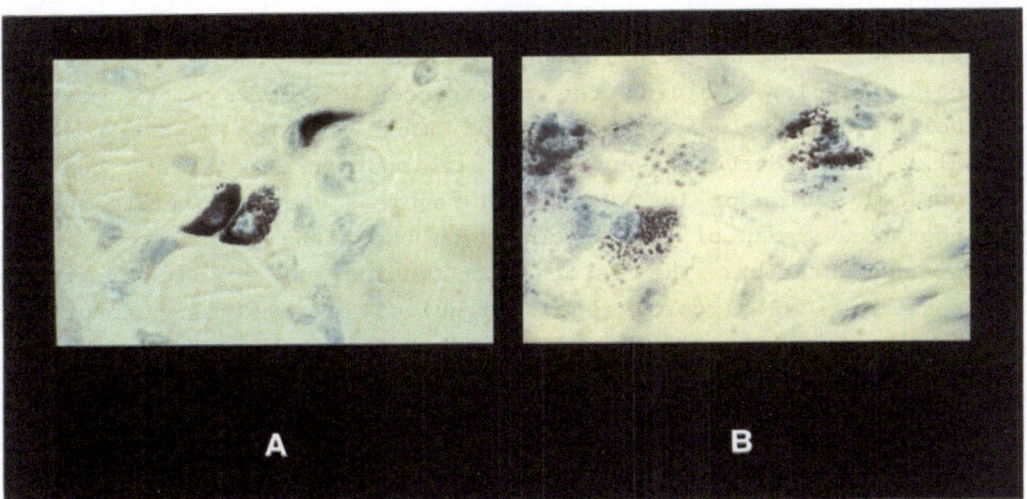

Figure 1.
Mast cell degranulation induced by lysoPS in CD-1 Swiss mice.
5 mg/kg of lysoPS were injected i.v. to mice. After 30 min,
the tongue was taken and stained with toluidine blue. (A),
untreated mice; (B) Mice treated with lysoPS. Magnification
1100x. Experimental details in Monastra et al. (1991).

al., 1988; Monastra et al., 1991). Four daily treatments with lysoPS (25 mg/kg, i.p.) induce extensive depletion of mast cell granules yielding the appearance of almost empty mast cells in histologic specimens of mouse tongue (Monastra et al., 1991). Structure-activity relationships demonstrate that optimal action of lysoPS in vivo requires the carboxylate and the amino group of serine linked to the alpha carbon atom in L configuration. The ester bond linking the acyl chain in position 1 of glycerol is also required for maximal potency. Reacylation into PS is not necessary for the action on mast cells.

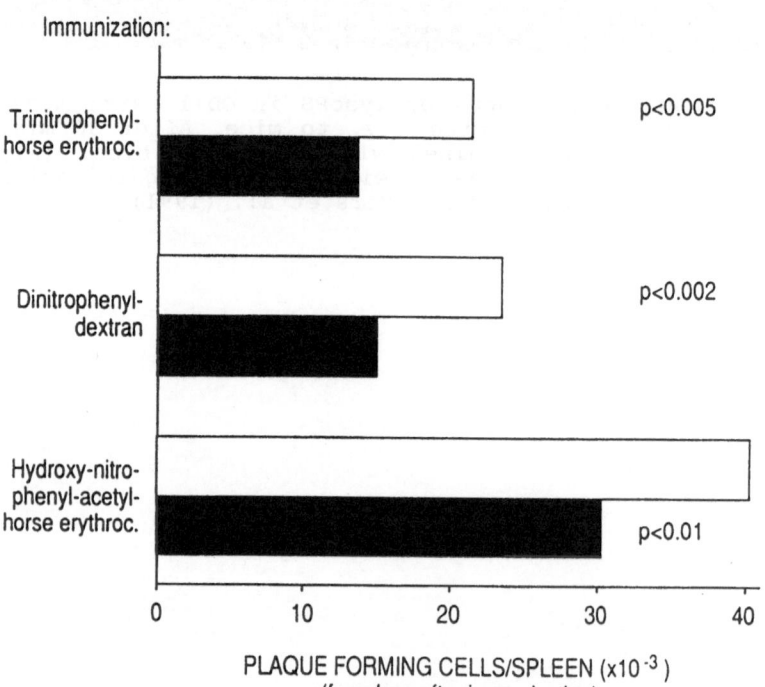

PLAQUE FORMING CELLS/SPLEEN $(\times 10^{-3})$
(four days after immunization)

Figure 2.
PS-induced decrease of primary humoral immune response in $(C57BL/10xDBA/2)F_1$ mice. PS (25 mg/kg i.v.) was given to mice 4h before the antigens (5×10^8 trinitrophenyl-horse erythrocytes i.v., 5×10^8 hydroxy-nitrophenyl-acetyl-horse erythrocytes i.v., 100 µg of dinitrophenyl-dextran i.p.). Four days after immunization the spleen was assayed in triplicate for the plaque-forming cells. Experimental details in Ponzin et al. (1989)

Since PS accumulates in the spleen, it was thought that it might affect the humoral immune response occurring in this organ through the internalization of antigens by the local macrophages and their presentation to lymphocytes (Ponzin et al., 1989). As shown in Fig. 2, the i.v. injection of phospholipid in mice reduces the T-cell dependent and independent antibody production. In this action PS is only active when given before antigen, suggesting that the phospholipid affects the early steps of immune response when processed antigen leads to lymphocyte activation. Tests on T-helper cell activity indicated that these cells may be one target of PS.

Further indications were obtained measuring the action of PS on the lipopolysaccharide-induced serum level of tumor necrosis factor in mice (Monastra and Bruni, 1992). When lipopolysaccharide was given to mice pretreated with PS in a regimen of 3 daily injections (i.p.), the production of tumor necrosis factor was inhibited (Fig. 3). A similar PS effect was observed in rabbits. Since the serum level of tumor necrosis factor likely reflects the lipopolysaccharide-induced macrophage activation (Beutler and Cerami, 1988) the action of PS has been attributed to an inhibitory effect on these cells. In agreement, PS shows high affinity for mononuclear phagocytes (Allen et al., 1988) that ingest the phospholipid by their scavenger receptor (Nishikawa et al., 1990).

Recently it has been proposed that tumor necrosis factor participates in the pathogenesis of autoimmune disease, particularly multiple sclerosis (Selmaj and Raine, 1988). This prompted us to test the action of PS on experimental models of this illness. The development of experimental autoimmune encephalomyelitis (EAE) in mice closely mimics the progress of multiple sclerosis in man. When EAE was induced in mice by the injection of lymphocytes sensitized to myelin basic protein, the continuous administration of PS (30 mg/Kg i.p., daily) inhibited the development of illness (Monastra G., Cross A., Bruni A. and Raine C., submitted). The action of PS was reversible since the suspension of treatment resulted in the appearance of EAE with a lag of approximately one week. Furthermore, when PS was given to mice showing the signs of EAE, the phospholipid prevented the recurrent relapses which are a prominent feature of EAE in mice. The efficacy of PS in this model of autoimmune disease has been confirmed in histological studies which showed the absence of perivascular and parenchymal inflammation in the spinal cord and in the brain of phospholipid-treated mice.

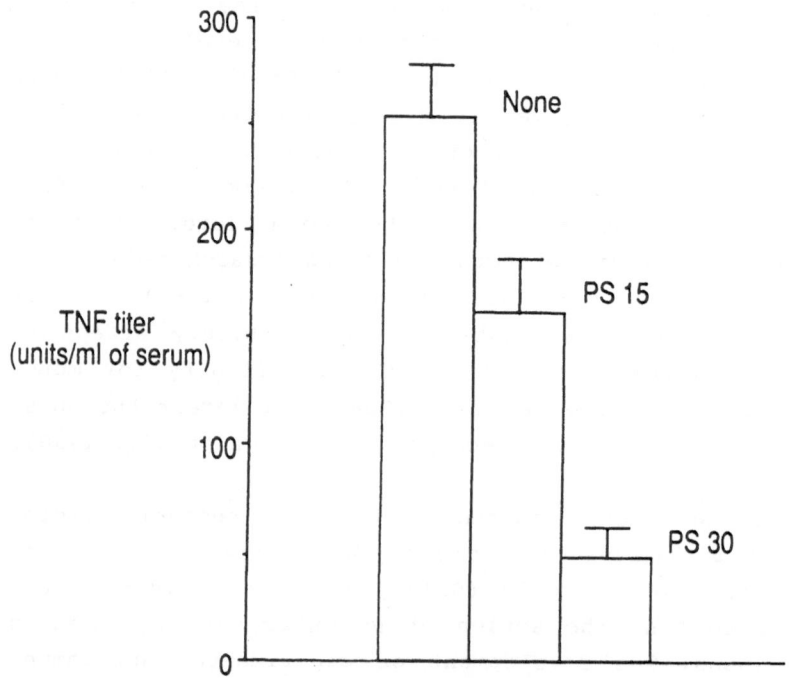

Figure 3.
Inhibition of endotoxin-induced serum level of tumor necrosis factor in CD-1 Swiss mice. Mice were treated daily for three days with the indicated doses (mg/kg i.p.) of PS. 90 min after the last phospholipid dose 5 µg of lipopolysaccharide from S. typhi was given i.p. The serum level of tumor necrosis factor was measured 60 min later by a biological assay (cytotoxic effect on L-929 murine fibrosarcoma cells). Details in Monastra and Bruni (1992).

IMMUNOMODULATION IN VITRO

As is evident from the above studies, PS and its main
metabolite lysoPS may have multiple effects on the immune
system. In accord, studies of PS and lysoPS in cultured cells
show that the phospholipids are active on mast cells, lymphocytes and macrophages. In mast cells the major effect is due
to lysoPS which behaves as an agonist, inducing degranulation
(Boarato et al., 1984). Mast cell stimulation has the same
structural requirements observed for the in vivo effect. For
reasons still to be understood, lysoPS induces histamine secretion in mouse mast cells by its own effect whereas in rat
mast cells the phospholipid acts synergistically with other
agonists. Among these are antigens and nerve growth factor
(Kolster et al., 1987). However, the signalling mechanism activated by lysoPS is the same in rat and mouse preparations
(Bellini et al., 1988; 1990). As shown in Fig. 4 for mouse
mast cells prelabeled with [^{3}H]arachidonate, the lysoPS-induced histamine release is associated with the formation of
labeled diacylglycerol and phosphatidate. The appearance of
diacylglycerol had a biphasic kinetics suggesting multiple
sources for the generation of this metabolite. Since under
these conditions hydrolysis of phosphoinositides is detected
(Bellini et al. 1988; 1990), it is likely that the first action of lysoPS causes mobilization of the small pool of these
phospholipids. The subsequent increase in [Ca^{2+}]i and the
stimulation of protein kinase C may then activate the phospholipases causing the mobilization of the larger phosphatidylcholine pool producing the amount of diacylglycerol required for sustained protein kinase C activation (reviewed by
Exton, 1990 and Billah and Anthes, 1990). The same sequence
of events has been observed upon activation of mast cell IgE
receptor (Gruchalla et al., 1990). In agreement, the kinetics
of histamine release under the lysoPS-induced stimulus shows
the same slow onset seen with activation of IgE receptor
(Fig. 4). At variance with mast cells, the action of serine
phospholipids on lymphocytes does not result from the ac-

tivation of a receptor-linked signalling pathway. Rather it is the consequence of PS or lysoPS incorporation into the membrane. The addition of PS on cultured lymphocytes is followed by the inhibition of mitogen-induced cell activation. In a mixed lymphocyte-macrophage population from mouse spleen, the phospholipid prevents the secretion of interleukin-2 induced by concanavalin A (Ponzin et al., 1989). In mononuclear cells from human blood, PS inhibits the DNA synthesis elicited by PHA or immunological stimuli (allogeneic lymphocytes, anti-CD3). Under these conditions PS prevents the expression of interleukin-2 receptor (Caselli et al., 1992).

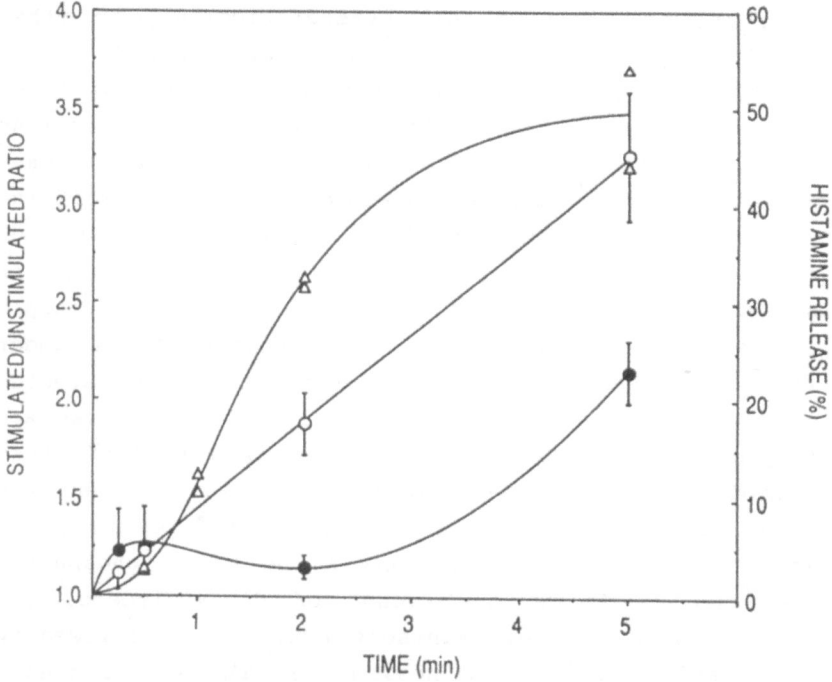

Figure 4.
Generation of diacylglycerol (●) and phosphatidate (○) during the lysoPS-induced histamine release in mouse mast cells. 10^6 [^{3}H]arachidonate-labeled mast cells were incubated with 1 μM of deoxy-lysoPS. The unstimulated value for diacylglycerol was 1454±155 dpm/10^5 cells, for phosphatidate 2631±342 (means ± S.E. n=3-5). The histamine secretion (△) was tested in parallel samples. Reproduced with permission from Bellini et al. (1990).

Attempts to clarify the action of PS revealed that the action of phospholipid is complex. Shortly after PS incorporation into the lymphocyte plasma membrane, the activation of protein kinase C decreases the phosphoinositide-dependent signalling system (Caselli E., Bellini F., Ponzin D., Baricordi O.R. and Bruni A., submitted). Furthermore, the continuous hydrolysis of PS in the incubation medium releases lysoPS and unsaturated fatty acids which together inhibit T cell function (Bellini F., Caselli E., Ponzin D. and Bruni A., submitted). In agreement, lymphocyte inhibition by PS is maximal with PS analogues containing unsaturated fatty acids. Also, a phospholipase A_2 active on PS is found in fetal calf serum which is a normal supplement in the medium used to study lymphocyte activation. The same phospholipase A_2 activity is also found in human serum (Bellini F., Caselli E., Ponzin D. and Bruni A., submitted). Inhibition of cell function by PS and lysoPS can be also demonstrated in macrophages. As shown previously (Gilbreath et al., 1986), the ingestion of liposomes containing PS by cultured mouse macrophages inhibits the microbicidal activity of these cells against leishmanias. Interestingly, we have found that PS does not affect the process of antigen degradation and presentation by these cells (Ponzin et al., 1989). It should be pointed out however that liposomes made exclusively with PS have been used in these experiments. Insufficient liposome uptake under these conditions may in part account for this results (Mietto et al., 1989).

GENERATION OF LYSOPHOSPHATIDYLSERINE

As described above most of the PS effect on the immune cells are the consequence of phospholipid hydrolysis by phospholipase A_2 producing lysoPS and unsaturated fatty acids. Since PS is distributed in the internal leaflet of plasma membrane, the generation of lysoPS in the intact cell does not result in lysophospholipid release into the external medium. Generation of lysoPS and its effect on contiguous cells are however possible upon cell damage and PS exposure to the extracellular medium. The formation of lysoPS has been studied in a population of rat leukocytes (Mietto et al., 1987). When these cells (67% polymorphonuclear and 33% mononuclear cells) were incubated with L-[^{14}C]serine, part of the label was incorporated into the lipid fraction. After 1 h of incubation at 37°C 65% was found in phosphatidylserine, 8.8% in phosphatidylethanolamine, 17.9% in sphingomyelin and 1.6% in lysophosphatidylserine. Chasing 2 h with cold serine it was found that the production of lysoPS during this time was 0.33 nmol/mg of leukocyte proteins. When these experiments were repeated with a preparation of leukocyte plasma membranes (0.25 mg protein/ml) the production of lysoPS in 1 h yielded in the surrounding medium a 0.3 μM solution of lysoderivative which showed biological activity on rat mast cells. In these studies it was also found that proteins added to the incubation medium prevented the lysoPS interaction with the inflammatory cells. However, the addition of tetradecanoylphorbolacetate restored the lysoPS uptake suggesting that activated cells are more sensitive to this phospholipid.

SUMMARY AND CONCLUDING REMARKS

When given by a parenteral route PS manifests pharmacological activity on the immune system. After conversion into lysoPS, the phospholipid causes mast cell degranulation in mice and rats, an effect followed by depletion of preformed mediators stored in these cells. In mice, PS inhibits the primary humoral immune response and prevents the lipopolysaccharide-induced serum level of tumor necrosis factor. Furthermore, the phospholipid inhibits the development of experimental autoimmune encephalomyelitis when the illness is induced in this animal by the intravenous injection of lymphocytes sensitized to myelin basic protein. In agreement with these in vivo observations, lysoPS activates cultured serosal mast cells to secrete histamine and PS inhibits DNA synthesis in human T lymphocytes stimulated through the antigen receptor. Moreover, PS inhibits the microbicidal activity of mouse macrophages against intracellular leishmanias (Gilbreath et al. 1986). Taken together these observations indicate that PS has inhibitory activity on lymphocyte and macrophage function, thereby producing immunosuppression. Since histamine has been shown to inhibit the synthesis of tumor necrosis factor occurring in macrophages (Vannier et al., 1991), a further contribution to the immunosuppressant effect may originate from the action of lysoPS on mast cells. However, as mast cell activation by serine phospholipids is only observed in certain rodent species (mouse, rat, gerbil, hamster), the contribution of mast cell depletion is expected to influence the action of PS only in these animals (Bruni, 1991). In all PS-induced effects there are indications that lysoPS formed in the blood stream or in tissues by a phospholipase A_2 plays a pivotal role. Thus, PS may be regarded as a precursor of the more active lysoderivative. Extending this concept, it seems likely that PS deacylation in cell membranes may serve the purpose of generating a metabolite, suitable to avoid exacerbation of immune response. This may occur in inflamed areas where many cells die and where our tests show that lysoPS is produced.

The action of PS in vivo may amplify the clinical perspectives of liposomes. These lipid structures are actively studied as a drug delivery system in cancer therapy. In this field considerable success has been obtained in increasing their circulation time by the inclusion of monosialoganglioside or phosphatidylinositol (Allen et al., 1989; Gabizon and Papahadjopoulos, 1988). Our data indicate that the inclusion of PS may contribute immunosuppressant activity to the phospholipid envelope. This property appears to be useful in diseases where certain cytokines (e.g. tumor necrosis factor) are responsible for pathogenesis and/or symptoms.

REFERENCES

Allen TM, Williamson P, Schlegel RA (1988) Phosphatidylserine
    as a determinant of reticuloendothelial recognition of
    liposome models of the erythrocyte surface. Proc Natl
    Acad Sci USA 85:8067-8071
Allen TM, Hansen C, Rutledge J (1989) Liposomes with prolong-
    ed circulation time: factors affecting uptake by
    reticuloendothelial and other tissues. Biochim Biophys
    Acta 981:27-35
Bellini F, Toffano G, Bruni A (1988) Activation of phospho-
    inositide hydrolysis by nerve growth factor and lysophos-
    phatidylserine in rat peritoneal mast cells. Biochim Bio-
    phys Acta 970:187-193
Bellini F, Viola G, Menegus AM, Toffano G and Bruni A (1990)
    Signalling mechanism in the lysophosphatidylserine-induc-
    ed activation of mouse mast cells. Biochim Biophys Acta
    1052:216-220
Beutler B, Cerami A (1988) Tumor necrosis, cachexia, shock,
    and inflammation: a common mediator. Ann Rev Biochem
    57:505-518
Billah MM, Anthes JC (1990) The regulation and cellular func-
    tions of phosphatidylcholine hydrolysis. Biochem J
    269:173-183
Boarato E, Mietto L, Toffano G, Bigon E and Bruni A (1984)
    Different responses of rodent mast cells to lysophospha-
    tidylserine. Agents Actions 14:613-618
Bruni A, Bigon E, Battistella A, Boarato E, Mietto L and
    Toffano G (1984) Lysophosphatidylserine as histamine re-
    leaser in mice and rats. Agents Actions, 14:619-625
Bruni A (1990) Phospholipid absorption and diffusion through
    membranes. In: Hanin I, Pepeu G (eds) Phospholipids.
    Plenum Press, New York, p 59-68
Bruni A (1991) Mast cell activation by serine phosholipids:
    influence on signal generation. J Immunol Res 3:1-3
Caselli E, Baricordi OR, Melchiorri L, Bellini F, Ponzin D,
    Bruni A (1992) Inhibition of DNA synthesis in peripheral
    blood mononuclear cells treated with phosphatidylserines
    containing unsaturated acyl chains. Immunopharmacology
    (in press).
Chang HW, Inoue K, Bruni A, Boarato E, Toffano G (1988)
    Stereoselective effects of lysophosphatidylserine in
    rodents. Br J Pharmac 93:647-653.
Elsbach P, Weiss J (1988) Phagocytosis of bacteria and phos-
    pholipid degradation. Biochim Biophys Acta 947:29-52
Exton JH (1990) Signaling through phosphatidylcholine break-
    down. J Biol Chem 265:1-4
Gabizon A, Papahadjopoulos D (1988) Liposome formulation with
    prolonged circulation time in blood and enhanced uptake
    by tumors. Proc Natl Acad Sci (USA) 85:6949-6953
Gilbreath MJ, Hoover DL, Alving CR, Swartz GM, Meltzer MS
    (1986) Inhibition of lymphokine-induced macrophage micro-
    bicidal activity against Leishmania major by liposomes:
    characterization of the physicochemical requirements for
    liposome inhibition. J Immunol 137:1681-1687

Gruchalla RS, Dinh TT, Kennerly DA (1990) An indirect pathway of receptor-mediated 1,2-diacylglycerol formation in mast cells. IgE receptor-mediated activation of phospholipase D. J Immunol 144:2334-2342

Kolster L, Jensen C, Bruni A, Mietto L, Toffano G and Norn S (1987) Effect of lysophosphatidylserine on the immunological histamine release. Biochim Biophys Acta 927:196-202

Mietto L, Boarato E, Toffano G, and Bruni A (1987) Lysophosphatidylserine-dependent interaction between rat leukocytes and mast cells. Biochim Biophys Acta 930:145-153

Mietto L, Boarato E, Toffano G and Bruni A (1989) Internalization of phosphatidylserine by adherent and non-adherent rat mononuclear cells. Biochim Biophys Acta 1013:1-6

Monastra G, Bruni A (1992) Decreased serum level of tumor necrosis factor in animals treated with lipopolysaccharide and liposomes containing phosphatidylserine. Lymphokine and Cytokine Res (in press)

Monastra G, Pege G, Zanoni R, Toffano G, Bruni A (1991) Lysophosphatidylserine-induced activation of mast cells in mice. J Lipid Mediators 3:39-50

Nishikawa K, Arai H and Inoue K (1990) Scavenger receptor-mediated uptake and metabolism of lipid vesicles containing acidic phospholipids by mouse peritoneal macrophages. J Biol Chem 265:5226-5231

Palatini P, Viola G, Bigon E, Menegus AM, Bruni A (1991) Pharmacokinetic characterization of phosphatidylserine liposomes in the rat. Br J Pharmacol 102:345-350

Ponzin D, Mancini C, Toffano G, Bruni A and Doria G (1989) Phosphatidylserine-induced modulation of the immune response in mice: effect of intravenous administration. Immunopharmacology 18:167-176

Selmaj KW, Raine CS (1988) Tumor necrosis factor mediates myelin and oligodendrocyte damage in vitro. Ann Neurol 23:339-346

Vannier E, Miller LC, Dinarello CA (1991) Histamine suppresses gene expression and synthesis of tumor necrosis factor alpha via histamine H_2 receptors. J Exp Med 174:281-284

DIRECT EFFECT OF PHOSPHATIDYLSERINE ON THE ACTIVITY OF THE POST-SYNAPTIC GABA$_A$ RECEPTOR COMPLEX

M.V. Rapallino, A. Cupello, P. Mainardi, G. Besio and C.W. Loeb
Clinica Neurologica
Università di Genova,
Via De Toni, 5
16132 Genova, ITALY

Neurotransmitter postsynaptic receptors, as integral membrane proteins, are located within the phospholipid bilayer of the acceptive membrane. Therefore it appears reasonable that the phospholipid environment of the plasma membrane influences and modulates the functional activity of these receptors. In addition, the phospholipid composition of the two leaflets of the bilayer may introduce a further element able to influence the function of these receptors.

Referring to GABA receptors, numerous studies have shown that the interaction of the neurotransmitter with its recognition site is influenced by treatment of the synaptic membranes with phospholipases or detergents (Giambalvo and Rosenberg, 1976; Lloyd and Davidson, 1979; Toffano et al., 1981). Binding of benzodiazepines to the GABA$_A$ receptor complex is also modified by synaptic membrane treatment with phospholipases (Ueno and Kuriyama, 1981). Even interactions between different sites within the receptor complex are modulated by the membrane phospholipid pool. In fact, Fujimoto and Okabayashi (1983) found that the stimulatory effect by Cl^- of diazepam binding to rat cerebellar membranes disappeared after both phospholipase A_2 and phospholipase C treatments.

Administration of phospholipids to animals can result in pharmacological effects. Our group (Loeb et al., 1982, 1984, 1986) and others (Toffano et al., 1984) have found that the administration of phosphatidylserine (PS) to rats simultaneously with GABA results in an antiepileptic effect, which is not shared by the administration of the same dose of plain GABA. The antiepileptic effect has been monitored by the EEG (Loeb et al., 1982, 1984) and a possible mechanism which may play a role in this central pharmacological effect of phosphatidylserine is a modulation of brain GABA$_A$ receptors activity, once the phospholipid gets into the brain.

To test this possibility we have used an experimental model involving the study of the permeation of labelled Cl^- across single plasma membranes, microdissected from the GABA acceptive Deiters' neurons in the rabbit

NATO ASI Series, Vol. H 70
Phospholipids and Signal Transmission
Edited by R. Massarelli, L. A. Horrocks,
J. N. Kanfer, and K. Löffelholz
© Springer-Verlag Berlin Heidelberg 1993

(Hydén et al., 1986). In this way we have investigated the effect of various concentrations of phosphatidylserine on both basal and GABA activated Cl⁻ permeation. The phospholipid was added in the medium bathing the extracellular side of the neuronal plasma membrane.

MATERIALS AND METHODS

1) Studies about the effect of PS on basal and GABA activated ^{36}Cl⁻ permeation; across single Deiters' membranes.

Throughout these experiments single plasma membranes from Deiters' neurons were prepared from adult male rabbits according to a free-hand dissection method described previously (Hydén et al., 1980; Rapallino et al., 1988; Rapallino et al., 1990). The free-hand dissected membranes were put over a 30 μm 0 hole on a gold-plated glass which was then put between two microchambers, M_1 and M_2 (Hydén et al., 1986). The membrane extracellular side always faced M_1, the M_1 medium being a low potassium buffer (NaCl 125 mM, Na acetate 30 mM, K acetate 2.3 mM; Hepes 10 mM, pH 7.4) containing ^{36}Cl⁻, 0.5 μCi/ml, added as Na^{36}Cl. The M_2 medium was a high potassium buffer mimicking the ionic composition of the intracellular fluid (NaCl 20 mM, Na acetate 6.5 mM, K acetate 112.5 mM; Hepes 10 mM, pH 7.4). In the experiments about the effects of exogenous phosphatidylserine on basal ^{36}Cl⁻ permeation, the membranes were preincubated with the phospholipid, at the various concentrations (10^{-9} to 10^{-3}M) studied, on the membrane extracellular side, for 3 minutes at 30°C. After the incubations, new M_1 medium containing PS, this time with the addition of ^{36}Cl⁻, was added and the incubations (30 seconds at 30°C) started.

In the case of the study of the effect of PS on GABA activated ^{36}Cl⁻ permeability, no preincubation was run. The M_1 medium, always containing ^{36}Cl⁻, 0.5 μCi/ml, contained in the various instances:
- a)- no drugs (control)
- b)- 10^{-5}M GABA
- c)- 10^{-5}M GABA plus various (10^{-13} to 10^{-3}M)
  concentrations of phosphatidylserine.

The incubations were performed for 30 seconds at 30°C. After the incubations, 10 μl aliquots were taken from M_1 and 2 μl ones from M_2 and counted for ^{36}Cl⁻ radioactivity in a liquid scintillation counter (Beckman LS 1801). The minimal level of counts in M_2 throughout the experiments was 393 dpm. Those in M_1 were always around 11,000 dpm.

Throughout the experiments, 10^{-3}M starting solutions of PS were prepared by suspending 2.5 mg in 3 ml of low potassium Hepes. The suspensions were sonicated 8 times (for 1 min at 30 sec intervals) at 0°C. A subsequent centrifugation for 20 min at 500 x g eliminated titanium contamination from the sonicator probe. The final PS concentrations were obtained by dilution with the buffer.

2) Studies about the effect of PS on the binding of GABA to
   Triton X-100 treated membranes from rat cerebellum.

Membranes were prepared from rat cerebellum and treated with Triton X-100 before the binding experiments. The cerebella were prepared from female Sprague-Dawley rats weighing around 250 g. They were frozen at -20°C and left overnight before the binding experiments. The tissue was then thawed and homogenized with 50 volumes of Tris/citrate 0.05 M, pH 7.1. The homogenates were then made 0.05% in Triton X-100 and incubated at 37°C for 30 minutes. After the treatment the suspensions were centrifuged at 50,000 x g for 10 min at 4°C. The pellets were resuspended in buffer and centrifuged again. After a further wash, the final pellet was resuspended in the buffer at 0.6 mg protein/ml. Buffer and [^{3}H]GABA were added to 1 ml samples of the resuspension to a final volume of 2 ml and a final concentration of the labelled neurotransmitter of 0.5 nM. When appropriate, part of the buffer was substituted by a suspension of phosphatidylserine (in the same buffer) so to give the desired final concentration, in the 10^{-8} to 10^{-3} M range. Non-specific binding was determined running two binding assays in the presence of an excess (0.1 mM) of cold GABA.

For the binding reactions, the suspensions were incubated for 5 minutes at 4°C. The incubations were interrupted by centrifugation at 50,000 x g. The membrane pellets were rapidly washed twice with 4 ml of cold (4°C) water and then counted. Specific binding data were obtained after subtraction of the appropriate non-specific binding values.

Materials

In our experiments we used γ-amino-butyric acid from Sigma, St. Louis, Mo, USA. Phosphatidylserine (1-stearoyl-2-oleyl) was kindly supplied by Fidia Research Laboratories. $^{36}Cl^-$ in NaCl solution, 3.7 mCi/gCl was from Amersham, U.K. γ-Amino-[2,3-^{3}H]butyric acid, spec. act. 80.8 Ci/mmol, was from New England Nuclear, U.S.A.

RESULTS

The issue whether the microchamber system is adequate for the study of $^{36}Cl^-$ permeation across single plasma membranes from Deiters' neurons has been taken up and discussed in a recent publication (Hydén et al., 1991). Several lines of evidence indicate the adequacy of the system, ruling out trivial leakages at the sides or via holes of the membrane preparation. Among these are the absence of passage of labelled inulin, the block of the permeation of Cl^- ions in the presence of a mixture of blockers of Cl^- channels and the permeation of labelled glucose from the outside to the inside of the membrane only in the presence of a downhill Na^+ gradient, outside → inside. If the gradient was reversed, no labelled glucose passed.

The effect of various concentrations of phosphatidylserine on basal $^{36}Cl^-$ permeation is reported in Figure 1. There is a decrease (by 32%) of the ion permeation from the outside to the inside of the membrane when $10^{-5}M$ PS is on the membrane extracellular side. That effect disappears at a PS concentration of $10^{-3}M$. The decrease of $^{36}Cl^-$ permeation was found even if the preincubation with PS was avoided (Figure 1).

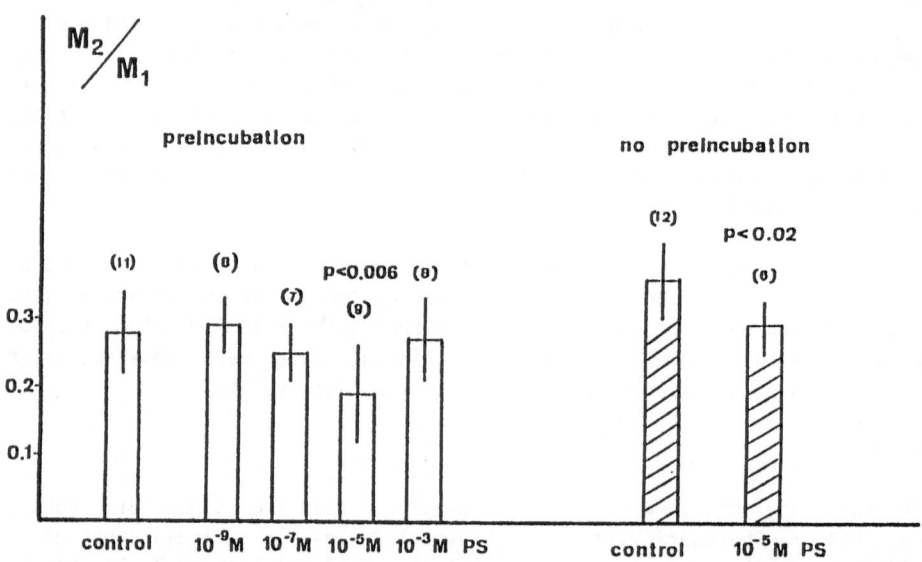

Fig. 1: Effect of phosphatidylserine on the basal $^{36}Cl^-$ passage across the Deiters' neuron membrane. The group of columns to the left represents experiments involving a 3 minute preincubation step either with plain buffer (control) or with PS. The group of columns to the right represents experiments without preincubation. The data are expressed as M_2/M_1 ratios, i.e. the ratios between $^{36}Cl^-$ concentration in the M_2 ("intracellular") and M_1 ("extracellular") chambers at the end of the incubations. The bars represent the standard deviations, whereas the numbers in parentheses represent the numbers of experiments. The p values refer to the comparison with the control.

The stimulation of $^{36}Cl^-$ passage across single Deiters' membranes by $10^{-5}M$ GABA (+28%) is reduced in the presence of phosphatidylserine from 10^{-11} to $10^{-5}M$ (Figure 2). Actually, the effect of GABA is completely erased in the 10^{-11} to $10^{-7}M$ range. Here again the PS effect disappears upon increasing its concentration to $10^{-3}M$.

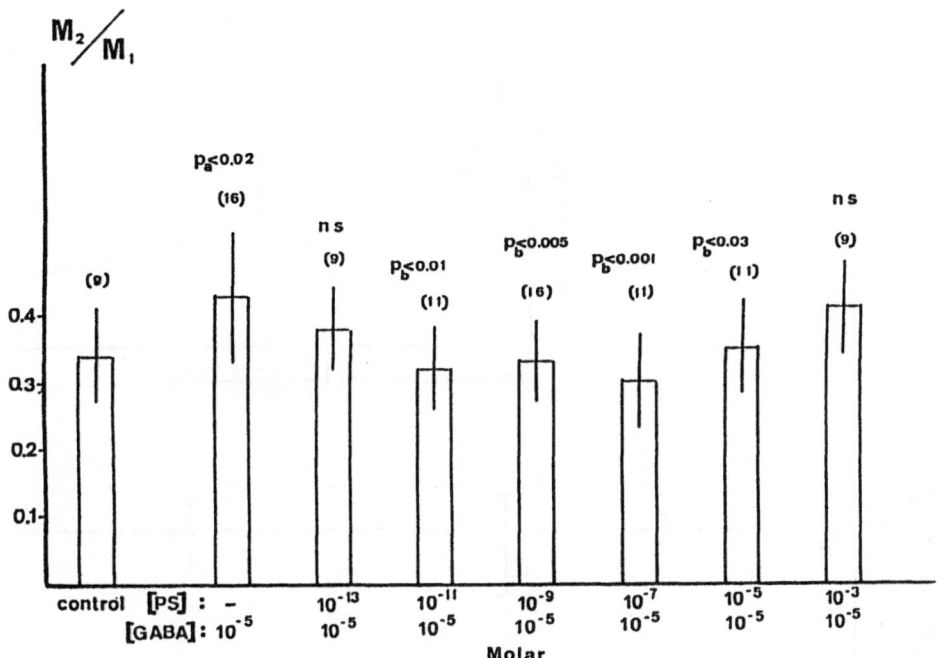

Fig. 2: Stimulation by GABA, $10^{-5}M$, of $^{36}Cl^-$ permeation across single Deiters' membranes and effect of various PS concentrations. The data are expressed as M_2/M_1 ratios. The absolute amounts of Cl^- ions passed across the membranes over 30 sec were $4.25 \pm 0.9 \times 10^{-7}$ mol (control) and $5.4 \pm 1.2 \times 10^{-7}$ mol ($10^{-5}M$ GABA). The bars and numbers within parentheses have the same meaning as in Figure 1. P_a refers to the comparison with the control, p_b to the comparison with the case in which $10^{-5}M$ GABA was present on the membrane extracellular side (second column from the left).

The effect of exogenous PS on the binding of labelled GABA to Triton treated rat cerebellum membranes is reported in Figure 3. The inset in the figure represents a typical Scatchard plot for the GABA binding which shows two components, a high affinity one ($K_D=8.9$ nM) and a low affinity one ($K_D=102$ nM). The binding data reported in Figure 3 involve the high affinity interaction, since GABA is 0.5 nM. The results show that PS in the 10^{-8} to 10^{-4}M range does not modify the binding of the neurotransmitter. Only the highest concentration used, 10^{-3}M, reduces the specific binding by 60%.

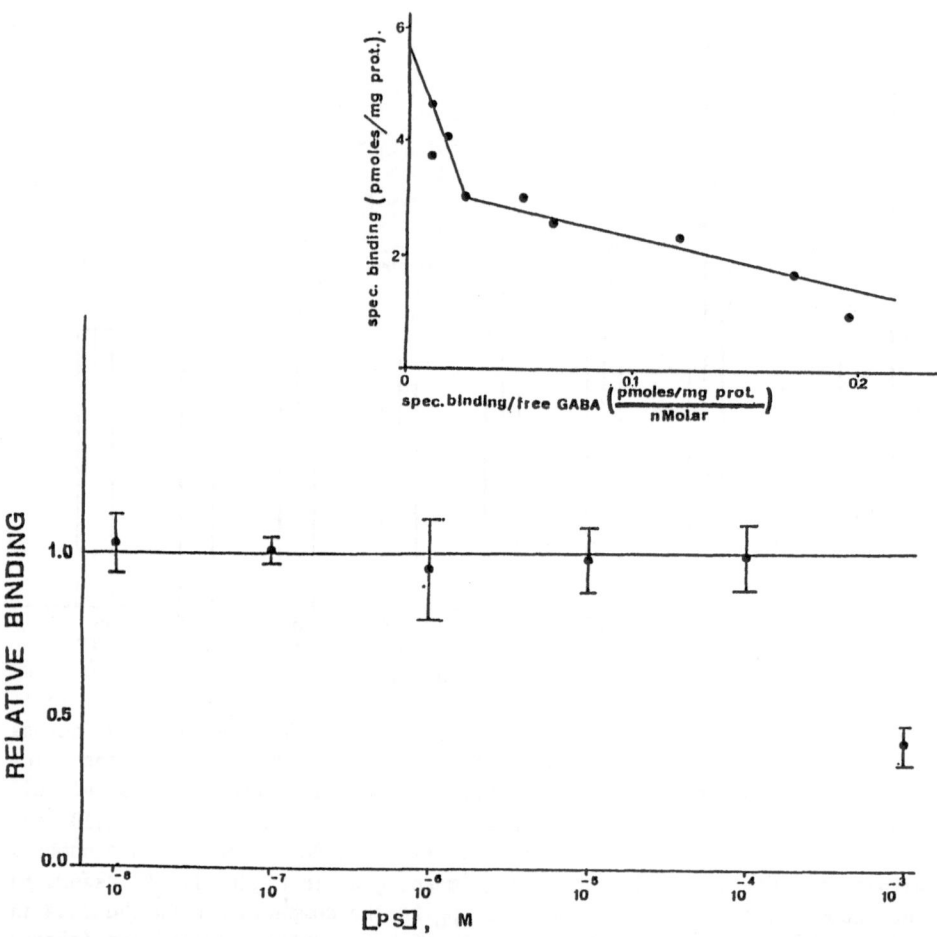

Fig. 3: Effect of various concentrations of phosphatidylserine on the specific binding of [^{3}H]GABA, 0.5 nM, to Triton X-100 treated membranes from rat cerebellum. The specific binding values are compared with the control value, where no PS was present, taken as 1.0. The curve is the average of 3 different experiments. In the inset is a typical Scatchard plot for GABA binding to such membrane preparations. The binding constants are: K_D = 8.9 nM and B_{Max} = 3.24 pmol/mg protein for the high affinity and K_D = 102 nM and B_{Max} = 2.44 pmol/mg protein for the low affinity component.

DISCUSSION

Phosphatidylserine, when is applied to GABA acceptive membranes at concentrations below 10^{-3}M, does not interfere with the interaction between the neurotransmitter and its recognition site (Figure 3). However, it interferes with both basal (at 10^{-5}M) and GABA-activated (at 10^{-11} to 10^{-5}M) Cl$^-$ "pores". This last action is very potent. It cannot be explained merely on the basis of an electrostatic interaction between the positive charges at the GABA activated Cl$^-$ channel mouth (Schofield et al., 1987) and the anionic polar head of PS. The very high affinity involved requires that, in addition to a purely electrostatic interaction, the hydrophobic tail of the lipid should find some sort of anchorage to the nerve membrane.

It is interesting to note that 10^{-3}M PS brings about effects abruptly different from those of the phospholipid concentrations immediately below, in all 3 cases studied here. In particular, at such a PS concentration the effects on basal and GABA-activated Cl$^-$ permeation disappear (Figures 1 and 2). In addition, 10^{-3}M in contrast with 10^{-4}M PS results in a remarkable inhibition of labelled GABA binding to rat cerebellum membranes (Figure 3). It is possible that at a phospholipid concentration as high as 10^{-3}M a large amount of aggregates are present, exerting detergent like effects on the membranes resulting in the abrupt changes observed.

Referring to the relevance of the effects we describe here to the pharmacological actions of phosphatidylserine, we found that after i.p. injection of PS to rats, at 740 mg/Kg, the phospholipid reaches an average brain concentration of 10^{-7}M (unpublished data). This PS level appears to be able to interfere with GABA post-synaptic action (Figure 2), however the effect found cannot explain the antiepileptic effect of the i.p. administration of PS in association with GABA (Loeb et al., 1982). The possible facilitation of PS on i.p. GABA passage to the blood stream and then from the blood to the brain (Loeb et al., 1988) appears to be a hypothesis to take into consideration.

REFERENCES

Fujimoto M, Okabayashi T (1983) Influence of phospholipase treatments on ligand binding to a benzodiazepine receptor GABA receptor-chloride ionophore complex. Life Sci 32:2393-2400

Giambalvo CT, Rosenberg T (1976) The effect of phospholipases and proteases on the binding of γ-amino butyric acid to junctional complexes of rat cerebellum. Biochim Biophys Acta 436:741-756

Hydén H, Lange PW, Larsson S (1980) S-100-glia regulation of GABA transport across the nerve cell membrane. J Neurol Sci 45:303-316

Hydén H, Cupello A, Palm A (1986) γ-Amino butyric acid stimulates chloride permeability across microdissected Deiters' neuronal membrane. Brain Res 379:167-170

Hydén H, Cupello A, Rapallino MV (1991) Regulation of inhibition by neuronal GABA$_A$ receptors interaction. Royal Society of Arts and Sciences, Goteborg, Biomedicine 1:1-78

Lloyd KG, Davidson L (1979) (^{3}H)GABA binding in brains from Huntington's chorea patients: altered regulation by phospholipid. Science 205:1147-1149

Loeb CW, Benassi E, Besio G, Maffini M, Tanganelli P (1982) Liposome entrapped GABA modifies behavioral and electrographic changes of penicillin-induced epileptic activity. Neurology 32:1234-1238

Loeb CW, Benassi E, Besio G, Bo GP, Mainardi P, Scotto PA, Faverio A (1984) Antiepileptic effect of γ-amino-butyric acid and phosphatidylserine in rats. IRCS Med Sci 12:465

Loeb CW, Besio G, Mainardi P, Scotto PA, Benassi E, Bo GP (1986) Liposome-entrapped γ-amino-butyric acid inhibits isoniazid-induced epileptogenic activity in rats. Epilepsia 27:98-102

Loeb CW, Marinari UM, Benassi E, Besio G, Cottalasso D, Cupello A, Maffini M, Mainardi P, Pronzato MA, Scotto PA (1988) Phosphatidylserine increases in vivo the synaptosomal uptake of exogenous GABA in rats. Exp Neurol 99:440-446

Rapallino MV, Cupello A, Hydén H (1988) Direct evidence for the presence of GABA$_A$ receptors on the Deiters'neurone plasma membrane cytoplasmic side. Brain Res 462:350-353

Rapallino MV, Cupello A, Mainardi P, Besio G, Loeb CW (1990) Effect of phosphatidylserine on the basal and GABA-activated Cl$^-$ permeation across single nerve membranes from rabbit Deiters' neurons. Neurochem Res 15:593-596

Schofield PR, Darlison MG, Fuijta N, Burt DR, Stephenson FA, Rodriguez H, Rhee LM, Ramachandran J, Reale V, Glencorse TA, Seeburg PH, Barnard EA (1987) Sequence and functional expression of the GABA$_A$ receptor shows a ligand-gated receptor super-family. Nature 328:221-227.

Toffano G, Aldinio C, Balzano M, Leon A, Savoini G (1981) Regulation of GABA receptor binding to synaptic plasma membrane of rat cerebral cortex: the role of endogenous phospholipids. Brain Res 222:95-102

Toffano G, Mazzari S, Zanotti A, Bruni A (1984) Synergistic effect of phosphatidylserine with γ-amino-butyric-acid in antagonizing the isoniazid-induced convulsions in mice. Neurochem Res 9:1065-1073

Ueno E, Kuriyama K (1981) Phospholipids and benzodiazepine recognition sites of brain synaptic membranes. Neuropharmacology 20:1169-1176

PHOSPHATIDYLSERINE REVERSES THE AGE-DEPENDENT DECREASE IN CORTICAL ACETYLCHOLINE RELEASE IN THE RAT.

G. Pepeu, F. Casamenti and C. Scali
Department of Preclinical and Clinical Pharmacology
University of Florence
Viale Morgagni 65
50134 Florence, Italy

In this paper the actions of the natural 1,2-diacyl-sn-glycerol-3-phosphoserine (PS) on brain cholinergic mechanisms in aging rats will be reviewed. Furthermore, preliminary results, obtained with a 1,3-diacyl-sn-glycerol-2-phosphoserine (PS-S1), will also be presented. PS-S1 represents an interesting tool for investigating the role and mechanism of action of serine phospholipids and may have therapeutic implications.

Age-associated central cholinergic hypofunction.

A hypofunction of brain cholinergic mechanisms is present in normal aging in man and animals, and is particularly pronounced in some of the age-related degenerative diseases such as senile dementia of Alzheimer and Alzheimer type (Pepeu, 1988). In normal aging, the neurons of the forebrain cholinergic nuclei show a decrease in size and loss of dendritic spines (Fisher et al., 1989) associated with a reduced density of the cortical cholinergic network (Geula and Mesulam, 1989). On the other hand, a large loss of forebrain cholinergic neurons is a feature of senile dementia of Alzheimer and Alzheimer's type (Price et al., 1982).

In the aging man the cholinergic hypofunction is indirectly demonstrated by an increased sensitivity to the amnesic effects of anticholinergic drugs (Sunderland et al., 1985). In old rodents, the cholinergic hypofunction is demonstrated by a decrease in acetylcholine (ACh) synthesis and release from potassium-depolarized brain slices (Gibson and Peterson, 1981; Gibson et al., 1981) and potassium-depolarized cortical, hippocampal and striatal synaptosomes (Meyer et al.,1984; Consolo et al., 1986). Similarly, endogenous ACh release from electrically stimulated cortical slices prepared from old rats is lower than that from slices prepared from young rats (Pedata et al., 1983). The decrease in evoked ACh release already begins in 14-month-old rats (Vannucchi and Pepeu, 1987) and by 18 months is 50% smaller than that found in 3-month-old rats. Extracellular ACh levels investigated by transverse microdialysis

NATO ASI Series, Vol. H 70
Phospholipids and Signal Transmission
Edited by R. Massarelli, L. A. Horrocks,
J. N. Kanfer, and K. Löffelholz
© Springer-Verlag Berlin Heidelberg 1993

in the cerebral cortex, hippocampus and striatum of 18-month-old rats is also markedly lower than in 2-month-old rats (Wu et al., 1988).

Evidence based on the findings that antimuscarinic agents disrupt cognitive functions (Rusted and Warburton, 1989), and on the neurochemical and behavioral effect of brain lesions affecting cholinergic pathways (Pepeu et al., 1990; Dunnett et al., 1887), indicate that cholinergic hypofunction may be responsible in part, at least, for the cognitive deficits associated with aging (Collerton, 1986), although in human pathology the specific functional consequences of the hypofunction remain unknown (Fibiger, 1991).

Assuming that cholinergic hypofunction plays an important pathogenetic role, in recent years much effort has been devoted to the search for drugs able to correct it, and through this action, improve cognitive deficits. The effort has been mostly aimed at developing less toxic anticholinesterase or directly acting cholinomimetic agents, with favorable pharmacokinetic properties, such as tacrine, metrifonate, heptylphysostigmine and AF102B (Giacobini and Becker, 1988). Unfortunately, the clinical results have been so far limited.

Following this research strategy, PS may also be considered a potentially useful drug, since it has been repeatedly shown that it acts, indirectly, on brain cholinergic mechanisms as will be described in the following paragraphs.

Effect of PS on age-associated cholinergic hypofunction.

Indirect evidence that PS may interact with central cholinergic mechanisms results from the finding that a single administration (75 mg/Kg i.p.) antagonizes the disrupting effect of scopolamine on spontaneous alternation (Pepeu et al., 1980), and on passive avoidance conditioned responses (Zanotti et al., 1986). Direct evidence was given by Casamenti et al., (1979) with the demonstration that PS (75-150 mg/Kg i.p.) stimulated ACh release from the cerebral cortex, investigated by the cortical cup technique. However these investigations were carried out on adult rats. The first demonstration that PS administration improved age-induced cholinergic hypofunction in the rat was given by Pedata et al (1985). They observed that a 30 day treatment of 24-month-old rats with PS 15 mg/kg i.p. restored ACh release from electrically stimulated cortical slices to adult values. Phosphatidylcholine at the same dose was inactive.

As shown by Vannucchi and Pepeu (1987) the improvement of ACh release from electrically stimulated cortical slices in aging rats can be seen after at least 7 days of daily i.p. injections of PS (15 mg/Kg) and lasts for 5

259

days after the interruption of the treatment. In young rats and in aging rats treated with PS there is no decrease in ACh content of the cortical slices at the end of the stimulation cycles. On the contrary, there is a 40% decrease, approximately, in the stimulated slices prepared from aging rats, since in aging rats "de novo" synthesis is unable to supply a sufficient amount of ACh for the evoked release, and ACh stores are progressively depleted. Confirmation that repeated PS administrations correct the age-associated impairment in ACh synthesis was obtained by Vannucchi et al., (1990) by incubating cortical slices taken from untreated ,and PS treated aging rats with tritiated choline. It was found that PS administration prevents the age-associated reduction in total evoked ACh release but does not correct the reduction in evoked tritiated ACh. This finding indicates that, in aging rats, PS is able to increase the availability of endogenous choline (Ch) for "de novo" ACh synthesis and release.

Taken together the "ex vivo" experiments demonstrate that a short term pretreatment of 16 to 24-month-old rats with PS (15 mg/Kg i.p.) markedly reduces the age-associated impairment of the cholinergic mechanisms. In PS-treated old rats, evoked ACh release from, and ACh synthesis and content in the stimulated slices are higher than in the untreated, and not significantly different from those found in young rats.

The question of whether the effect of PS on age-related cholinergic hypofunction can also be seen "in vivo", has been answered by the work of Casamenti et al., (1991) in which basal ACh and Ch output from the cerebral cortex of 3- and 19-month-old freely moving rats was measured by microdialysis. In the untreated aging rats, ACh and Ch output were 39 and 16% lower, respectively, than in young rats. PS administration (15 mg/Kg i.p. daily for 8 days) to the old rats markedly attenuated the decrease in ACh release, and slightly increased Ch output. No effect was detected in young rats treated with PS, or in old rats treated with phosphatidylcholine and phosphoserine.

Experiments with PS-S1

PS-S1 is 1,3-dipalmitoyl-sn-glycerol-2-phosphoserine), which differs from PS in the stereochemistry of the glycerol backbone and the symmetric fatty acids. In order to ascertain whether PS-S1 is also active on the age-related cholinergic hypofunction, ACh and Ch output from the cerebral cortex were investigated in 23-month-old male Wistar rats by transverse microdialysis according to the procedure described by Casamenti et al., (1991). Briefly, the microdialysis tubing (AN 69 membrane, Dasco, Italy) was covered with Super-Epoxy glue along the whole of its length except for 8 mm, which corresponds to the length of the parietal cortex and was stereotactically inserted transversely under ketamine anaesthesia.

Table 1.   EFFECT OF PS-S1 ADMINISTERED DAILY FOR 30 DAYS ON BASAL AND
K⁺-EVOKED ACh OUTPUT FROM THE CEREBRAL CORTEX OF 3 AND 23-MONTH-OLD RATS

AGE (mo)	DOSE mg/Kg i.p.	N° rats	ACh OUTPUT	
			Basal pmol/20min	Evoked % increase
			(mean + S.E.M.)	
3	-	5	5.1 ± 0.7 **	143.0 ± 36.3 **
23	-	8	2.8 ± 0.2 (-43% vs 3 mo) **	79.7 ± 19.9 (-56% vs 3 mo) #
23	15	12	3.4 ± 0.2 (-34% vs 3 mo) **	150.3 ± 24.5 (+88% vs 23 mo) #
23	30	10	3.0 ± 0.2 (-42% vs 3 mo)	155.5 ± 24.5 (+95% vs 23 mo)

Student's t-test, two-tailed; difference statistically significant: **
P<0.01 vs 3 mo; # P<0.05 vs 23 mo

On the following day the microdialysis tubing was perfused at a constant
flow rate (2.0 μl/min) with Ringer solution containing 7 μM physostigmine
sulfate.   When the concentration of KCl was raised to 100 mM the NaCl
content was correspondingly reduced. After a 1 h equilibration period the
perfusate was collected every 20 min and ACh and Ch content of the
ialysate assayed by HPLC with an electrochemical detector with Ag/AgCl
reference electrode and platinum working electrode at +250 mV (Damsma et
al.,   1987).   1,3-Dipalmitoyl-sn-glycerol-2-phosphoserine (PS-S1) was
kindly supplied by Fidia S.p.A. Abano Terme, Italy and was suspended in
Tris buffer by sonication.

Two experiments were performed.   In the first, shown in Table 1, two
groups of 23-month-old rats were injected daily for 30 days with either 15
or 30 mg/kg of PS-S1. Controls were injected with an equal volume of Tris
buffer.   Basal release was calculated as the mean of 4-5 samples. 100 mM
KCl was then added to the superfusing fluid for two collection periods.
The maximum increase in the evoked ACh release was found in the second
sample.   Then the release returned rapidly to the basal level.

In young rats the K⁺-evoked release was 143% larger than basal release.

In the untreated 23-month-old rats both basal and evoked ACh release were
significantly smaller than in the 3-month-old rats. In old rats receiving

Table 2.   EFFECT OF PS-S1 (15 mg/Kg i.p. daily for 8 days) ON BASAL AND K+-EVOKED ACh OUTPUT FROM THE CEREBRAL CORTEX OF 3 AND 23- MONTH-OLD RATS

AGE mo	N° rats	ACh OUTPUT (pmol/20 min + S.E.M.)		
		Basal	Evoked	% Increase
3	5	5.2 + 0.7	12.3 + 1.8	136
			**	**
23	8	2.9 + 0.2	5.3 + 0.8	82
		(-43% vs 3 mo)	(-54% vs 3 mo)	
			*	
23	4	2.7 + 0.1	9.2 + 2.8	140
		(-48% vs 3 mo)	(-25% vs 3 mo)	

Student's t-test, two-tailed; difference statistically significant: ** P<0.01, * P<0.05 vs 3 mo

PS-S1 (15 and 30 mg/Kg i.p.) for 30 days, the basal release was as low as in untreated 23-month-old rats. On the contrary the percent increase in evoked release was as large as in the 3-month-old rats. However the absolute values of the evoked release in old rats receiving 15 mg/Kg PS-S1 were still 37% lower, and in those receiving 30 mg/Kg were 41% lower than in young rats.

Ch levels were 41.5 ± 4.8 pmol/20 min in 3 month-old rats and 28.8 ± 3.2 in untreated old rats with a statistically significant 31% decrease. A 22% and 7% decrease was found in the old rats receiving 15 and 30 mg/Kg of PS-S1 respectively.   In all groups of rats K^+ depolarization brought about an increase in Ch output which was never statistically significant.

In the second experiment, PS-S1 was given to 23-month-old rats at the dose of 15 mg/Kg i.p. for 8 days. The results are shown in Table 2. The treatment did not increase the basal ACh release in aging rats but enhanced the K^+-evoked release.   The percent increase of the evoked ACh release over the basal release was as large in treated aging as in young rats.   The Ch efflux was 30% lower in the untreated aging than in the young rats.   In the treated old rats it was only 14% lower than in the young rats.   No statistically significant increase in Ch efflux occurred after K^+ depolarization.

These preliminary experiments show that PS-S1 has no effect on basal ACh release in aging rats, while PS increases basal ACh release in aging rats, as demonstrated by Casamenti et al., (1991). On the other hand, PS-S1 treatment, similarly to PS treatment, slightly increases basal Ch efflux.

Furthermore, in both experiments, the percent increase of the evoked release in treated aging rats was as large as in young rats. It appears that this effect is due more to a restoration by PS-S1 of the ability of cholinergic neurons to respond to K+ depolarization than to an increase in ACh stores. A comparison between PS-S1 and PS with regard to this action is impossible, since PS influence on K^+ depolarization has not yet been investigated.

The mechanism by which PS stimulates ACh synthesis and release in aging brain has not yet been clarified. Both 1,2 and 1,3 phosphatidylserines are able to activate protein kinase C (Lee and Bell, 1989). Comparing the present results with those obtained, under the same conditions, by Casamenti et al., (1991), the only effect that PS and PS-S1 have in common seems to be the ability to partly restore Ch efflux in aging rats. It may be pertinent to mention that protein kinase C activation accelerates choline phospholipid hydrolysis, possibly by activating phospholipase D or phospholipase C (Löffelholz, 1989). On the other hand, the changes in protein kinase C activity occurring during aging are restored by repeated PS administration (Calderini et al., 1986). However, the increase in Ch efflux in aging rats is associated with an increase in basal ACh release only after PS treatment, making it difficult to propose a relationship between the two events. Finally, it is still a matter of debate as to whether the pharmacological effects exerted by PS, and presumably also by PS-S1, result from a direct action on brain cells or, indirectly, from an interaction with unidentified peripheral factors (see Bruni et al., this volume).

Conclusions

The finding that the treatment with PS restores ACh release from the cerebral cortex in aging rats, as demonstrated by "ex vivo" and "in vivo" experiments, suggests that the age- associated cholinergic hypofunction may be corrected. An improvement in cholinergic function in aging rats has been obtained by Fisher et al., (1987) following intracerebro-ventricular infusion of nerve growth factor (NGF). They found that the increase in choline acetyltransferase and ACh release was associated with an improvement in the age-associated impairment of the behavioral performance. Similarly PS treatment improves the acquisition of a passive avoidance conditioned response (Corwin et al., 1985; Vannucchi et al., 1990) and the performance in a water maze (Zanotti et al., 1989) in aging rats.

To which extent the findings obtained in aging rats can be extrapolated to aging man needs to be fully investigated. However, the administration of PS orally for 3 months to Alzheimer's patients (Amaducci, 1988), and to elderly subjects affected by age-associated memory impairment (Crook et al., 1991) has been shown to improve some of the cognitive deficits. Investigations on PS-S1 are at a very preliminary stage and more experiments are needed in order to define its pharmacological and biochemical properties and potential therapeutic applications.

Acknowledgment

The investigations reported in this work were supported by CNR grants 88.01827.04, 89.02737.04 and 91.00416. PF40

References

Amaducci L (1988) Phosphatidylserine in the treatment of Alzheimer's disease: results of a multicenter study. Psychopharmacol Bull 24:130-134

Calderini G, Bellini F, Bonetti A, Galbiati E, Rubini R, Zanotti A, Toffano G (1986) Pharmacological properties of phosphatidylserine in the aging brain: biochemical aspects and therapeutic potential. In: Horrocks LA, Freysz L, Toffano G (eds) Phospholipid research in the nervous system. Biochemical and moleceular pharmacology. Liviana Press, Padova, p 233

Corwin J, Dean RL, Bartus RT, Rotrosen J, Watkins DL (1985) Behavioral effects of phosphatidylserine in the aged Fisher 344 rats: amelioration of passive avoidance deficits without changes in psychomotor task performance. Neurobiol Aging 8:11-15

Crook TH, Tinkleberg J, Yesavage J, Petrie W, Nunzi MG, Massari DC (1991) Effects of phosphatidylserine in age-associated memory impairment. Neurology 41:644-649

Casamenti F, Mantovani P, Amaducci L, Pepeu G (1979) Effect of phosphatidylserine on acetylcholine output from the cerebral cortex of the rat. J Neurochem 32:529-533

Casamenti F, Scali C, Pepeu G (1991) Phosphatidylserine reverses the age-dependent decrease in cortical acetylcholine release: a microdialysis study. Eur J Pharmacol 194:11-16

Collerton D (1986) Cholinergic function and intellectual decline in Alzheimer's disease. Neuroscience 19:1-28

Consolo S, Wang JX, Fiorentini F, Vezzani A, Ladinsky H (1986) In vivo and in vitro studies on the regulation of cholinergic neurotransmission in striatum, hippocampus and cortex of aged rats. Brain Res 374:212-218

Damsma G, Lammerts van Bueren D, Westerink BHC, Horn AS (1987) Determination of acetylcholine in the femtomole range by means of HPLC, a post-column enzyme reactor, and electrochemical detection. Chromatographia 24:827-831

Dunnett SB, Whishaw IQ, Jones GH, Bunch ST (1987) Behavioral, biochemical and ihistochemical effects of different neurotoxic amino acids injected into nucleus basalis magnocellularis of rats. Neuroscience 20:653-669

Fibiger HC (1991) Cholinergic mechanisms in learning, memory and dementia: a review of recent evidence. Trends Neurol Sci 14:220-223

Fisher W, Victorin K, Bjorklund A, Williams LR, Varon S and Gage FH (1987) Amelioration of cholinergic neuron atrophy and spatial memory impairment in aged rats by nerve growth factor. Nature 329:65-68

Fisher W, Gage FH, Bjorklund (1989) Degenerative changes in forebrain cholinergic nuclei correlate with cognitive impairments in aged rats. European J Neurosci 1:34-40

Geula C, Mesulam MM (1989) Cortical cholinergic fibers in aging and Alzheimer's disease: a morphometric study. Neuroscience 33:469-476

Giacobini E, Becker R (1988) Current research in Alzheimer therapy. Taylor and Francis, New York, Philadelphia, Washington and London

Gibson GE, Peterson C (1981) Aging decrease oxidative metabolism and the release and synthesis of acetylcholine. J Neurochem 37:78-984

Gibson GE, Peterson C, Jenden DJ (1981) Brain acetylcholine synthesis declines with senescence. Science 213:674-676

Lee MH, Bell RM (1989) Phospholipid functional groups involved in protein kinase C activation, phorbol ester binding, and binding to mixed micelles. J Biol Chem 264:14797-14805

Löffelholz K (1989) Receptor regulation of choline phospholipid hydrolysis. Biochem Pharmacol 38:1543-1549

Meyer EM, St Onge E, Crew FT (1984) Effect of aging on rat cortical presynaptic cholinergic processes. Neurobiol Aging 5:315-317

Pedata F, Slavikova Y, Kotas A and Pepeu G (1983) Acetylcholine release from rat cortical slices during postnatal development and aging. Neurobiol Aging 6:337-339

Pedata F, Giovannelli G, Spignoli G and Pepeu G. (1985) Phosphatidylserine increases acetylcholine release from cortical slices in aged rats. Neurobiol Aging 6:337-339

Pepeu G (1988) Acetylcholine and brain aging. Pharmacol Res Commum 20:91-97

Pepeu G, Gori G, Bartolini L (1980) Pharmacologic and therapeutic perspectives on dementia: an experimental approach. In: Amaducci L, Davison AN, Antuono P (eds) Aging of the brain and dementia. Raven, New York, p 271

Pepeu G, Di Patre PL, Casamenti F (1990) Spontaneous and drug stimulated recovery of cortical cholinergic function after lesion of the nucleus basalis. In: Steriade M, Biesold D (eds) Brain cholinergic systems. Oxford University Press, Oxford New York Tokyo Toronto, p 357

Price DL, Whitehouse PJ, Struble RG, Coyle JT, Clark AW, Delong MR (1982) Alzheimer's disease and Down's syndrome. In: Synex FM, Merril CR (eds) Alzheimer's disease, Down's syndrome, and aging. Ann NY Acad Sci 396:145-164

Rusted JM, Warburton DM (1989) Cognitive models and cholinergic drugs. Neuropsychobiol 21:31-36

Sunderland T, Tariot P, Murphy DL, Weingartner H, Mueller EA, Cohen RM (1985) Scopolamine challenges in Alzheimer's disease. Psychopharmacology 87:247-249

Vannucchi MG, Pepeu G (1987) Effect of phosphatidylserine on acetylcholine release and content in cortical slices from aging rats. Neurobiol Aging 8:403-407

Vannucchi MG, Casamenti F, Pepeu G (1990) Decrease in acetylcholine release from cortical slices in aged rats: investigations into it reversal by phosphatidylserine. J Neurochem 55:819-825

Wu CF, Bertorelli R, Sacconi M, Pepeu G, Consolo S (1988) Decrease of brain acetylcholine release in aging freely-moving rats detected by microdialysis. Neurobiol Aging 9:357-361

Zanotti A, Valzelli L, Toffano G (1986) Reversal of scopolamine amnesia by phosphatidylserine in rats. Psychopharmacology 90:274-275

Zanotti A, Valzelli L, Toffano G (1989) Chronic phosphatidylserine treatment improves spatial memory and passive avoidance in aged rats. Psychopharmacology 99:316-321

CONTROL OF PHOSPHATIDYLCHOLINE METABOLISM

Dennis E. Vance, Grant M. Hatch, Haris Jamil, Tomoko Nishimaki-Mogami, Rockford W. Samborski, Lilian Tijburg and Amandip K. Utal
Lipid and Lipoprotein Research Group and Department of Biochemistry
University of Alberta
Edmonton, Alberta T6G 2S2 Canada

The biosynthesis and catabolism of phosphatidylcholine (PC) are tightly coupled processes. In the past decade significant progress has been made in unraveling control mechanisms for regulation of PC biosynthesis (Vance, 1990; Tijburg et al., 1989). In contrast the mechanisms that regulate PC catabolism are poorly understood. Renewed interest in regulation of PC catabolism has been sparked by the discovery that PC is a source of diacylglycerol, a second messenger that activates protein kinase C (Löffelholz, 1989; Exton, 1990). The purpose of the present chapter is to review the recent progress in these two areas of research with emphasis on the more recent contributions from the authors' laboratory.

Regulation of Phosphatidylcholine Biosynthesis

Over the past decade four major mechanisms have been identified for the regulation of PC biosynthesis. A summary of the current status of these mechanisms follows.

Regulation by Supply of Fatty Acids

The most striking results in various cells lines have been obtained by treatment of cells in culture with exogenous fatty acids (Vance, 1990).

NATO ASI Series, Vol. H 70
Phospholipids and Signal Transmission
Edited by R. Massarelli, L. A. Horrocks,
J. N. Kanfer, and K. Löffelholz
© Springer-Verlag Berlin Heidelberg 1993

For example, treatment of HeLa cells with 0.35 mM sodium oleate results in a 5- to 10-fold stimulation of PC biosynthesis. This is apparently due to translocation of CTP:phosphocholine cytidylyltransferase (CT) from cytosol where it is inactive to membranes where it is active. In this instance the supply of CDP-choline made via the CT reaction apparently limits the rate of PC biosynthesis in the HeLa cells. The membrane to which CT translocates is probably the endoplasmic reticulum (ER) (Vance and Vance, 1988; Terce et al., 1988). If the oleate is removed by addition of albumin to the cell medium, the rate of PC biosynthesis returns to normal and CT translocates from the membranes to the cytosol. The specificity for the translocation of CT has been studied (Cornell and Vance, 1987). A negative charge is not required since oleoyl alcohol is also a potent mediator of CT binding to membranes. Diacylglycerol will also promote CT binding. The mechanism for the fatty acid effect is not known. One possibility is simply a dilution of the concentration of PC in the membranes somehow attracts the enzyme to the membrane where CT is activated. One major problem is the lack of evidence that free fatty acids have a physiologically relevant role in modulation of PC biosynthesis (Vance, 1990).

Cyclic AMP and Okadaic Acid Inhibition of Phosphatidylcholine Biosynthesis

The second mechanism involves the inhibition of PC biosynthesis in hepatocytes by cAMP-analogues, glucagon and okadaic acid (Vance, 1990). There is good agreement in a variety of different cells that short term incubations (0.5-3h) with cAMP analogues cause an inhibition of PC biosynthesis. We showed a loss of CT activity from microsomes when cells were incubated with chlorophenylthio-cyclic AMP (CPT-cAMP) (Pelech et al., 1981), but failed to see an effect on CT translocation into cytosol of hepatocytes treated with glucagon (Pelech et al., 1984). After the successful purification of CT (Feldman and Weinhold, 1987), it became possible to determine if CT were an in vitro substrate for cAMP-dependent protein kinase (Sanghera and Vance,

1989). We were able to detect up to 0.2 moles of ^{32}Pi incorporated per mole of CT on serine residues and this coincided with a decreased binding of CT to membranes. Whether this occurred in intact hepatocytes remained an open question which was answered after an antibody was obtained to a synthetic peptide derived from the cDNA sequence for CT (Kalmar et al., 1990). This peptide when linked to KLH (keyhole limpet hemocyanin) generated an antibody in rabbits that quantitatively immunoprecipitated CT from hepatocyte cytosol (Jamil et al., 1991). Subsequent studies showed that CT was a protein phosphorylated on approximately 7 peptides and the state of phosphorylation was unchanged when hepatocytes were incubated with CPT-cAMP. This result strongly suggested that the inhibition of PC biosynthesis in hepatocytes incubated with cAMP analogues or glucagon was not due to a direct effect of the cAMP-dependent kinase on CT.

The emphasis of the research then shifted to other potential mechanisms that might alter PC biosynthesis. It is well known that cAMP causes an inhibition of fatty acid biosynthesis and it was conceivable that this would result in a decreased supply of diacylglycerol which could limit PC biosynthesis via CDP-choline:1,2-diacylglycerol cholinephosphotransferase. This hypothesis was tested and the results showed a nearly 50% reduction in diacylglycerol in cAMP-treated hepatocytes compared to control incubations. Correlation studies showed that the cAMP effect could be reversed by the addition of oleic acid which caused an increase in the concentration of diacylglycerol and the rate of PC biosynthesis without an effect on the translocation of CT. In these experiments, the correlation between the rate of PC biosynthesis and diacylglycerol levels in hepatocytes was r^2 = 0.93 (Jamil et al., 1991). The data strongly suggest that the rate of PC biosynthesis is inhibited in hepatocytes incubated with cAMP analogues or glucagon because of an inhibition of fatty acid biosynthesis and consequent reduction of diacylglycerol levels.

In an alternative approach to understand the role of CT phosphorylation in regulation of PC biosynthesis, we have utilized okadaic acid, a potent inhibitor of protein phosphatases 1 and 2A (Cohen and Cohen, 1989; Cohen et al., 1990). In an initial study we showed that addition of okadaic acid to liver postmitochondrial supernatants inhibited the translocation of CT from cytosol to microsomes (Hatch et al., 1990). Subsequent studies showed an inhibition of PC biosynthesis in hepatocytes treated with okadaic acid and there was less CT activity associated with the membrane fraction (Hatch et al., 1991a). The mechanism for the okadaic acid effect was explored further. Significant differences in the level of phosphorylation of CT were observed in hepatocytes treated with okadaic acid (Hatch et al., 1991b). However, knowing the linkage between diacylglycerol and the cAMP inhibition of PC biosynthesis, we also investigated the levels of diacylglycerol in okadaic acid treated hepatocytes. Various time course studies suggested that okadaic acid, which causes an inhibition of acetyl-CoA carboxylase and fatty acid synthesis (Cohen et al., 1990), inhibits PC biosynthesis by causing a reduction in the supply of diacylglycerol rather than an effect on CT phosphorylation (Hatch et al., 1991b). The function of phosphorylation of CT is unknown and requires further investigation.

Thus, the present evidence supports the hypothesis that cAMP and okadaic acid inhibit fatty acid biosynthesis which lowers the level of diacylglycerol. The supply of this substrate for the cholinephosphotransferase reaction thus limits the rate of PC biosynthesis.

Phosphatidylcholine Feedback Inhibits Its Own Biosynthesis

Feedback inhibition of a pathway by the final biosynthetic product is well known in many metabolic pathways but not often observed in phospholipid biosynthetic pathways (Vance, 1990). Recent studies from our laboratory have provided convincing evidence for a potent feedback control mechanism in PC biosynthesis (Yao et al., 1990; Jamil et al., 1990). We had been investigating the role of PC biosynthesis in the secretion of lipoproteins in hepatocytes and found that cells depleted of choline were defective in the secretion of very low density lipoproteins but all other hepatic proteins were secreted normally (Vance, 1990). In the course of these studies we were curious what effect choline deficiency had on CT since we knew that the rate of PC biosynthesis was decreased by 70 to 80 %. We found a 2-fold increase in CT on the membranes of the choline deficient cells (Yao et al. 1990). We explored several ideas to explain the increased CT binding to cellular membranes and eliminated concentration of fatty acids or a change in the state of phosphorylation of CT. However, we noted a very strong correlation between the distribution of CT and the concentration of PC in the cell membranes. When choline was added to the deficient cells, the levels of PC rose over a 2 hour period by 15% and this coincided with a release of CT from membranes to the cytosol (r^2 = 0.98) (Jamil et al., 1990). The translocation could have also been due to a rise in the aqueous precursors of PC (e.g., choline, phosphocholine, CDP-choline). However, when PC levels were increased by the addition of methionine which promoted the methylation of phosphatidylethanolamine, the release of CT into the cytosol was also observed. Similarly, addition of low levels of lyso-PC to the cells resulted in uptake and acylation to form PC and this correlated with release of CT into the cytosol (Jamil et al., 1990). Taken together, these data show for the first time that PC is a potent feedback inhibitor of its own biosynthesis.

Treatment of various types of cells with phospholipase C results in enhanced translocation of CT from cytosol to membranes (Sleight and Kent, 1983) and specifically the ER in Krebs II cells (Tercé et al., 1988). It has been speculated that the translocation of CT is due to a deficiency of PC in the cellular membranes but direct evidence for this hypothesis has not yet been presented. In light of the results from the choline deficiency experiments, a decrease in the levels of PC after phospholipase C digestion would be an attractive hypothesis.

Diacylglycerol as a Modulator of CT Translocation and PC Biosynthesis

As described above, the supply of diacylglycerol as a substrate for the cholinephosphotransferase reaction can directly control the rate of PC biosynthesis. On the other hand, in vitro studies have shown that an increase in the concentration of diacylglycerol can enhance the binding of CT to membranes (Cornell and Vance, 1987). Recent evidence now suggests that at least in HeLa cells, the concentration of diacylglycerol can regulate the binding of CT to membranes where it is activated and the activity of CT appears to limit the rate of PC biosynthesis in these cells (Utal et al., 1991).

Diacylglycerol has been implicated as a mediator of CT translocation as a result of studies on the mechanism by which PC biosynthesis is stimulated by the tumor promoter, tetradecanoyl phorbol acetate (TPA). Incubation of HeLa cells with this compound causes a 3- to 4-fold increase in the rate of PC biosynthesis and a translocation of CT from cytosol to membranes (Vance, 1990; Utal et al., 1991). Over the years several studies investigated the mechanism by which TPA causes CT translocation and eliminated changes in the levels of fatty acids and changes in the state of phosphorylation of CT (Vance, 1990; Watkins and Kent, 1990; Utal et al., 1991). We considered the idea that a change in the level of diacylglycerol might be responsible for the translocation of CT and the increased rate of PC biosynthesis. Various different approaches suggested a strong correlation between the concentration of diacylglycerol, CT translocation and PC biosynthesis (Utal et al., 1991). The working model is that TPA activates protein kinase C which in turn activates within minutes a phospholipase C that degrades PC on the plasma membrane. The diacylglycerol generated migrates to the ER where it enhances the binding and activation of CT. Thus, the rate of PC biosynthesis is accelerated. A direct phosphorylation of CT by protein kinase C does not appear to be involved since pure CT is not a substrate for protein kinase C in vitro (Jamil and Vance, unpublished results). Moreover, the state of phosphorylation of CT in HeLa cells is unchanged after treatment with TPA (Watkins and Kent, 1990; Utal et al., 1991). In addition, HeLa cells, downregulated for protein kinase C by incubation with TPA for 24 h, showed a stimulation of PC biosynthesis and CT translocation when incubated with the soluble diacylglycerol, dioctanoylglycerol.

Thus, it can be concluded that diacylglycerol can modulate PC biosynthesis by two different mechanisms: 1) as a substrate for the cholinephosphotransferase reaction as demonstrated in the studies with cAMP and okadaic acid; 2) as a mediator of CT translocation and activation in HeLa cells treated with TPA.

Regulation of PC catabolism

There is a paucity of information about the mechanisms that regulate the rate of PC catabolism. The problem is considerably more complex than understanding the regulation of PC biosynthesis. One common difficulty between biosynthetic and catabolic studies is that many of the substrates and enzymes are membrane components. However, there the comparison ends. There are four different types of phospholipases (A_1, A_2, C and D) and several lysophospholipases. One or more of these enzymes is found in virtually all cell types and in most cellular compartments (cytosol, ER, mitochondria, lysosomes, etc.). In addition, phospholipid catabolism is complicated by the various types of molecular species of the phospholipids and that the fatty acyl species of PC can be remodeled by deacylation-reacylation reactions. On top of this complexity, it is now almost certain that PC serves as a precursor of diacylglycerol for activation of protein kinase C and the phospholipases C and/or D that degrade PC are regulated by GTP-binding proteins (Löffelholz, 1989; Exton, 1990). The present discussion will be limited to recent experiments that show the choline-deficient model to be a powerful system in which to investigate PC catabolism. Secondly, we discuss evidence that PC catabolism is indeed regulated and not merely a constitutive process with regulation only on the biosynthetic side.

We had considered for a number of years how to study the control of PC catabolism in cultured hepatocytes. We made very little progress in pulse-chase studies with choline because the radioactivity slowly accumulated in PC owing to the large pool of phosphocholine (Vance, 1990). When catabolism was studied after a 24 h chase period, we experienced two additional problems. First, the catabolic products (glycerophosphocholine, phosphocholine and choline) had a high radioactive background making it difficult to detect the possible increase in labeled catabolites. Secondly, the viability of the primary hepatocytes is limited and cell death begins to complicate any experiments. These problems dissapated when choline-deficient hepatocytes were utilized for pulse-chase studies (Tijburg et al., 1991a). When these cells were pulsed with labeled choline, there was a rapid incorporation of label into PC because of the depleted level of phosphocholine in the cells. In addition, there was minimal radioactivity in the catabolites and the hepatocytes were viable.

We therefore undertook a series of experiments with choline-deficient hepatocytes prelabeled with [^{3}H]choline and chased $\pm$ unlabeled choline. Addition of the unlabeled choline enhanced the degradation of PC by approximately 2-fold (Tijburg et al., 1991a). The major product of catabolism was glycerophosphocholine for both choline-deficient and -supplemented cells and glycerophosphocholine formation was enhanced 2-fold in the choline-supplemented cells. Comparable results on the formation of glycerophosphocholine were obtained when [^{3}H]methionine or [^{3}H]glycerol were used as labeled precursors (Tijburg et al., 1991a). Assays of microsomal, mitochondrial and cytosolic fractions from choline-deficient and -supplemented rat livers showed a 1.4-fold increase in the formation of glycerophosphocholine in the microsomes from choline-supplemented hepatocytes. These results demonstrated that the rate of PC catabolism could be regulated in rat hepatocytes.

In a subsequent study the specificity of choline for stimulation of the formation of glycerophosphocholine was investigated (Tijburg and Vance, 1991). Supplementation of the cells with monomethylethanolamine had a minimal effect on the rate of PC catabolism and the formation of glycerophosphocholine. In contrast, supplementation of the choline-deficient hepatocytes with dimethylethanolamine increased the catabolism of PC by 1.6-fold compared to unsupplemented hepatocytes. This effect was accompanied by a 2.5-fold increase in the production of labeled glycerophosphocholine. Thus, it appears that formation of phosphatidyldimethylethanolamine from its corresponding base mimics the effect of the synthesis of PC from choline in increasing PC catabolism, whereas the effect of monomethylethanolamine is much less pronounced.

As mentioned previously another complication for studies on the regulation of PC catabolism is the remodelling of the molecular species. Prelabeling of choline-deficient hepatocytes with [^{3}H]choline facilitated studies on the transformation of molecular species of PC that occurs in hepatocytes. Analysis of the molecular species of PC by HPLC showed that, at the end of the radioactive pulse, approximately 75% of the label was incorporated into palmitate-containing species and only 16% of the species contained stearate (Tijburg et al., 1991b). During the chase period ± choline there was a redistribution of label and after 12 h approximately 56% of the total radioactivity was in palmitate-containing species and 37% was recovered in stearate-containing species. There was a precursor-product relationship between palmitate-containing species and stearate-containing species with arachidonate or linoleate on the *sn*-2 position. The relative rates of the deacylation reactions in choline-supplemented cells were increased with respect to the reacylation reactions compared to choline-deficient cells. This agrees with the previous findings that the formation of glycerophosphocholine was enhanced in choline-supplemented cells (Tijburg et al., 1991a). It seems likely that the regulation of deacylation and reacylation of lysoPC plays an important role in the regulation of PC concentrations in the cell as well as the composition of the molecular species.

The availability of the choline-deficient hepatocytes for rapid labeling of PC has also allowed us to study for the first time the effect of phosphorylation reactions on the rate of PC catabolism (Hatch and Vance, 1991). In pulse-chase experiments with [^{3}H]choline there was a 60 to 75% enhanced formation of glycerophosphocholine when hepatocytes were incubated with okadaic acid or CPT-cAMP. When both compounds were incubated with the cells, there was an additive effect on glycerophosphocholine formation. When PC in the hepatocytes was labeled with [^{3}H]methionine, an enhanced degradation of PC was observed in the presence of okadaic acid or CPT-cAMP that was similar to that observed when cells were labeled with [^{3}H]choline (Hatch and Vance, 1991). In vitro assays of phospholipase-lysophospholipase activity in crude mitochondrial fractions showed a 28 to 47% increase in catabolism of PC when the hepatocytes were preincubated with okadaic acid or CPT-cAMP compared to control cells. Thus, it appears that the rate of PC catabolism can be regulated by phosphorylation/dephosphorylation mechanisms that involve cAMP-dependent protein kinase and protein phosphate phosphatases. The enzymes and pathways involved remain to be elucidated.

Why should CPT-cAMP and okadaic acid stimulate PC catabolism and at the same time inhibit formation of PC as discussed in an earlier section of this chapter? We do not know but have speculated that it may involve another mechanism to generate fatty acids for supply of energy under short term conditions (Hatch and Vance, 1991). In the longer term decreased formation of PC could not be tolerated since PC is essential for cellular life. Incubations of hepatocytes for 6 or more hours with CPT-cAMP did reverse the inhibition of PC biosynthesis and stimulated PC formation (Pelech et al., 1982).

Future Directions

In this short review we have presented recent results primarily from our own work which shed light on the mechanisms that cells use to maintain levels of PC. During the 1980s we have seen very significant developments in the enzymology and regulation of the biosynthesis of PC. We can anticipate that our understanding of the regulatory features of PC biosynthesis should become much more sophisticated during the 1990s. The regulation of PC catabolism is almost a new subject for this decade. Our level of understanding of catabolic pathways and the nature of the enzymes involved is almost primitive. We have only just begun to probe regulatory mechanisms. The challenges are large but should yield new and hopefully unexpected insights in the future.

Acknowledgements - We are grateful to Sandi Ungarian for expert technical assistance. Our research was supported by a grant from the Medical Research Council of Canada. D.E.V. is a Medical Scientist of the Alberta Heritage Foundation for Medical Research. G.M.H. was supported by Fellowships from the Heart and Stroke Foundation of Canada and the Alberta Heritage Foundation for Medical Research.

References

Cohen, P. and Cohen, P.T.W. (1989) Protein phosphatases come of age. J. Biol. Chem. 265:21435-21438.

Cohen, P., Holmes, C.F.B. and Tsukitani, Y. (1990) Okadaic acid: a new probe for the study of cellular regulation. Trends Biochem. Sci. 15:98-102.

Cornell, R. and Vance, D.E. (1987) Translocation of CTP:phosphocholine cytidylyltransferase from cytosol to membranes in HeLa cells: stimulation by fatty acid, fatty alcohol, mono- and diacylglycerol. Biochim. Biophys. Acta. 919:26-36.

Exton, J.H. (1990) Signaling through phosphatidylcholine breakdown. J. Biol. Chem. 265:1-4.

Feldman, D.A. and Weinhold, P.A. (1987) CTP:phosphorylcholine cytidylyltransferase from rat liver: isolation and characterization of the catalytic subunit. J. Biol. Chem. 262:9075-9081.

Hatch, G.M., Lam, T.-S., Tsukitani, Y. and Vance, D.E. (1990) Effect of NaF and okadaic acid on the subcellular distribution of CTP:phosphocholine cytidylyltransferase activity in rat liver. Biochim. Biophys. Acta 1042: 374-379.

Hatch, G.M., Jamil, H. Utal, A.K. and Vance, D.E. (1991b) Feedforward regulation of phosphatidylcholine biosynthesis: Okadaic acid inhibits phosphatidylcholine biosynthesis by reducing 1,2-diacylglycerol content in isolated rat hepatocytes. submitted.

Hatch, G.M., Tsukitani, Y. and Vance, D.E. (1991a) The protein phosphatase inhibitor, okadaic acid, inhibits phosphatidylcholine biosynthesis in isolated rat hepatocytes. Biochim. Biophys. Acta 1081:25-32.

Hatch, G.M. and Vance, D.E. (1991) CPT-cAMP and okadaic acid enhance phosphatidylcholine catabolism in choline-deficient rat hepatocytes. Biochem. Cell Biol., In Press.

Jamil, H., Utal, A.K. and Vance, D.E. (1991) Cyclic AMP-induced inhibition of phosphatidylcholine biosynthesis is caused by a decrease in cellular diacylglycerol levels in cultured rat hepatocytes. Submitted.

Jamil, H., Yao, Z. and Vance, D.E. (1990) Feedback regulation of CTP:phosphocholine cytidylyltransferase translocation between cytosol and endoplasmic reticulum by phosphatidylcholine. J. Biol. Chem. 265:4332-4339.

Kalmar, G.B., Kay, R.J., Lachance, A., Aebersold, R. and Cornell, R.B. (1990) Cloning and expression of rat liver CTP:phosphocholine cytidylyltransferase: An amphipathic protein that controls phosphatidylcholine synthesis. Proc. Natl. Acad. Sci. USA 87:6029-6033.

Löffelholz, K. (1989) Receptor regulation of choline phospholipid hydrolysis: A novel source of diacylglycerol and phosphatidic acid. Biochem. Pharmacol. 38:1543-1549.

Pelech, S.L., Pritchard, P.H., Sommerman, E.F., Percival-Smith, A. and Vance, D.E. (1984) Glucagon inhibits phosphatidylcholine biosynthesis via the CDP-choline and transmethylation pathways in cultured rat hepatocytes. Can. J. Biochem. Cell Biol. 62:196-202.

Pelech, S.L., Pritchard, P.H. and Vance, D.E. (1981) cAMP analogues inhibit phosphatidylcholine biosynthesis in cultured rat hepatocytes. J. Biol. Chem. 256:8283-8286.

Pelech, S.L., Pritchard, P.H. and Vance, D.E. (1982) Prolonged effects of cyclic AMP analogues of phosphatidylcholine biosynthesis in cultured rat hepatocytes. Biochim. Biophys. Acta 713:260-269.

Sanghera, J.S. and Vance, D.E. (1989) CTP:phosphocholine cytidylyltransferae is a substrate for cAMP-dependent protein kinase in vitro. J. Biol. Chem. 264:1215-1223.

Sleight, R. and Kent, C. (1983) Regulation of phosphatidylcholine biosynthesis in mammalian cells:III. Effects of alterations in the phospholipid compositions of chinese hamster ovary and LM cells on the activity and distribution of CTP:phosphocholine cytidylyltransferase. J. Biol. Chem. 258:836-839.

Tercé, F., Record, M., Ribbes, G., Chap, H. and Douste-Blazy, L. (1988) Intracellular processing of cytidylyltransferase in Krebs II cells during stimulation of phosphatidylcholine synthesis. Evidence that a plasma membrane modification promotes enzyme translocation specifically to the endoplasmic reticulum. J. Biol. Chem. 263:3142-3149.

Tijburg, L.B.M., Geelen, M.J.H. and van Golde L.M.G. (1989) Regulation of the biosynthesis of triacylglycerol, phosphatidylcholine and phosphatidylethanolamine in the liver. Biochim. Biophys. Acta 1004:1-19.

Tijburg, L.B.M., Nishimaki-Mogami, T. and Vance, D.E. (1991a) Evidence that the rate of phosphtidylcholine catabolism is regulated in cultured rat hepatocytes. Biochim. Biophys. Acta, in press.

Tijburg, L.B.M., Samborski, R.W. and Vance, D.E. (1991b) Evidence that remodeling of the fatty acids of phosphatidylcholine is regulated in isolated rat hepatocytes and involves both the sn-1 and sn-2 positions. Biochim. Biophys. Acta, in press.

Tijburg, L.B.M. and Vance, D.E. (1991) Head group specificity in the regulation of phosphatidylcholine catabolism in rat hepatocytes. Biochim. Biophys. Acta, in press.

Utal, A.K., Jamil, H. and Vance, D.E. (1991) Diacylglycerol signals the translocation of CTP:phosphocholine cytidylyltransferase in HeLa cells treated with 12-O-tetradecanoylphorbol 13-acetate. Submitted.

Vance, D.E. (1990) Phosphatidylcholine metabolism: masochistic enzymology, metabolic regulation, and lipoprotein assembly. Biochem. Cell Biol. 68:1151-1165.

Vance, J.E. and Vance, D.E. (1988) Does rat liver Golgi have the capacity to synthesize phospholipids for lipoprotein secretion? J. Biol. Chem. 263:5898-5909.

Watkins, J.D. and Kent. C. (1990) Phosphorylation of CTP:phosphocholine cytidylyltransferase in vivo:Lack of effect of phorbol ester treatment in HeLa cells. J. Biol. Chem. 265:2190-2197.

Yao, Z., Jamil, H. and Vance, D.E. (1990) Choline deficiency causes translocation of CTP:phosphocholine cytidylyltransferase from cytosol to endoplasmic reticulum in rat liver. J. Biol. Chem. 265:4326-4331.

THE CONVERSION OF ETHANOLAMINE CONTAINING COMPOUNDS TO CHOLINE DERIVATIVES AND ACETYLCHOLINE

R. Massarelli, C. Andriamampandry*, N.E. Haidar, M. Andriamampandry
M. Carrara**, G. Sorrentino***, I.N. Singh*, L. Freysz, A. Hubsch
and J.N. Kanfer*
Centre de Neurochimie-Cronenbourg
CNRS
23 rue du Loess
67037 Strasbourg Cedex
France

One of the main issues in the understanding of the biochemistry and pathology of the brain concerns the mechanisms by which choline (Cho) is supplied to nerve cells. In brain tissue Cho, originating from endogenous synthesis in the liver (methylation of phosphatidylethanolamine) or as free base from the diet (Pardridge and Oldendorf, 1977; Cornford et al., 1978; Wecker and Trommer, 1984; see also Löffelholz, this volume), can be taken up from the blood circulation through the blood brain barrier. The brain itself is, however, capable of producing Cho from endogenous sources (Mozzi et al., 1979; Blusztajn et al., 1979; Crews et al., 1980; Zeisel, 1985; Blusztajn et al., 1987). In contrast to the fluctuating plasma levels, which are directly dependent upon diet, the brain tissue seems to have a homeostatic mechanism that maintains Cho at a steady state level. At low plasma concentrations of Cho there is more Cho exiting the brain than entering it producing a negative arterio-venous difference (AVD). A high plasma Cho level "forces" its uptake in the brain and raises its concentration (Dross and Kewitz, 1972; Tucek, 1978; Klein et al., 1991).

Whenever Cho is taken up in nerve cells or in the brain (Jope, 1979; Tucek, 1984; Massarelli et al., 1986; Tucek, 1985; Massarelli et al., 1974; Barker and Mittag, 1975; Haubrich et al., 1975; Simon et al., 1976) or in other organs (Martin, 1968; Acara and Rennick 1972; Haga and Noda, 1973; Yamamura and Snyder, 1973; Gardiner and Gwee, 1974; Zeisel et al., 1980; Sheard and Zeisel, 1986; Lerner, 1989), it is immediately phosphorylated and, in cholinergic cells, acetylated. Phosphocholine (PCho) may be eventually converted into phosphatidylcholine (PtdCho) from

*Department of Biochemistry, University of Manitoba, Winnipeg, Manitoba, Canada; **Department of Pharmacology, University of Padua, Italy; ***Institute of Neurological Sciences, University of Naples, Italy

NATO ASI Series, Vol. H 70
Phospholipids and Signal Transmission
Edited by R. Massarelli, L. A. Horrocks,
J. N. Kanfer, and K. Löffelholz
© Springer-Verlag Berlin Heidelberg 1993

which free Cho can be released via the activation of phospholipase D or by other catabolic processes. The degradation of Cho phospholipids appears to contribute to the negative A-V difference in Cho (AVδ) (Blusztajn and Wurtman, 1981).

Dross and Kewitz (1972) postulated that free ethanolamine (Etn), phosphoethanolamine (PEtn) and phosphatidylethanolamine (PtdEtn) may all be converted in the brain to the corresponding choline derivatives. In this chapter we will review these findings and will present the experimental evidence showing the existence of these reactions in clonal cell cultures derived from human neuroblastoma and the possibility that labelled acetylcholine (AcCho) may be found after incubation of cholinergic cells with [^{3}H]Etn. Data will be summarized indicating the actual predominance of the conversion of PCho in nervous tissue, over that of PtdCho and finally the presence, in rat brain, of at least two enzymatic activities responsible for the conversion of PEtn to PCho.

The conversion of water-soluble Etn derivatives in the nervous tissue

The conversion of PtdEtn to PtdCho by the stepwise methylation of the former was observed in rat brain and in neuronal and glial cell cultures derived from chick embryo brain cortices (Mozzi and Porcellati, 1979; Blusztajn et al., 1979; Crews et al., 1980; Mozzi et al., 1980, 1981a,b; Dainous et al., 1982). With the exception of the work of the group of Kewitz (Kewitz and Pleul, 1977) such a reaction had not been observed previously; it was assumed instead that the capability of the brain, and the nervous tissue in general, to synthetise Cho by the methylation of phospholipids as the liver does, was irrelevant. It was subsequently supposed that the negative AVδ observed in the brain derived essentially from the production of Cho via the catabolism of phospholipids (Blusztajn and Wurtman, 1981).

The experiments on neuronal cultures indicated that the methylation pathway exists in the nervous tissue, at a much lower activity than in the liver, but it was also found that methyl derivatives of PEtn and free Etn could be obtained after incubation with [^{3}H]Etn (Andriamampandry et al., 1989a). Other experiments performed in the presence of unlabeled or tritiated MeEtn and Me$_2$Etn led to the synthetic pathway which is sketched in scheme 1. The results showed that in chick neurons the preferential methylation pathway is via the conversion of PEtn.

In rat neuronal cultures, however, the conversion of PtdEtn to PtdCho was as efficient as that of PEtn to PCho. Moreover in these neurons it is

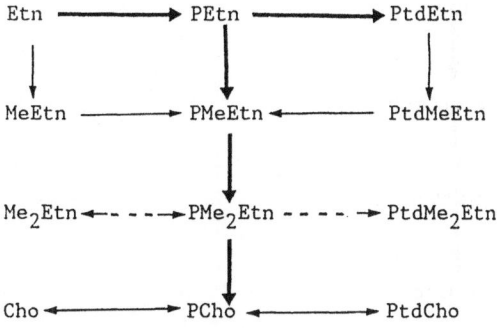

Scheme 1.   Conversion of [³H]Etn to various phosphorylated and methylated products by cultured chick brain neuronal cells (Andriamampandry et
al., 1991).

possible that free Etn may be directly converted into free Cho according
to scheme 2.

More recently, experiments were performed on clonal cell lines from human
neuroblastoma which have the characteristics of being catecholaminergic
(LA-N-1) or cholinergic (LA-N-2) and to be morphologically differentiated
by retinoic acid. The basic characteristics of these cells such as
enzymatic activities (Seeger et al., 1977; Singh et al., 1990), receptor
function (Sorrentino et al., in press) and choline metabolism (Blusztajn
et al., 1985; Singh et al., 1991) have already been studied.

This model was used to answer three questions: a) whether human neuronal
cells (even if tumoral in origin) could also be capable to methylate water
soluble derivatives of Etn, b) whether a differentiating agent such as
retinoic acid might influence this conversion and c) whether radioactive
AcCho might eventually be formed from radioactive Etn.

Experimental approach

The extraction, separation and identification of the various monomethylEtn
(MeEtn), dimethylEtn (Me₂Etn) and Cho derivatives have been previously
described (Andriamampandry et al., 1989a).

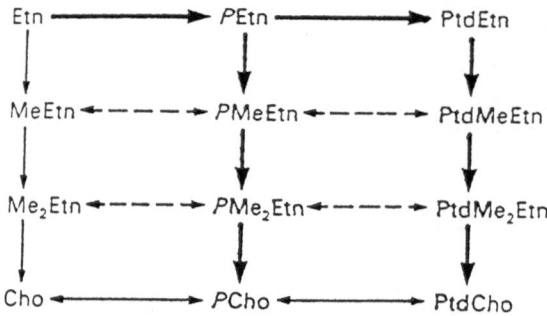

Scheme 2. Conversion of [³H]Etn to various phosphorylated and meth-ylated products by cultured chick brain neuronal cells (Andriamampandry et al., 1989)

Cell cultures

Human neuroblastoma clonal cell lines LA-N-1 and 2 were obtained from Dr. R. Seeger, University of California, Los Angeles, and maintained in culture with Leibovitz' F15 medium supplemented with 15% foetal calf serum. Cell replication was performed using trypsin in a Ca^{2+} free PBS medium.

Incubations were performed adding [³H]Etn (the amount of radioactivity and the specific activity is specified in the Legends to the Figures) to the growth media.

RESULTS AND DISCUSSION

The results indicate that both LA-N clones have the capability to convert Etn compounds into Cho containing compounds (Table 1). Quite interestingly the addition of retinoic acid to the growth medium influenced this conversion differently in the two clones.

Comparing Figure 1A (LA-N-2) with Figure 2A (LA-N-1) the incorporation of free Etn in the former clone after RA treatment decreased significantly after 60 and 120 minutes of incubation whereas in the latter it is remarkably increased. The labelling in MeEtn and Me2Etn derivatives increased in RA treated LA-N-2 cells (Fig. 1B) but not in the corresponding derivatives of LA-N-1 cells (Fig. 2B - this was not

TABLE 1. LAN-2 labelling of Cho and AcCho (in pmoles) from Etn and Cho precursors

		Cho	AcCho
[³H]Etn	Exp 1	11.8 ± 2.3	9.9 ± 1.6
	Exp 2	8.1 ± 0.8	8.0 ± 0.7
[³H]Chρ	Exp 1	75.6 ± 9.3	112.5 ± 12.8
	Exp 2	84.4 ± 8.5	82.0 ± 7.8

Cells were incubated 12 hours with the precursors and the amount synthesized in pmoles was calculated on the basis of the specific activity present in the growth medium. N=4; mean ± SD for both separate experiments.

significant after 60 min of incubation - and Fig. 2C). Similarly the conversion to free Cho in LA-N-1 cells increased slightly (Fig. 1C) while in LA-N-1 neurons a similar increase in labelling was not statistically significant (Fig. 2D).

The incorporation of the label into PEtn after RA treatment was strongly stimulated in LA-N-2 cells (Fig. 3) and remained at a higher level in the other derivatives, including PCho, even if not significant (Fig. 3B, C and D). The labelling of PEtn was increased in RA treated LA-N-1 cells after 60 min of incubation (Fig. 4) but not after 90 min, nor there was any significant difference in the labelling of the other metabolites (Fig. 4B, C and D).

In both cell types the incorporation of Etn into PtdEtn was severely decreased by the treatment with RA (Figs. 5A and 6A) as well as the incorporation into PtdCho and sphingomyelin (Figs. 5B, C and 6B, C).

The results confirm the existence of mechanisms for the incorporation of Etn into Cho containing compounds in these human cell lines similar to those found in chick and rat primary neuronal cultures and in the rat brain. The results also indicate that the differentiation induced by RA influences differently the incorporation and metabolism of [³H]Etn in the two cell lines. It might be suggested that the observed differences were due to the different nature of these neurons (LA-N-2 are cholinergic and LA-N-1 cells catecholaminergic). But, at present, this remains only a speculation.

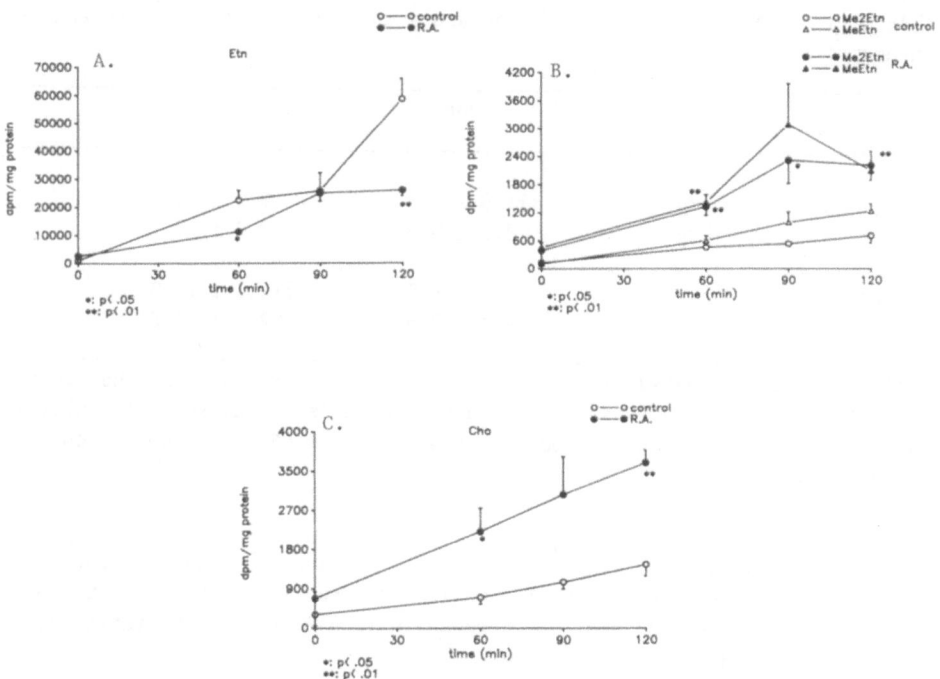

Fig. 1. Incubation of LA-N-2 neuroblastoma cholinergic cell line with [³H]Etn (5 μCi/ml of Leibovitz F15 growth medium: specific activity: 19 μCi/nmol; the label and the incubation conditions were the same in all LA-N-2 experiments). The Figures represent the appearance of the label as function of time in free Etn (A), MeEtn and Me₂Etn (B) and Cho (C) in cells grown in the presence of retinoic acid (10^{-5} for 8 days) or in the presence of the vehicle only as the control. Various compounds were extracted and separated by bidimensional TLC on silica gel plates as described in Andriamampandry et al. (1989a). Asterisk represents significance by the two-tail Student's t-test. Mean of 3 or 4 values ± SD.

Further experiments have shown that the incubation with radioactive Etn leads to the appearance of radioactive AcCho (Fig. 7) in parallel with that of free Cho (Fig. 7B). The labelling in both free Cho and AcCho was greater when the cells were grown in the presence of RA.

To ascertain the relative importance of the labelling of AcCho from Etn, LA-N-2 cells have been incubated in parallel with radioactive Etn or Cho and the results shown in Table 1 indicate that 10% of AcCho may be labelled with Etn. These experiments also showed that, in these cells, the ratio of the labelling of Cho to AcCho is close to one regardless of the radioactive precursor used, Cho or Etn.

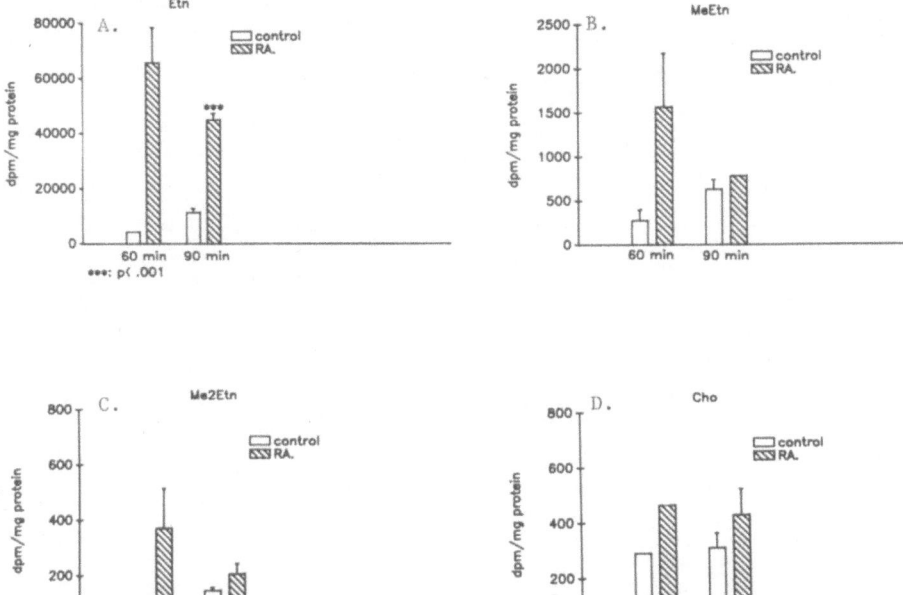

Fig. 2. Incubation of LA-N-1 neuroblastoma cathecholaminergic cell line with [³H]Etn (2.5 μCi/ml of Leibovitz F15 growth medium; specific activity 29.5 μCi/nmol. The label and the incubation conditions were the same in all LA-N-1 experiments). The Figures represent the appearance of the label as function of time in free Etn (A), MeEtn (B), Me2Etn (C) and Cho (D). Samples were treated as in the legend of Figures 1. Bars represent the average of 2 (without SD) or 3 values ± SD. Significance was measured according to Student's t test.

The following tentative conclusions may be drawn:
(a) human neurons, as previously shown for rat and chick neurons, have the capability to convert Etn into Etn and Cho containing derivatives with some differences between the avian and the mammalian (rat) neurons (compare scheme 1 and 2). The preferential path of conversion is the methylation of the water soluble phospho-derivatives.

(b) in human neuronal cell lines the differentiation induced by RA increased the incorporation of free [³H]Etn into the catecholaminergic LA-N-1 cells whereas, in the cholinergic LA-N-2 neurons, an increased incorporation of the label was observed into the Etn methylated products PCho, free Cho and AcCho.

288

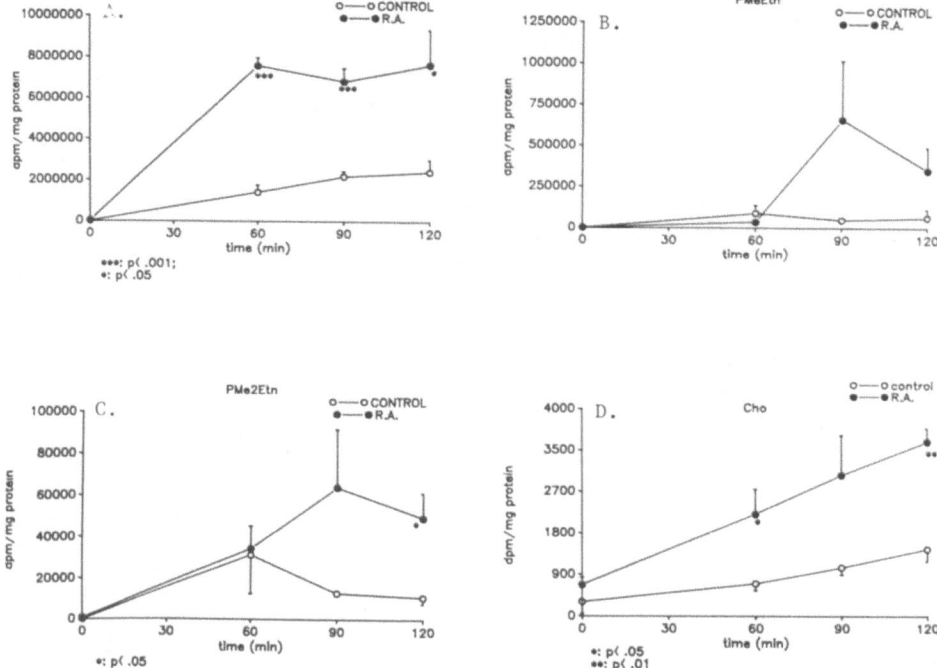

Fig. 3. Incubation of LA-N-2 cells with [³H]Etn. Appearance of the label in PEtn (A), PMeEtn(B), PMe2Etn (C), Cho (D). Incubations were performed and samples were treated as in the legend of Figure 1. Each point represents the average of 3 or 4 values ± SD.

(c) in LA-N-2 cells the Cho needed for AcCho synthesis may derive from the increased methylation of PEtn since the corresponding labelling into free Etn and its methylated derivatives is not affected by the RA treatment. Moreover the differentiation induced by RA reduces the synthesis of PtdEtn and of its methylated derivatives.

(d) in the case of LA-N-1 neurons the phosphorylation of Etn seems reduced by RA with a corresponding strong increase in the labelling of free Etn.

(e) AcCho may be synthetized from [³H]Etn up to 10% of corresponding labelling with Cho after 12 hours of incubation. The apparent rate of AcCho synthesis from Etn parallels that from Cho.

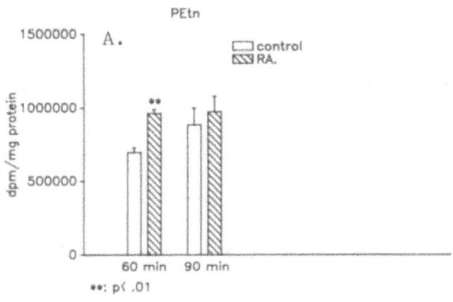
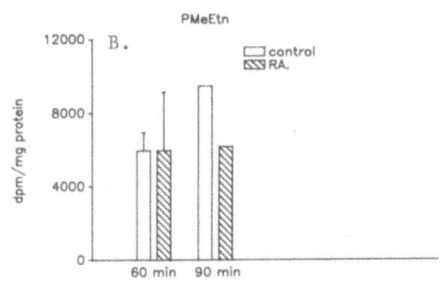
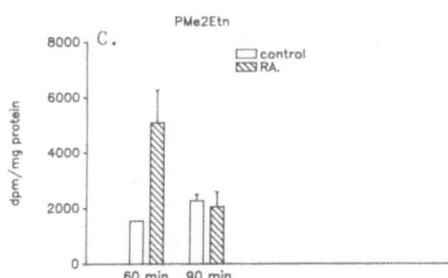
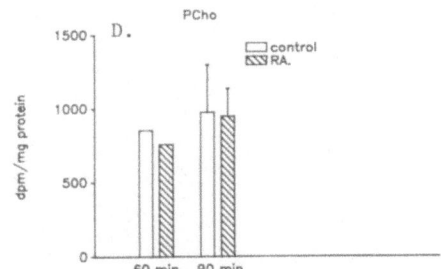

Fig. 4. Incubation of LA-N-1 cells with [³H]Etn. Appearance of the label in PEtn (A), PMeEtn (B), PMe₂Etn (C) and PCho (D). Incubations were performed as in the legend of Figure 1 and samples were treated as in the legend of Figure 2. Bars are the average of 2 or 3 values ± SD. Significance was measured with Student's t-test.

Modulation of the conversion reactions of Etn

Three different pharmacological agents have been used in an attempt to modulate these reactions in chick primary neurons (Andriamampandry et al., 1991): a catecholaminergic agonist (isoproterenol), an inhibitor of PtdEtn methylation in plants (2-hydroxyethylhydrazine, 2-HEH) and the growth of cells in the presence of exogenous gangliosides under conditions known to stimulate neuritogenesis and neurotrophism. A summary of the results is presented in scheme 3 and indicates that in chick primary neurons:

(a) the conversion of free Etn to free Cho is not affected by any of the treatments.

(b) the methylation of PEtn to PCho is decreased in the presence of isoproterenol (the effect is reversed by propanolol) and of gangliosides but it is increased by the addition of 2-HEH.

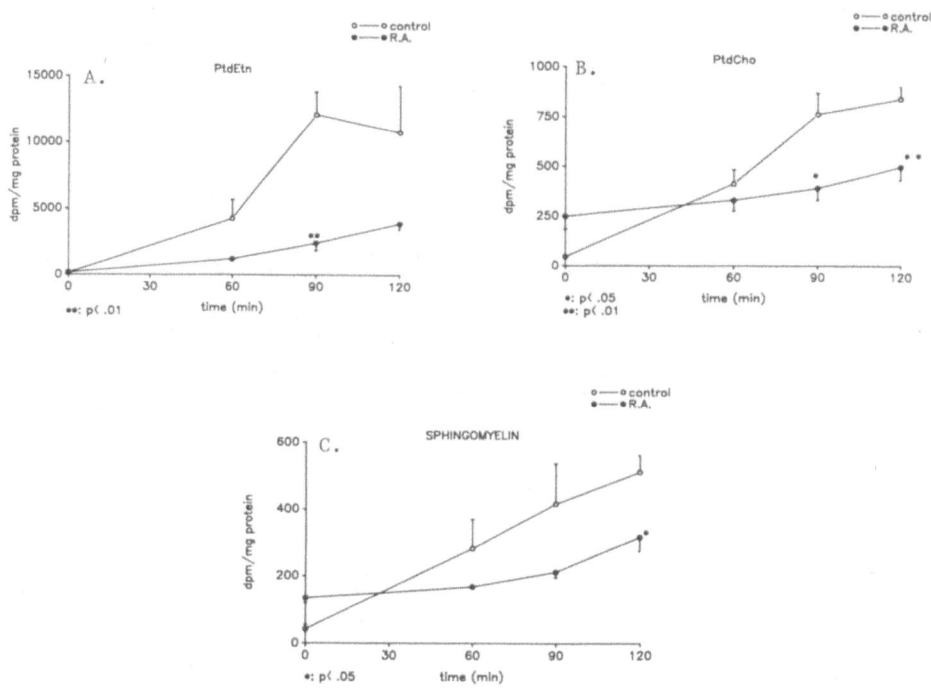

Fig. 5. Incubation of LA-N-2 cells with [³H]Etn. Appearance of the label in PtdEtn (A), PtdCho (B) and sphingomyelin (C). The experimental conditions were as explained in the legend for Figure 1.

(c) the methylation of PtdEtn to PtdCho is increased in the presence of isoproterenol and gangliosides and inhibited, as expected from the results obtained in yeast (Nikawa and Yamashita, 1983), by 2-HEH.

These observations support the finding that, in chick neurons, the methylation of free Etn is nearly undetectable or unimportant from a functional point of view and indicate that the main conversions are made from PEtn and PtdEtn. These two mechanisms appear to be inversely related since the inhibition of the one stimulates the other and vice-versa.

Conversion of Etn in vivo

It was possible, after intraventricular injection of [³H]Etn in the rat brain to identify radioactive MeEtn, Me₂Etn and Cho containing compounds

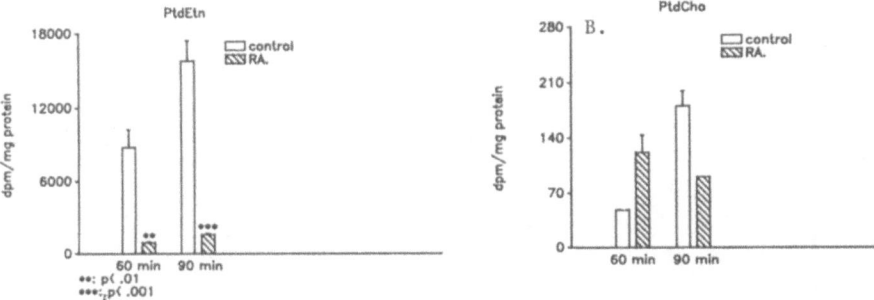

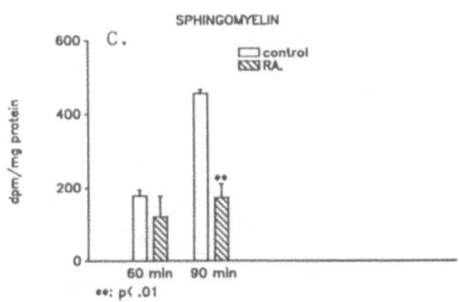

Fig. 6. Incubation of LA-N-1 cells with [³H]Etn. Appearance of the label in PtdEtn (A), PtdCho (B) and sphingomyelin (C). The experimental conditions were as explained in the legends of Figures 1 and 2.

as a function of time (Andriamampandry et al., 1989a). The results showed that, in the rat brain and in rat neuronal cultures, the preferential pathways of conversion of Etn to Cho are through the methylation of both PEtn and PtdEtn. Moreover the methylation of free Etn, even if of much lower importance than the other two, could not be excluded.

Further experiments revealed an interesting aspect of these pathways in the rat brain. There was a general decrease of the methylation activities as a function of the rat age (Andriamampandry et al., 1989b). This perhaps is a consequence of the decreased metabolism of Etn since this phenomenon had been shown in humans and rats using NMR spectroscopy (Brenton et al., 1985).

Why should the methylation of Etn compounds decrease as a function of age? A possibility to be considered might be that, in these experiments where [³H]Etn was injected intraventricularly, the permeation of the ventricular epithelia decreases in the brains of older rats. This did not appear to be the case when the results were recalculated on a percent

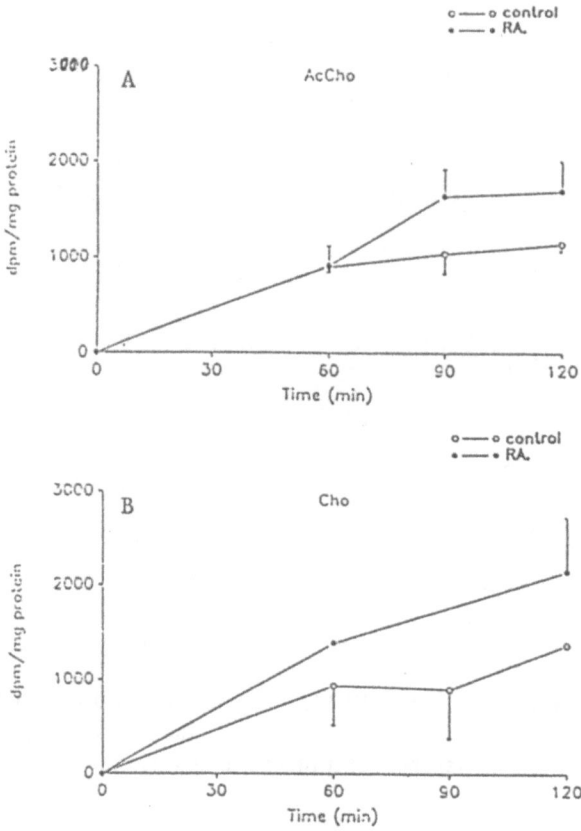

Fig. 7. Incubation of LA-N-2 cells with [³H]Etn. Appearance of the label in AcCho (A) and Cho (B). This was a separate experiment from that presented in Figures 1, 3 and 5. The incubation conditions were identical but the separation was attained by monodimensional TLC on cellulose plates using butanol/ethanol/acetic and water 100/20/33/17 (per volume) as developmental system.

basis. It appeared rather that the label from [³H]Etn might be increasingly incorporated into some other metabolites as a function of age. Preliminary experiments indicate that this might be reflected in the protein fraction (perhaps as a part of the protein-glycan-PI anchor; see Low, this symposium).

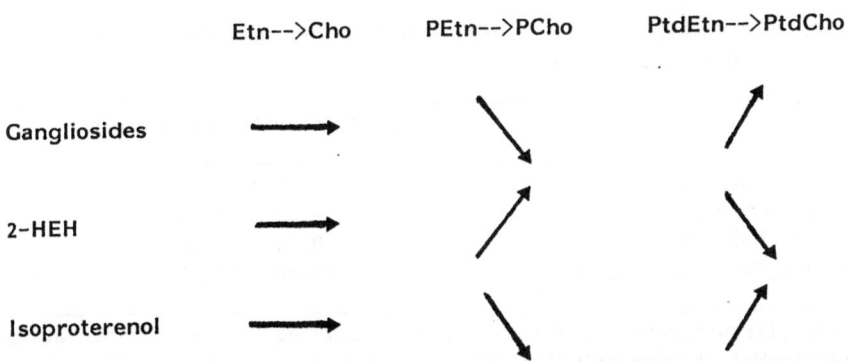

Scheme 3. Effect of pharmacological agents on the methylation conversion mechanisms. Experimental conditions are detailed in Andriamampandry et al., 1991.

Presence of a cytosolic enzymatic activity(ies) methylating PEtn to PCho

Any labelling experiment suffers from limitations inherent with the specific activity of the precursor being used and with the time factor which may obscure a degradation phenomenon occurring simultaneously with an apparent appearance of synthesis. The presence of a methyltransferase enzymatic activity would support the reality of the Etn conversion reactions observed in cell cultures and in the rat brain. Considering that the results of the labelling experiments indicated that the main route was that of PEtn to PCho it was presumed that any enzymatic activity(ies) catalyzing these reactions should be present in the cytoplasm. Such activities were found in the supernatant of a rat brain homogenate (Table 2) using PEtn or the phosphorylated monomethylated and dimethylated derivatives as substrates (Andriamampandry et al., 1990).

An ammonium sulfate fractionation of such supernatant showed at least one enzymatic activity precipitated between 20 and 40% of saturation and some remaining activity at higher concentrations of ammonium sulfate when PEtn (Table 3) and PMeEtn were the substrates. Another enzymatic activity was precipitated at a higher concentration of ammonium sulfate (50-80%) when PMe2Etn was the substrate. It appears then that the rat brain has, at least, two enzymatic activities capable of converting PEtn to PCho. In more recent experiments the enzymatic activity that catalyzes the conversion of PMe2Etn to PCho has been isolated to homogeneity.

TABLE 2. Distribution of phosphoethanolamine N-methyltransferase(s) activity(ies) in the rat brain

Fraction	(pmol product/mg protein/hr)
whole homogenate	421.20 ± 39.45
P2 (20,000 g pellet)	0.90 ± 0.007
P3 (105,000 g pellet)	0.70 ± 0.02
S3 (105,000 g supernatant)	738.60 ± 51.20

The N-methyltransferase activities were estimated following incubation with [³H]AdoMet (S-adenosylmethionine) and phosphoethanolamine. The values are the average of 3 determinations ± SD and have been adjusted for the controls (boiled enzyme).

CONCLUSION

This work started as the study of one of the possible ways that the cholinergic neuron might have developed to supply Cho for the synthesis of AcCho via the methylation of Etn and, notably, of PtdEtn. The unusual behavior of nerve cells to preferentially convert PEtn to PCho has opened a new and previously unsuspected development.

Since the initial report of preliminary data on the conversion of water soluble Etn compounds in chick neurons and glial cells (Massarelli et al., 1982) the work on the production of Cho from Etn leads to the following conclusions:

(a) labelled MeEtn, Me₂Etn and Cho containing compounds can be found after incubation with tritiated Etn in neural cells, including glial cells (Dainous et al., 1982; Massarelli et al., 1986).

b) There appears to be some species-specific differences between mammals (rat neuronal cultures, rat brain and human tumoral neurons) and avians (chick).

c) apparently inoperant pathways in chick neurons (such as the conversion of PtdEtn) may be made functionally operant by pharmacological means; suggesting that when one route of methylation is inhibited the other alternative pathway continues at an increased rate suggesting a necessity for the cell to maintain the process of methylation.

(d) AcCho may be labelled from radioactive Etn indicating that this base contributes some of the Cho necessary for AcCho synthesis.

TABLE 3. Distribution of N-methyltransferase activities in the cytosol of a rat brain homogenate following an ammonium sulphate precipitation

PEtn as substrate	Sp.Act.	Tot. Act.	% Recovery
S3	7.66	2105.22	100.00
0-20%	0.00	0.00	0.00
20-30%	6.13	442.21	21.00
30-40%	33.11	1798.02	85.40
40-50%	0.00	0.00	0.00
50-80%	2.25	214.60	10.20
PMeEtn as substrate			
S3	7.09	1946.91	100.00
0-20%	0.00	0.00	0.00
20-30%	12.21	880.95	45.24
30-40%	18.50	1004.72	51.82
40-50%	0.00	0.00	0.00
50-80%	1.22	116.36	8.31
PMe₂Etn as substrate			
S3	7.70	2116.48	100.00
0-20%	0.00	0.00	0.00
20-30%	0.00	0.00	0.00
30-40%	0.00	0.00	0.00
40-50%	0.00	0.00	0.00
50-80%	26.65	2541.48	120.01

The N-methyltransferase activities were estimated following incubations with [^{3}H]AdoMet and PEtn, PMeEtn or PMe₂Etn. The values are the average of 3 determinations and have been adjusted for the controls (boiled enzymes).

(e) ageing reduces the methylation processes presumably reflecting a decreased activity of the methyltransferases.

This short review of the data concerning the conversion of Etn compounds to Cho and its derivatives raises a number of questions. To enumerate them would be tedious, but there is a common ground for all of them: what is the physiological relevance of these conversion mechanisms? The observed labelling of Cho derivatives from Etn supports the hypothesis that these pathways are only one of the essential means to supply Cho to the brain. The adaptability that the cell shows when the main route of conversion is blocked by pharmacological means may indicate the need of the cell for these biochemical pathways. It remains to be established why this should be so.

ACKNOWLEDGMENTS

Supported in part from grants from the Medical Research Council of Canada (JNK) and NATO (860804 to RM).

REFERENCES

Acara M, Rennick B (1972) Renal tubular transport of choline: modifications caused by intrarenal metabolism. J Pharmacol Exper Ther 182:1-13

Andriamampandry C, Freysz L, Kanfer JN, Dreyfus H, Massarelli R (1989a) Conversion of ethanolamine, monomethylethanolamine and dimethylethanolamine to choline containing compounds by neurons in culture and the rat brain. Biochem J 264:555-562

Andriamampandry C, Freysz L, Kanfer JN, Massarelli R (1989b) The conversion of ethanolamine into choline in rat neuronal cultures and in the rat brain: correlation between acetylcholine synthesis and age, in Pharmacological interventions on central cholinergic mechanisms in senile dementia (Alzheimer's disease), Kewitz, Thomsen, Bickel (eds), Zuckschwerdt Verlag

Andriamampandry C, Massarelli R, Freysz L, Kanfer JN (1990) A rat brain cytosolic N-methyltransferase(s) activity converting phosphorylethanolamine into phosphorylcholine. Biochem Biophys Res Commun 171:758-763

Andriamampandry C, Freysz L, Kanfer JN, Dreyfus H, Massarelli R (1991) Effects of monomethylethanolamine, dimethylethanolamine, gangliosides, isoproterenol and 2-hydroxyethylhydrazine on the conversion of ethanolamine to methylated products by cultured chick brain neurons. J Neurochem 56:1845-1850

Barker LA, Mittag TW (1975) Comparative studies of substances and inhibitors of choline transport and choline acetyltransferase. J Pharmacol Exper Ther 192:86-94

Blusztajn J, Zeisel SH, Wurtman RJ (1979) Synthesis of lecithin (phosphatidylcholine) from phosphatidylethanolamine in bovine brain. Brain Res 179:319-327

Blusztajn JK, Liskovitch M, Richardson UI (1987) Synthesis of acetylcholine from choline derived from phosphatidylcholine in human neuronal cell line. Proc Natl Acad Sci (USA) 84:5474-5477

Blusztajn JK, Wurtman RJ (1981) Choline biosynthesis by a preparation enriched in synaptosomes from rat brain. Nature 290:417-418

Brenton DP, Garrod PJ, Krywawysh S, Reynolds EO, Bachelard HS, Cox DW, Morris PG (1985) Phosphoethanolamine is a major constituent of phosphomonoester peak detected by 31NMR in newborn brain. Lancet 115

Cornford EM, Braun L, Oldendorf WH (1978) Carrier-mediated blood-brain barrier transport of choline and certain analogs. J Neurochem 30:299-308

Crews FT, Hirata F, Axelrod J (1980) Identification and properties of methyltransferases that synthesize phosphatidylcholine in rat brain synaptosomes. J Neurochem 34:1491-1498

Dainous F, Freysz L, Mozzi R, Dreyfus H, Louis JC, Porcellati G, Massarelli R (1982) Synthesis of choline phospholipids in neuronal and glial cell cultures by the methylation pathway. FEBS Lett 146:221-223

Dross K, Kewitz H (1972) Concentration and origin of choline in the rat brain. Naunyn Schmiedeberg's Arch Pharmacol 274:91-106

Gardiner JE, Gwee MC (1974) The distribution in the rabbit of choline administered by injection or infusion. J Physiol 239:459-476

Haga T, Noda H (1973) Choline uptake systems of rat brain synaptosomes. Biochim Biophys Acta 291:564-575

Haubrich DR, Wang PFL, Clody DE, Wedeking PW (1975) Increase in rat brain acetylcholine induced by choline or deanol. Life Sci 17:975-980

Jope RS (1979) High affinity choline transport and acetylCoA production in brain and their roles in the regulation of acetylcholine synthesis. Brain Res 180:313-344

Kewitz H, Pleul O (1977) Synthesis of choline from ethanolamine in rat brain. Proc Natl Acad Sci (USA) 3:2181-2195

Klein J, Koppen A, Löffelholz K (1991) Uptake and storage of choline by rat brain: influence of dietary choline supplementation. J Neurochem 57:370-375

Lerner J (1989) Choline transport specificity in animal cells and tissue. Comp Biochem Physiol C: Comp Pharmacol Toxicol 93:1-9

Martin K (1968) Concentrative accumulation of choline by human erythrocytes. J Gen Physiol 51:497-516

Massarelli R, Sensenbrenner M, Ebel A, Mandel P (1974) Kinetics of choline uptake in mixed neuronal-glial and exclusively glial cultures. Neurobiology 4:414-418

Massarelli R, Dainous F, Freysz L, Dreyfus H, Mozzi R, Floridi A, Siepi D, Porcellati G (1982) The role of choline phospholipids and choline neosynthesis in the regulation of acetylcholine metabolism. in Basic and clinical aspects of molecular neurobiology, AM Giuffrida-Stella, G Gombos, B Benzi, HS Bachelard (eds), Menarini Foundation, pp 147-155

Massarelli R, Mykita S, Sorrentino G (1986) The supply of choline to the brain in The Astrocytes, S Federoff, A Vernadakis (eds), Plenum Press, New York, pp 155-178

Mozzi R, Porcellati G (1979) Conversion of phosphatidylethanolamine to phosphatidylcholine in rat brain by the methylation pathway. FEBS Lett 100: 363-366

Mozzi R, Andreoli V, Porcellati G (1980) Involvement of S-adenosyl-methionine in brain phospholipid metabolism. In: Cavallini D, Gaull GE, Zappia V (eds) Natural Sulfur Compounds: Novel Biochemical and Structural Aspects. Plenum Press, New York, pp 41-54

Mozzi R, Siepi D, Andreoli V, Piccinin GL, Porcellati G (1981b) N-Methylation of phosphatidylethanolamine in different brain areas of the rat. Bull Molec Biol Med 6:6-15

Mozzi R, Siepi D, Andreoli V, Porcellati G (1981b) The synthesis of choline plasmalogen by the methylation pathway in rat brain. FEBS Lett 131:115-118

Nikawa J, Yamashita S (1983) 2-hydroxyethylhydrazine as a potent inhibitor of phospholipid methylation in yeast. Biochim Biophys Acta 751:201-209

Pardridge WM, Oldendorf WH (1977) Transport of metabolic substrates through the blood-brain barrier. J Neurochem 28:5-12

Seeger RC, Rayner S, Banerjee A, Chung H, Lang WE, Neusteim HB, Benedict WF (1977) Morphology, growth, chromosomal pattern and fibrinolytic activity of two new human neuroblastoma cell lines. Cancer Res 37, 1364-1371

Sheard NF, Zeisel SH (1986) An in vitro study of choline uptake by intestine from neonatal and adult rats. Ped Res 20:768-772

Simon JR, Atweh S, Kuhar MJ (1976) Sodium-dependent high affinity choline uptake: a regulatory step in the synthesis of acetylcholine. J Neurochem 26:909-922

Singh IN, Sorrentino G, McCartney DG, Massarelli R, Kanfer JN (1990) Enzymatic activities during differentiation of the human neuroblastoma cells, LAN-1 and LAN-2. J Neurosci Res 25:476-485

Singh IN, Sorrentino G, Massarelli R, Kanfer JN (1991) The metabolic fate of [³H-methyl]choline in cultured human neuroblastoma cell lines, LAN-1 and LAN-2. Molec Chem Neuropathol 14:53-65

Sorrentino G, Singh IN, Hubsch A, Kanfer JN, Mykita S, Massarelli R (1992) Muscarinic binding sites in a catecholaminergic human neuroblastoma cell line. Neurochem Res, in press

Tucek S (1978) Acetylcholine synthesis in neurons. Chapman and Hall, London

Tucek S (1984) Problems in the organization and control of acetylcholine synthesis in brain neurons. Prog Biophys Mol Biol 44:1-46.

Tucek S (1985) Regulation of acetylcholine synthesis in the brain. J Neurochem 44:11-24

Wecker L, Trommer BA (1984) Effects of chronic (dietary) choline availability on the transport of choline across the blood-brain barrier. J Neurochem 43:1762-1765

Yamamura HI, Snyder SH (1973) High affinity transport of choline into synaptosomes of rat brain. J Neurochem 21:1355-1374

Zeisel SH, Story DL, Wurtman RJ, Brunengraber H (1980) Uptake of free choline by isolated perfused rat liver. Proc Natl Acad Sci (USA) 77:4417-4419

Zeisel SH (1985) Formation of unesterified choline by rat brain. Biochim Biophys Acta 835:331-343

PHOSPHOLIPID METABOLISM IN A HUMAN CHOLINERGIC CELL LINE - POSSIBLE
INVOLVEMENT OF THE BASE EXCHANGE ENZYME ACTIVITIES

I. Singh, D. McCartney, R. Massarelli* and J. N. Kanfer
Department of Biochemistry
and Molecular Biology
University of Manitoba
Winnipeg, Canada R3E 0W3

Considerable effort has been expended to unravel the details of the
mechanisms that allow a cell to respond in a predetermined manner to the
arrival of a ligand at the external surface of its plasma membrane.
Although it is well accepted that the proteins residing in these membranes
function as receptors or channels, it is reasonable to assume that the
lipids present in this lipid enriched environment also play a role in
these mechanisms. This is illustrated by the polyphosphoinositides which
are cleaved by a specific phospholipase C and are the source of the
phosphorylated inositols responsible for the release of calcium from
intracellular stores (Hokin, 1985). A decade ago it was fashionable to
investigate the involvement of the phospholipid N-methyl transferases that
were proposed to be important for signal transmission across biological
membranes (Hirata and Axelrod, 1980).

The latest additions to this catalogue are the major mammalian
phospholipid, phosphatidylcholine, and the several related phospholipases
catalyzing specific hydrolytic reactions such as phospholipases A_2, C
and D (Billah and Anthes, 1990; Exton, 1990).

Numerous model systems, predominantly cultured mammalian cells, have been
employed to investigate these problems and we have utilized the human
neuroblastoma cell designated LA-N-2. These intact cells have been
employed to investigate: (a) the effect of the phorbol ester, TPA, on
phospholipid metabolism, (b) the possibility that intact phosphocholine
can exit the cell, (c) the probability that phospholipase A_2 activation
triggers phospholipase D activity, (d) the regulation of the serine base
exchange activity *in vitro*.

*CNRS Centre de Neurochimie
67037 Strasbourg-Cronenbourg
France

NATO ASI Series, Vol. H 70
Phospholipids and Signal Transmission
Edited by R. Massarelli, L. A. Horrocks,
J. N. Kanfer, and K. Löffelholz
© Springer-Verlag Berlin Heidelberg 1993

MATERIALS AND METHODS

Materials

[9,10^{3}H]Myristic acid (39.3 Ci/mmol), [^{14}C-methyl]choline chloride (53 mCi/mmol), and L-[(U)-^{14}C]serine (52.1 mCi/mmol) were purchased from DuPont-New England Nuclear, Boston, USA. [2-^{14}C]Ethan-1-ol-2-amine hydrochloride was purchased from Amersham Canada Limited, Ontario, Canada. Falcon T-25 flasks, Leibovitz's L-15 medium, and heat-inactivated fetal bovine serum were obtained from Flow Labs. Serine-free L-15 medium was prepared in this laboratory.

Cell culture and labeling and protocols

The human neuroblastoma cell line, LA-N-2 (passage 81) was obtained from Dr. Robert Seeger, University of California, Los Angeles. The maintenance and growth conditions for this cell line were as described (Singh et al., 1990).

Analyses of radiolabeled lipid and water-soluble components

The water-soluble components, phosphocholine (PCho), glycerophosphocholine (GroPCho), CDP-choline and choline (Cho) were separated on silica gel G60 plates with a solvent system composed of 1.2% NaCl/CH$_3$OH/concentrated NH$_4$OH (50/50/5, v/v/v) (Yavin, 1976) employing the corresponding authentic radioactive compounds as standards. The plates were air-dried and processed for radioautograms using Kodak X-O MAT film.

The insoluble residue remaining after the lipid extraction was the material which represented the labeled proteins. Each experiment was conducted on at least three separate occasions with different batches of the LA-N cells.

Measurement of base exchange enzyme in vitro

The preparation of the rat brain membranes enriched in the base exchange enzymes and the assay procedure were as previously (Kobayashi et al., 1988) except that the Azolectin microdispersion was omitted from the incubations. The reaction mixture contained from 150 μg to 190 μg particulate protein, 5 mM CaCl$_2$, 50 mM HEPES pH 8.0, either 0.48 mM [^{14}C]ethanolamine (0.5 - 1.0 x 10^4 dpm/nmol) or 0.38 mM L-[^{14}C]serine (2 - 3 x 10^4 dpm/nmol) or 0.38 mM [^{14}C]choline (2 - 3 x 10^4 dpm/nmol) in a total incubation volume of 0.24 ml at 37°C for 20 min. The incorporation into lipids was carried out as previously described (Kobayashi et al., 1988). The activities varied with individual membrane preparations from 8 to 20 nmol/mg protein for the choline base exchange enzyme activity, from 52 to 95 for the ethanolamine base exchange enzyme activity and from 9 to 28 for the L-serine base exchange activity.

The various standard lipids were from Serdary. The data presented is from a representative experiment that had been repeated on a least three separate occasions with freshly prepared membranes varying somewhat in their absolute activities.

Protein determination
Protein was determined (Lowry et al., 1951) using bovine serum albumin as the standard.

Statistical analysis
Statistical significance was assessed using Student's t-test.

Results

Tumor-promoting phorbol esters such as 12-0 tetra-decanoylphorbol 13-acetate (TPA) are known to stimulate the metabolism of membrane phospholipids (Weinstein, 1981) including phosphatidylcholine (PtdCho) and other phospholipids (Guy and Murray, 1982; Lockney et al., 1984; Mufson et al., 1981; Nishino et al., 1983; Kiss and Anderson, 1989). It was of interest to examine the influence of TPA on phospholipid metabolism in these LA-N-2 cells.

TPA on choline prelabeled cells
       The protocol for these studies was as follows:

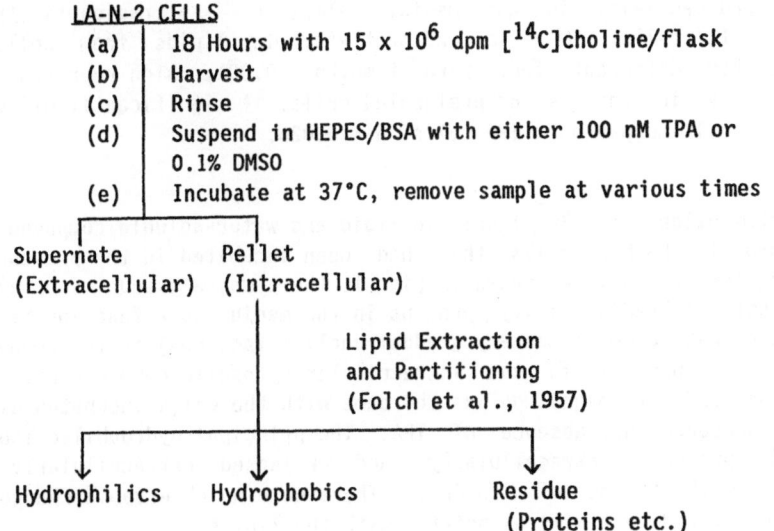

LA-N-2 CELLS
  (a) | 18 Hours with 15 x 10^6 dpm [^{14}C]choline/flask
  (b) | Harvest
  (c) | Rinse
  (d) | Suspend in HEPES/BSA with either 100 nM TPA or
       | 0.1% DMSO
  (e) | Incubate at 37°C, remove sample at various times

Supernate   Pellet
(Extracellular) (Intracellular)

                 Lipid Extraction
                 and Partitioning
                 (Folch et al., 1957)

Hydrophilics   Hydrophobics   Residue
                                          (Proteins etc.)

SCHEME 1

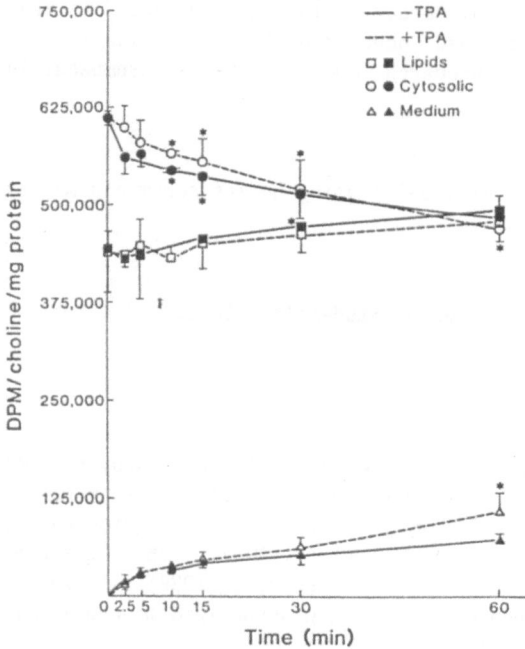

Figure 1.   Time-course of the release of radioactivity from
[^{14}C-methyl]choline prelabeled LA-N-2 cells.   Values represent means ±
S.D. of three separate experiments, analyzed in triplicate in the absence
[0.1% DMSO alone] (filled) or presence (open) of 100 mM TPA.
△-----△ radioactivity in the medium cells; o-----o radioactivity in
cytosol   of   cells;   □-----□ radioactivity in lipids from cells. *
significantly different from zero time, p < 0.001, which represents the
radioactivity in the 24 hr prelabeled cells. ** significantly different
from DMSO cells compared to TPA cells, p < 0.025.

The distribution of ^{14}C label in lipid and water-soluble compounds was
determined in LA-N-2 cells that had been incubated in the presence of
[^{14}C]Cho for 24 hr. As shown in Fig. 1, there was a gradual increase in
the amount of radioactivity present in the medium as a function of time
and there was a complementary gradual decline (p<0.005) in the amount of
radioactivity present in the intracellular cytosolic compartment. This
phenomenon and its magnitude was the same with the cells incubated either
in the presence or absence of TPA. The principal hydrophilic labeled
compound   present   intracellularly   and   released extracellularly was
[^{14}C]PCho with traces of GroPCho.   There were no observable changes in
the amount of radioactivity associated with the lipids.

Electropermeabilization

This observation suggested that [14C]PCho may be released intact directly from the cytosolic compartment into the medium rather than arising directly from phosphatidylcholine by a phospholipase C hydrolysis. We were unsuccessful in detecting such phospholipase C activity in these LA-N-2 cells (data not shown). To test the possibility of a direct release of phosphocholine, electropermeabiliztion experiments were performed to introduce phospho[14C]choline into the LA-N-2 cells. As shown in scheme 2, there was 5- to 12- fold more phosphocholine released from electropermeabilized cells incubated at 37°C than from either of the two separate controls. These results support the likelihood that phosphocholine can be released intact from an intracellular compartment into the extracellular medium and provides a mechanism for the observations shown in Fig. 1. Caution should be exercised in concluding that the appearance of PCho extracellularly is a demonstration of phospholipase C activity.

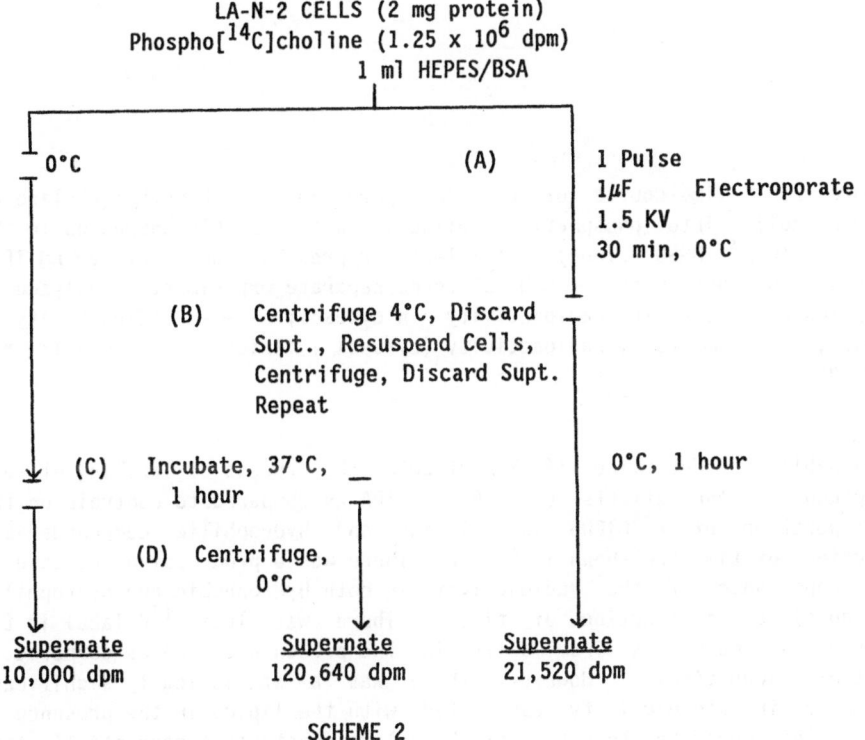

SCHEME 2

Choline labeling of LA-N-2 cells

Since there were no demonstrable effects of TPA upon the metabolism of choline containing phospholipids of prelabeled cells it seemed

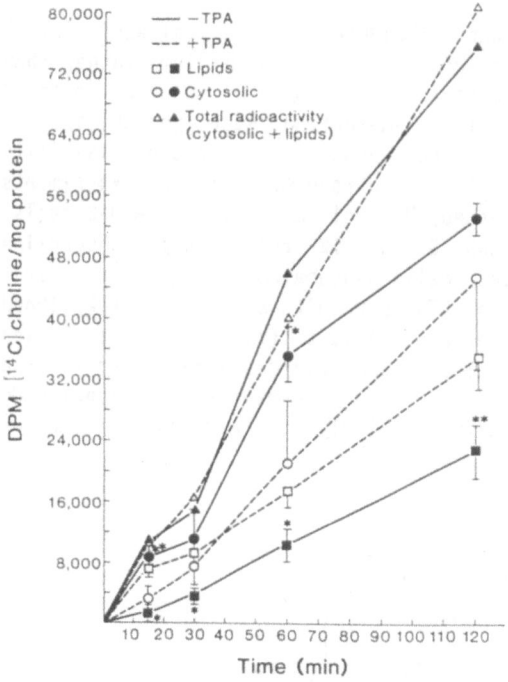

Figure 2. Time-course of the incorporation of [^{14}C-methyl]choline by LA-N-2 cells into phosphatidylcholine and water-soluble compounds in the absence (0.1% DMSO alone) (filled) or presence (open) of 100 nM TPA. Values represent means ± S.D. of three separate experiments, analyzed in triplicate. o-----o radioactivity in cytosol; Δ-----Δ radioactivity in lipids; □-----□ total radioactivity (cytosol + lipids); * p < 0.001; ** p < 0.005

reasonable to determine if TPA affected the process of labeling of such compounds. The effects of 100 nM TPA as compared to controls on the incorporation of [^{14}C]Cho into lipids and hydrophilic compounds as a function of time are shown in Fig. 2. There was a progressive increase in the appearance of the radioactivity in both hydrophobic and hydrophilic compounds as a function of time. There was less ^{14}C label in the hydrophilic compounds from cells in the presence of TPA as compared to control incubations. However, there was a statistically significant increase in radioactivity associated with the lipids in the presence of TPA. This could be the result of channeling substrate preferentially into the intermediates of the *de novo* pathway. However, there was no difference in the labeling of CDP choline in the presence or absence of TPA. The sum of radioactivity in hydrophilic and hydrophobic compounds

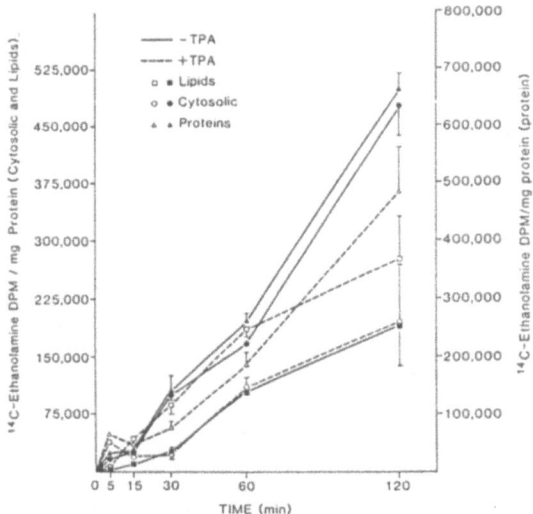

Figure 3. Time-course of the incorporation of [^{14}C]ethanolamine by LA-N-2 cells into phospholipids and water-soluble compounds in the absence (0.1% DMSO alone) (filled) or presence (open) of 100 nM TPA. Values represent means ± S.D. of triplicate incubations from a single experiment. Similar results were obtained when the same experiment was repeated twice. o-----o radioactivity in cytosol from cells; □-----□ radioactivity in lipids from cells; Δ-----Δ radioactivity in proteins from cells.

present in cells labeled in the presence or absence of TPA was identical (Fig. 2).

Ethanolamine labeling of LA-N-2 cells
It seemed useful to learn if these effects of TPA were restricted only to the incorporation of [^{14}C]choline into lipid and water-soluble compounds; therefore, similar experiments were undertaken with [^{14}C]ethanolamine. The results of labeling of the lipids, the hydrophilic compounds and the proteins by [^{14}C]ethanolamine are presented in Fig. 3. TPA caused a marked stimulation of the appearance of [^{14}C]ethanolamine in phosphatidylethanolamine accompanied by a decrease in the hydrophilic compounds as compared to the results obtained with control cells. There was a considerable amount of radioactivity associated with the proteins and this increased linearly with time in the absence or presence of TPA. However, there was less radioactivity associated with the proteins in the presence of TPA.

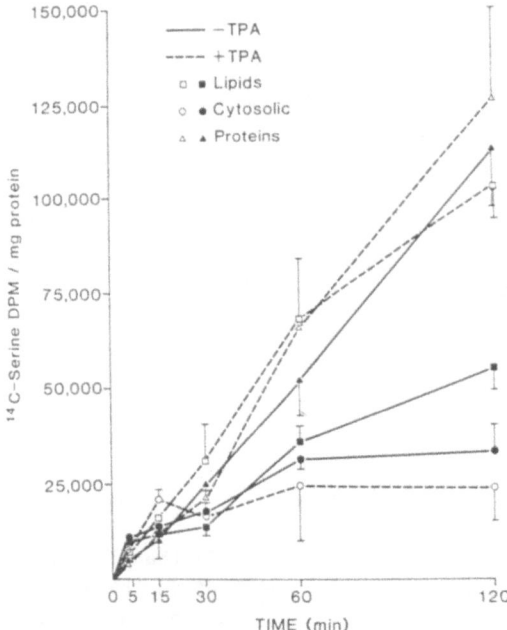

Figure 4. Time-course of the incorporation of [^{14}C]serine by LA-N-2 cells into phosphatidylserine and water-soluble compounds in the absence (0.1% DMSO alone) (filled) or presence (open) of 100 nM TPA. Values represent means ± S.D. of three separate experiments, analyzed in triplicate incubations. o-----o radioactivity in cytosol; □-----□ radioactivity in lipids; Δ-----Δ radioactivity in proteins in cells.

Serine labeling of LA-N-2 cells

Since TPA had similar effects on both choline and ethanolamine incorporation it seemed reasonable to determine if this also applied to serine. The results of labeling of the lipids, the hydrophilic compounds and the protein by [^{14}C] L-serine is presented in Fig. 4. TPA caused a marked stimulation of [^{14}C] L-serine incorporation into phosphatidyl-serine, with some phosphatidylethanolamine appearing at 120 min, accompanied by a reduction in labeling of hydrophilic compounds as compared to control cells. As expected, this amino acid was also incorporated into protein, but there were no differences between cultures labeled in the presence or absence of TPA.

Retinoic acid grown cells

Previous reports demonstrated that LA-N-2 cells propagated in the presence of 10^{-5} M retinoic acid (RA) underwent morphological differentiation with accompanying changes in certain enzymatic activities (Singh et al.,

TABLE 1. Comparison of incorporation of [^{14}C]choline, [^{14}C]ethanolamine and L-[^{14}C]serine into various classes of cellular compounds of retinoic acid and control LAN-2 cells. (All values expressed as dpm/mg protein/120 minutes) (a = p<0.025, RA vs. Ctrl; b = p<0.05 RA vs. Ctrl; c = p<0.005 Ctrl vs. TPA; d = p<0.005 RA vs. Ctrl; e = p<0.001 RA vs. Ctrl; f = p<0.001 Ctrl vs. TPA; g = p<0.01 Ctrl vs. TPA

	Cytosol	Lipids	Proteins
[^{14}C]choline labeled cells			
Control - TPA	126,127 + 9,347	17,172 + 1,326	
Control + TPA	73,835 + 3,574^c	28,943 + 2,105^c	
RA treated - TPA	69,691 + 20,997^a	28,442 + 6,058^b	
RA treated + TPA	59,195 + 8,017	20,107 + 3,480	
[^{14}C]ethanolamine labeled cells			
Control - TPA	206,955 + 9,721	94,801 + 1,021	250,184 + 23,415
Control + TPA	82,373 + 11,973^f	352,173 + 24,030^f	123,032 + 8,396^c
RA treated - TPA	203,355 + 8,614	55,687 + 8,033^d	57,649 + 5,310^e
RA treated + TPA	151,612 + 10,313	56,641 + 9,535	63,383 + 7,433
[^{14}C]serine labeled cells			
Control - TPA	20,344 + 1,276	30,246 + 8,323	68,124 + 1,374
Control + TPA	11,049 + 529^c	58,190 + 6,541^g	115,411 + 2,517^f
RA treated - TPA	21,291 + 722	15,297 + 1,423^b	18,503 + 11,049^e
RA treated + TPA	19,259 + 3,599	13,254 + 2,459	19,929 + 2,536

1990). Therefore, it seemed reasonable to determine whether TPA exerted a similar influence on the labeling with L-[^{14}C]serine, [^{14}C]choline and [^{14}C]ethanolamine of LA-N-2 cells differentiated with RA as with those propagated in the absence of retinoic acid. These results are presented in Table 1, which shows that the retinoic acid grown cells as compared to control cells: (a) were unresponsive to TPA, (b) had a reduced capacity to incorporate labeled choline, ethanolamine or serine into their characteristic compounds.

The base exchange activities
The results obtained thus far implicated the base exchange enzyme activities as being responsible for the TPA-enhanced incorporation of choline, ethanolamine and serine into their corresponding phospholipids. This was based upon (a) inability to detect enhanced labeling of the precursors of the *de novo* pathway and (b) the serine base exchange activity is the sole mechanism available to mammals for phosphatidylserine production (Folch et al., 1957; Kanfer, 1980, 1989). TPA is known to liberate free fatty acids due to phospholipase A_2 activation (Weinstein et al., 1979) to increase diacylglycerols due to phospholipid hydrolysis and perhaps to affect free sphingosine levels.

Effects of fatty acids
The results of incubating these rat brain membranes in the presence of varying concentrations of sodium oleate revealed that the choline base exchange enzyme activity was inhibited in a dose-dependent manner. In contrast the serine and ethanolamine base exchange enzyme activities were stimulated at intermediate, 0.1 to 0.5 mM, concentrations of sodium oleate (Fig. 5, right panel). Identical results were obtained with the same concentrations of sodium linolenate. Sodium arachidonate at concentrations above 0.5 mM inhibited both the L-serine and ethanolamine base exchange enzyme activities, but at 0.1 to 0.25 mM concentrations the serine base exchange activity was doubled and the ethanolamine base exchange enzyme activity increased by 65%. However, the choline base exchange enzyme activity was inhibited in a dose-dependent manner by all concentrations of arachidonate employed (Fig. 5, left panel). Na palmitate was without any effect on these activities.

Effects of sphingosine
The rat brain membrane preparation was incubated with varying concentrations of sphingosine to determine if this compound could modulate the activity of the base exchange enzymes. There was a doubling of the L-serine base exchange enzyme activity at 0.2 mM sphingosine and only a 25% increase of the ethanolamine base exchange enzyme at 0.1 mM sphingosine. Higher concentrations of sphingosine inhibited the choline base exchange enzyme completely, the ethanolamine base exchange enzyme by 85% and the L-serine base exchange enzyme by about 30% (Fig. 6).

Effects of diacylglycerol and nucleotides
1,2-Diolein from 0.05 to 5 mM did not affect either of the three base exchange enzyme activities. The presence of 10 mM GTP reduced the choline base exchange enzyme activity by 30%, the ethanolamine base exchange enzyme activity by 25% and the serine base exchange enzyme activity by 47%. However, the presence of ATP, CTP, GTP, UTP, ADP, cAMP, GTPS, cGMP, or GDP at the same concentration did not affect these three enzyme activities.

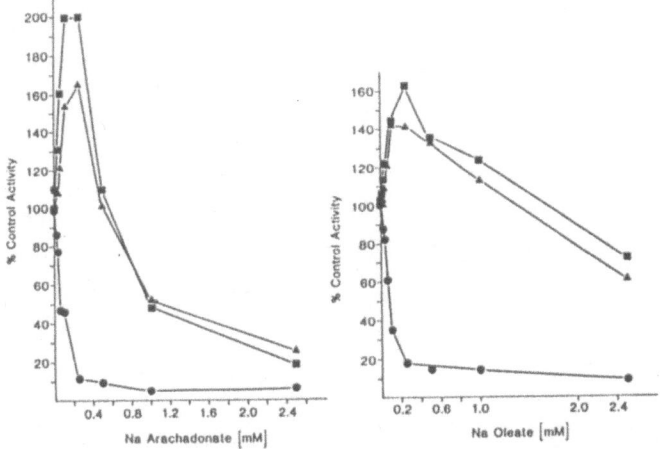

Figure 5. The effect of varying sodium oleate (right) and sodium arachidonate (left) concentrations upon the individual base exchange enzyme activities of rat brain membranes enriched in the endoplasmic reticulum. The L-serine base exchange enzyme activity is shown as squares, that of the choline base exchange enzyme as circles and that of the ethanolamine base exchange enzyme as triangles.

TPA and phospholipase D activation

PLD of mammalian tissues is a latent enzyme and the most efficient activator is oleic acid (Kanfer, 1989), which is liberated through PLA_2 hydrolysis of phospholipids. TPA has been shown to activate PLD activity of cultured cells (Dennis et al., 1991), however, the possibility that this is an indirect effect has not been excluded. We have employed mellitin, a presumed PLA_2 activator to liberate oleate which in turn can activate PLD. A convenient measure of PLD activity with intact cells is the appearance of phosphatidylethanol as a reflection of the "transphosphatidylation" activity of this enzyme (Kanfer, 1989).

The protocol employed for these investigations was:

LA-N-2 cells
(a) [^{3}H]myristic acid, 1.5 Ci/ml for 18 hours labeling
(b) Removed medium, rinse 2 X
(c) Incubate 1 hour with either EtoH alone or mellitin alone or EtoH + mellitin or TPA alone or TPA + mellitin
(d) Harvest cells, wash, lipid extraction
(e) TLC lipids with ethyl-acetate/isooctane/acetic acid/water (13:2:3:10) solvent
(f) Radioautography, scrape and count

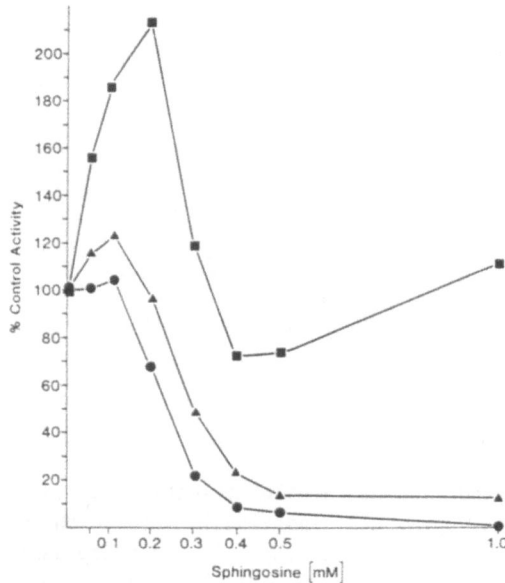

Figure 6. The effect of varying concentrations of sphingosine upon the individual base exchange enzyme activities of rat brain membranes enriched in the endoplasmic reticulum. The L-serine base exchange activity is shown as squares, that of the choline base exchange enzyme as circles and that of the ethanolamine base exchange enzyme as triangles.

The results summarized in Table 2 show that about 0.7% of the total radioactivity is recovered as PA from samples without any additions and 0.06% as PtdEt. Incubation in the presence of EtOH increases the PtdEt at least 4-fold and in the presence of both EtOH and mellitin, this is increased 9-fold. In contrast, TPA does not augment the PtdEt formation. As expected both mellitin and TPA seem to increase the amount of radioactivity associated with DAG and FFA.

DISCUSSION

These results suggest the following sequence of events.
(1) Agonist binding to receptor activates a G protein (G - ?, G_{pla}).
(2) G protein stimulates phospholipase A_2 resulting in the release of unsaturated fatty acids as $C_{18:1}$, $C_{20:4}$, etc.
(3) The $C_{18:1}$ stimulates phospholipase D (Weinstein et al., 1979) activity resulting in formation of phosphatidic acid.

(4) The phosphatidic acid produced is rapidly hydrolyzed by phosphatidic acid phosphatase to 1,2-*diacylglycerol*.

(5) The $C_{20:4}$ and $C_{18:1}$ activate the L-serine/ethanolamine base exchange enzyme (Suzuki and Kanfer, 1985) activity producing *phosphatidylserine*.

(6) 1,2-Diacylglycerol elevates the free sphingosine levels.

(7) Sphingosine stimulates the L-serine base exchange enzyme (Taki and Kanfer, 1978) activity, producing *phosphatidylserine*.

Therefore, PKC activity can be increased because:

(a) $C_{20:4}$ activates PKC α and γ (Nishizuka, 1988)

(b) *1,2 DAG* and *phosphatidylserine* activate the Ca^{2+} phospholipid-dependent protein kinase C (Nishizuka, 1988)

Schematically this would be

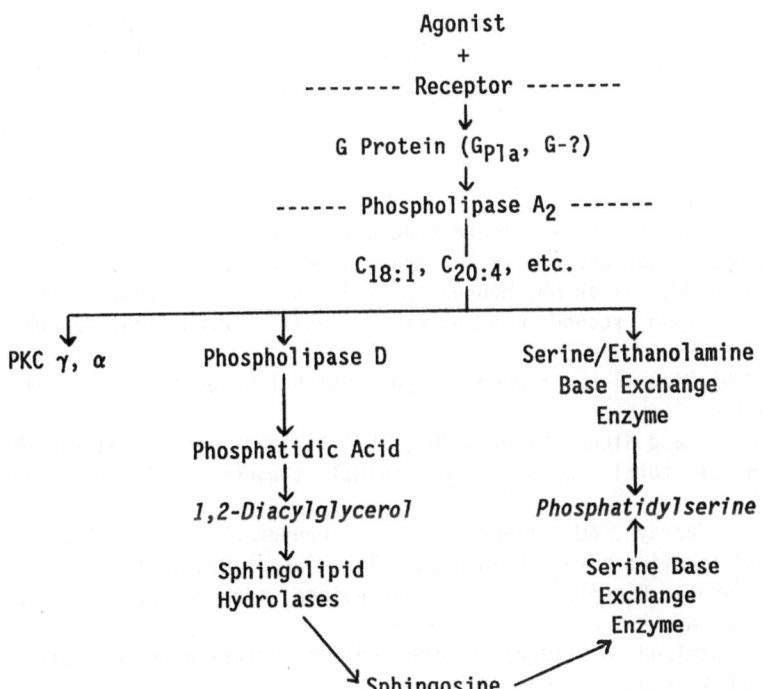

TABLE 2. The percentage distribution of radioactivity LA-N-2 in cells prelabeled with [^{3}H]myristic acid after challenge with various agents

	Percent of Radioactivity				
	PtdCho + PtdEtn + PtdSer	PtdOH	PtdEt	MG	DAG + FFA
1. No additions	89.1	0.69	0.06	1.24	8.5
2. + Ethanol (EtOH) (0.5% = 85 mM)	85.35	0.55	0.23	1.11	12.2
3. + Mellitin (50μg/ml)	80.95	0.58	0.10	1.55	16.45
4. + TPA (100 nM)	82	0.59	0.09	1.7	15.28
5. EtOH + Mellitin	78.3	0.54	0.55	1.8	18.8
6. EtOH + TPA	83.5	0.52	0.28	1.7	13.65

REFERENCES

Billah MM and Anthes JC (1990) The regulation and cellular functions of phosphatidylcholine hydrolysis. Biochem J 269:281-291

Dennis EA, Rhea SG, Billah MM, Hannun YA (1991) Role of phospholipases in generating lipid second messengers in signal transduction. FASEB J 5:2068-2077

Exton JH (1990) Signalling through phosphatidylcholine breakdown. J Biol Chem 265:1-4

Folch J, Lees M and Sloane Stanley GH (1957) A simplified method for the isolation of total lipides from animal tissues. J Biol Chem 226:497-509

Guy GR and Murray AW (1982) Tumor promoter stimulation of phosphatidylcholine turnover in HeLa cells. Cancer Res 42:1980-1985

Hokin LE (1985) Receptors and phosphoinositide-generated second messengers. Ann Rev Biochem 54:205-235

Hirata F and Axelrod J (1980) Phospholipid methylation and biological signal transmission. Science 209:1082

Kanfer JN (1980) The base exchange enzymes and phospholipase D of mammalian tissues. Can J Biochem 12:1370

Kanfer JN (1989) Phospholipase D and the base exchange enzyme. In: Vance D (ed), Phosphatidylcholine Metabolism, CRC Press, Boca Raton, pp. 65

Kiss Z and Anderson WB (1989) Phorbol ester stimulates the hydrolysis of phosphatidylethanolamine in leukemic HL-60, NIH 3T3, and baby hamster kidney cells. J Biol Chem 264:1483-1487

Kobayashi M, McCartney DG and Kanfer JN (1988) Developmental changes and regional distribution of phospholipase D and base exchange enzyme activities in rat brain. Neurochem Res 13:771-776

Lockney MW, Golomb HM and Dawson G (1984) Phorbol ester tumor promoters specifically stimulate choline phospholipid metabolism in human leukemic cells. Biochim Biophys Acta 796:384-392

Lowry OH, Rosebrough NJ, Farr AL and Randall RJ (1951) Protein measurement with the Folin phenol reagent. J Biol Chem 193:265-275

Mufson RA, Okin E and Weinstein IB (1981) Phorbol esters stimulate the rapid release of choline from prelabelled cells. Carcinogenesis 2:1095-1102

Nishino H, Fujiki H, Terada M, and Sato S (1983) Enhanced incorporation of radioactive inorganic phosphate into phospholipids of HeLa cells by tumor promoters. Carcinogenesis 8:1943-1945

Nishizuka Y (1988) The molecular heterogeneity of protein kinase C and its implications for cellular regulation. Nature 334:661-665

Singh IN, Sorrentino G, McCartney DG, Massarelli R and Kanfer JN (1990) Enzymatic activities during differentiation of the human neuroblastoma cells, LA-N-1 and LA-N-2. J Neurosci Res 25:476-485

Suzuki T and Kanfer JN (1985) Purification and properties of an ethanolamine-serine base exchange enzyme of rat brain membranes. J Biol Chem 260:1394-1399

Taki T and Kanfer JN (1978) A phospholipid serine base exchange enzyme. Biochim Biophys Acta 528:309-317

Weinstein IB, Lee L-S, Fisher PB, Mufson A and Yamasaki H (1979) Action of phorbol esters in cell culture: Mimicry of transformation, altered differentiation, and effects on cell membranes. J Supramol Struct 12:195-208

Weinstein IB (1981) Current concepts and controversies in chemical carcinogenesis. J Supramol Struct Cell Biochem 17:99-120

Yavin E (1976) Regulation of phospholipid metabolism in differentiating cells from rat brain hemispheres in cultures: patterns of acetylcholine, phosphocholine and choline phosphoglyceride labeling from [methyl-^{14}C]choline. J Biol Chem 251:1392-1397

DISPOSITION OF CHOLINE FOR THE BRAIN

H. Kewitz and O. Pleul
Department of Clinical Pharmacology
Klinikum Steglitz
Free University of Berlin
Hindenburgdamm 30
1000 Berlin 45
Germany

Choline keeps a crucial position in the metabolism and in the function of mammalian brain because it is a cornerstone for the synthesis of both acetylcholine, one of the most significant transmitters in the brain, and phosphatidylcholine, one of the backbones in biological membranes. The concentration of choline, which in rat brain is not higher than 28 nmol/g, is amazingly low and the turnover time of 45 sec is very high at least in comparison with the very large amount of 20,000 nmol/g of choline containing phospholipids with a turnover time of 17 hrs and also with regard to the very high turnover of acetylcholine depending on functional demands.

Scheme of Choline-metabolism in brain

Figures: nmoles/g

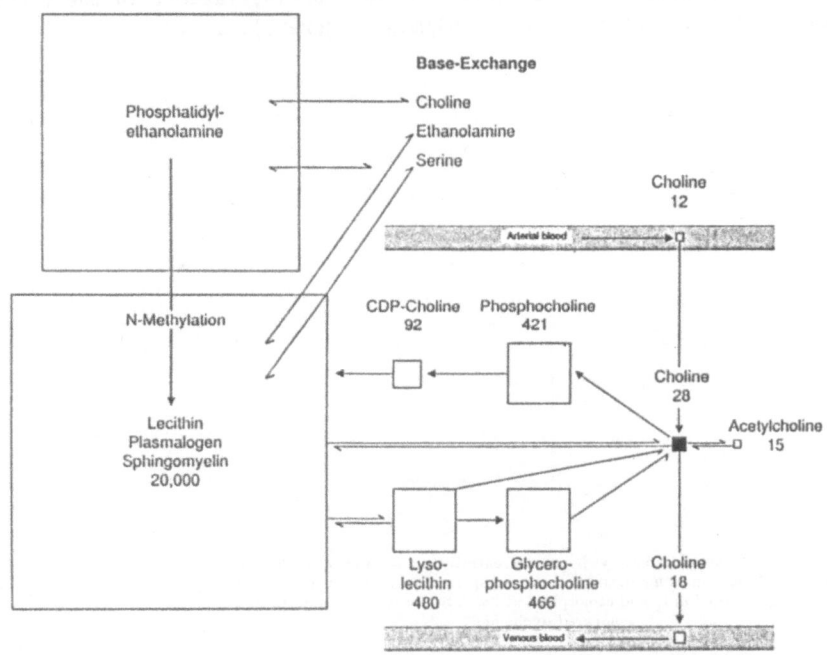

NATO ASI Series, Vol. H 70
Phospholipids and Signal Transmission
Edited by R. Massarelli, L. A. Horrocks,
J. N. Kanfer, and K. Löffelholz
© Springer-Verlag Berlin Heidelberg 1993

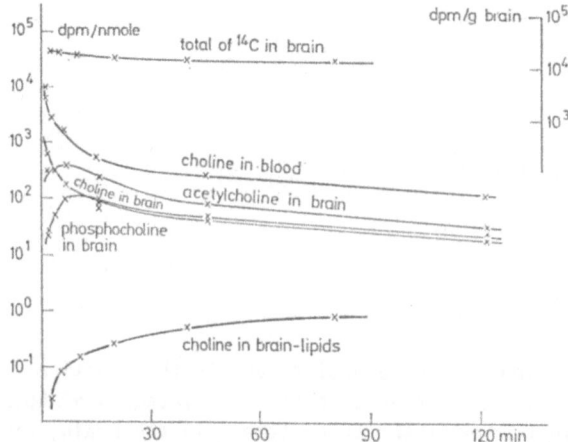

Fig.1. Specific radioactivities of choline, acetylcholine, phosphorylcholine, and choline in lipids in the brain and of choline in the blood of rats after i. v. injection of ^{14}C-labelled choline

In addition Figure 1 shows that choline is contained in a virtually unique precursor pool for both acetylcholine and phosphocholine as well since both show the same specific radioactivity as choline at their maximum where the curves are crossing.

2. After separation of the brain from the circulation, as by decapitation, choline accumulates immediately in the brain, fastest in the first few minutes, at a velocity of 20.7 nmol/min (Figure 2).

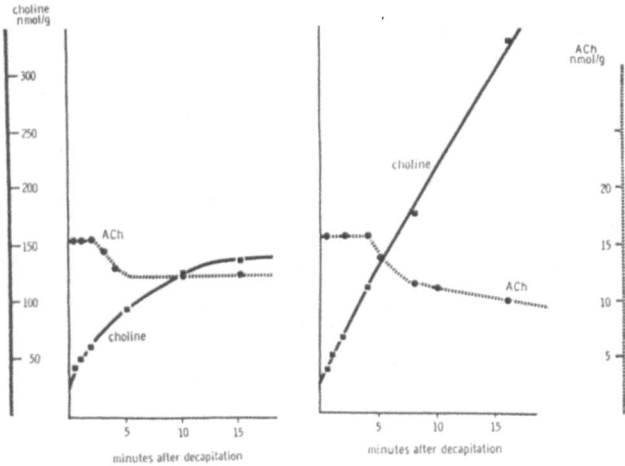

Fig. 2.Changes of choline- and acetylcholine-concentrations in rat brain after decapitation. Left: Brains remaining in skull at room temperature (approx. 20°C). Right: Dissected brains stored in humid atmosphere at 38°C. Scale: Acetylcholine 10-fold compared to choline

The sizes of the squares on the scheme reflect exactly those proportions in which the concentration of choline is related to the various choline containing intermediates. The scheme also shows the very complex metabolic interconnections including the indirect links to phosphatidylethanolamine via the transfer of methyl groups at the N-position and also via base exchange. It is quite impressive to see the many inputs which the choline pool receives from different sources including those not shown on the scheme, namely de novo synthesis of choline by stepwise transmethylation of unesterified ethanolamine and of phosphoethanolamine to phosphocholine as described in the foregoing papers given by Massarelli and by Kanfer. Not shown on the scheme is the formation of additional choline which eventually may be liberated from lysophosphatidylcholine transported from the blood to the brain, a mechanism suggested by Illingworth and Portman (1972) to serve the supply of choline to the brain. Considering all these different supplements of choline to such a small pool and regarding the requirement to keep the concentration of choline constant at the optimal level in order to meet the demands both, the impulse-directed need for acetylcholine and the metabolically-dependent requirement for phosphatidylcholine, it seems quite obvious that a whole system of control mechanisms must exist. Almost nothing is known so far about such provisions but it might be conceivable to regard the outflow of choline from the brain into the venous blood as one link in the chain of events to keep the choline concentration constant at an optimal level. We suggest that choline may be produced in excess to the needs and that the surplus does overflow into the blood circulation. The next speaker, Konrad Löffelholz, will elaborate at that aspect in more detail.

Choline may originate from blood-stream, from the metabolism or from preformed compounds coming from the blood. From experiments to explore the disposition of choline three findings have led to the hypothesis that choline might be formed in the brain either by N-transmethylation of ethanolamine from the donor S-adenosylmethionine or by liberation from circulating lysophosphatidylcholine transported through the blood-brain barrier into the brain.

1. After i.v. administration of tracer amounts of labelled choline the specific activity of choline remained 6 to 7 times higher in the blood than in the brain, although choline is transported very efficiently and fast through the blood-brain barrier (Dross and Kewitz, 1966; Kewitz et al., 1973).

From those results it became clear that unlabelled or low labelled choline must be formed in the brain and continuously mixed with the labelled fraction provided from the blood.

The largest proportion, namely 65%, originated from the degradation of glycerophosphocholine (Table 1) (Kewitz et al., 1972).

This result has suggested that there may be a net outflow of choline from the brain via the circulation.

Table 1. Concentrations ± SD of phosphocholine (PCh) and glycerophosphocholine (GPCh) in the brain of rats after decapitation (N = 4 at each time interval)

Minutes after decapitation	PCh nmol/g	GPCh nmol/g
0.5	283 ± 12	580 ± 11
4.0	272 ± 23	539 ± 6
8.0	301 ± 11	471 ± 37
16.0	288 ± 24	383 ± 11

This has been confirmed by the third finding which showed that the concentration of choline in the arterial blood was lower than in the venous outflow (Table 2) (Dross and Kewitz, 1972).

These findings have been confirmed by others (Choi et al., 1975; Aquilonius et al., 1975) and gave rise to the idea that the outflow of choline via the circulation arises from newly formed choline within the brain.

It was known at that time that choline can be synthesized de novo by stepwise N-methylation of phosphatidylethanolamine in the liver using S-adenosylmethionine as the donor (Bremer and Greenberg, 1961). In the meantime this reaction has been shown to occur also in the brain (Mozzi and Porcellati, 1979; Crews et al., 1979; Blusztajn et al., 1979; Blusztajn and Wurtman, 1981; Hattori et al., 1984) and worked out in detail by Massarelli (1991).

On the other side Illingworth and Portman (1972) suggested that lysophosphatidylcholine may be transported from the blood into the brain to deliver choline, which could be easily split off. In order to test this hypothesis a series of experiments have been performed using labelled choline, methionine, ethanolamine or lysophosphatidylcholine, which were administered intravenously and the time courses of the specific radioactivities of choline and its metabolites in blood, brain, liver, and skeletal muscle were followed. As an indicator for the formation of choline in the brain, the ratio between specific radioactivities of choline in the blood and in the brain has been used under steady state conditions (Kewitz and Pleul, 1975; Kewitz et al., 1975).

When choline is formed in the brain, this ratio becomes smaller after labelled methionine, ethanolamine, or lysophosphatidylcholine than after labelled choline, because additional sources of choline will also contain the label.

From the results given table 3, one can see that the ratio was significantly reduced from 6.8 at 3 hrs after labelled choline to 0.67 after labelled methionine, to 0.79 after ethanolamine and to 1.5 after labelled lysophosphatidylcholine.

Table 2. Arterio-venous choline difference, nmol/ml.
lower line: means with SD

Arterial blood	Venous blood		Jugularis vein minus
femoral artery	femoral vein	jugular vein	femoral artery
13.0	12.4	20.2	7.2
12.4	12.7	19.8	7.4
12.1	13.9	16.6	4.5
12.4	10.7	20.8	8.4
12.1	12.5	21.9	9.8
12.6	13.3	19.4	6.8
12.4	13.7	20.1	7.7
11.9	12.7	17.3	5.4
11.4		18.8	7.4
		22.0	
		18.3	
		17.7	
12.3 ± 0.5	12.7 ± 1.0	19.4 ± 1.7	7.2 ± 1.5

Thus, the reply to the questions asked before was very clear. Choline is provided in the brain by both de novo synthesis and transport in the form of lysophosphatidylcholine.

Table 3. Ratios of specific radioactivities (± SD) of choline in blood versus choline in brain 3 and 6 h after i.v. tracer doses of labelled choline, methionine, ethanolamine or lysophosphatidylcholine (lysoPC) in rats

Precursor	3 h	6 h
choline	6.8 (± 0.9)	4.4 (± 0.4)
methionine	0.67 (± 0.05)*	0.98 (± 0.2)**
ethanolamine	0.79 (± 0.1)*	0.76 (± 0.1)**
lysoPC	1.52 (± 0.17)*	

* $p < 0.005$
** $p > 0.001$

The results in Table 4 may help to conclude whether de novo synthesis of choline in the brain is in accord with the well known transmethylation reaction.

Table 4. Ratios of specific radioactivities of choline versus metabolites in the brain of rats 3 h after i.v. bolus administration of tracer doses of labelled precursors

metabolite pecursor	acetyl- choline	phospho- choline	phosphatidyl- choline	glycerophospho- choline
choline	1.1	0.8	10.0	7.3
methionine	1.3	2.0	52.0	22.9
ethanolamine	2.2	1.6	57.0	21.6
lysophosphatidyl- choline	4.4	0.8	9.0	7.5

The results are the ratios between specific radioactivities of choline and of some metabolites in the brain 3 hours after an intravenous bolus injection of tracer doses of various precursors. The two extremes which could have been expected by the transfer of the label due to methylation of ethanolamine could either be a tenfold increase of the ratio between specific activities of choline and phosphatidylcholine or no change in comparison to the experiment with labelled choline. The tenfold increase would have meant that transmethylation had occurred on the level of unesterified ethanolamine and no change would have meant transmethylation would have occurred at the level of phosphatidylethanolamine. Neither of those extremes took place but the ratio showed a fivefold increase which means that almost 50% of the methylation may have occurred at unesterified

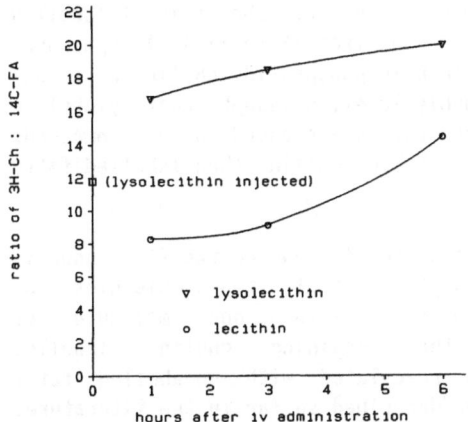

Figure 3. Time course of ratios between ³H and ¹⁴C in lysophosphatidylcholine (lysolecithin) and phosphatidylcholine (lecithin) of rat brain after i.v. injection of doubly labelled lysophosphatidylcholine.

ethanolamine and 50% at phosphatidylethanolamine whereas according to these figures no methylation of phosphoethanolamine may have taken place. Considering lysophosphatidylcholine as the precursor, the only considerable change obtained was an increase of the ratio between choline and acetylcholine from 1.1 to 4.4. Such a change is very unusual and could be explained either by an artifact or by the compartmentalization of choline in so far, that newly formed choline was not available for the synthesis of acetylcholine.

In order to get some additional information and since Pardridge and coworkers (1979) have argued that due to the very high affinity to albumin, 99.8% of the lysophosphatidylcholine in the plasma is tightly bound to protein, which means that 0.2% only would be available in its free and diffusible form, we decided to repeat in rats those experiments which had been performed in squirrel monkeys by Illingworth and Portman (1972). Thus, we injected doubly-labelled lysophosphatidylcholine intravenously and followed in phosphatidylcholine and lysophosphatidylcholine the ratio between ³H in the choline moiety to ¹⁴C in the fatty acid chain over time.

The results in rats (Figure 3) were different from those of Illingworth and Portman in monkeys. First we found in the brain of rats only 0.2% of the administered radioactivity instead of 1.0% in the brain of monkeys according to Illingworth and Portman. In rats the ratio of labelled choline to labelled fatty acids was raised from 12 to 17 in lysophosphatidylcholine and was reduced from 12 to 8 in phosphatidylcholine at 1 hour after the administration of the doubly labelled lysophosphatidylcholine with an initial ratio of 12. Thereafter ratios increased in both compounds which means that both integrated more labelled choline than labelled fatty acids.

The decrease of the ratio in phosphatidylcholine during the first hour is in accordance with the assumption that two molecules of lysophosphatidylcholine were condensed into one molecule of phosphatidylcholine. Concomitantly the recaining choline, labelled glycerophosphocholine may have been reacylated with unlabelled fatty acids, a reaction which has not been described so far in the literature. From these results, both mechanisms may be used for the formation of choline in the brain. The postulated transmethylation of unesterified ethanolamine (Kewitz, 1975) has been shown by Massarelli and coworkers at this meeting.

The transport of lysophosphatidylcholine from the blood into the brain cannot be excluded although the quantity is very low and an artifact due to unbound lysophosphatidylcholine in the short time-span after the bolus injection cannot be excluded. Nevertheless, it seems correct that a very small fraction of the doubly-labelled lysophosphatidylcholine had reached the brain and underwent metabolic reactions in these experiments.
In conclusion it is clear now that choline is taken up by the brain from the circulation but it is also synthesized de novo by N-methyltransferase and S-adenosylmethionine either from phosphatidylethanolamine or from phosphoethanolamine or from unesterified ethanolamine. Evidence has been obtained for the last two reactions in cultured human nerve cells by R. Massarelli and coworkers recently and presented at this symposium. Choline may also be formed by liberation from lysophosphatidylcholine, small proportions of which cannot be excluded to be transported from the blood into the brain. It is of utmost interest to study the mechanisms to keep the very small choline pool in the brain at a constant level.

Considering a turnover time of the choline pool of almost 45 seconds, a very risky situation could be produced by a sudden reduction or a drastic increase of choline within one minute due to rapid changes of the blood circulation or of choline concentrations in the blood. Therefore, several mechanisms may be postulated for that control and among others, one could be based on the principle of an excessive production of choline in combination with overflow mechanisms by which the surplus would be eliminated via blood circulation and which leads to the negative arteriovenous difference of choline across the brain.

References

Aquilonius SM, Ceder G, Lying-Tunell U, Malmlund HO, Schuberth J (1975) The arteriovenous difference of choline across the brain of man. Brain Res 99:430-433

Blusztajn JK, Wurtman RJ (1981) Choline biosynthesis by a preparation enriched in synaptosomes from rat brain. Nature 290:417-418

Blusztajn JK, Zeisel SH, Wurtman RJ (1979) Synthesis of lecithin (phophatidylcholine) from phosphatidylethanolamine in bovine brain. Brain Res 179:319-327

Bremer J, Greenberg DM (1961) A methyl transferring enzyme system of microsomes in the biosynthesis of lecithin (phosphatidylcholine). Biochim Biophys Acta 46:205-216

Choi RL, Freeman JJ, Jenden DJ (1975) Kinetics of plasma choline in relation to turnover of brain choline and formation of acetylcholine. J Neurochem 24:735-741

Crews FF, Hirata F, Axelrod J (1980) Identification and properties of methyltransferases that synthesize phosphatidylcholine in rat brain synaptosomes. J Neurochem 34:1491-98

Dross K, Kewitz H (1966) Der Einbau von i.v. zugeführtem Cholin in das Acetylcholin des Gehirns. Naunyn-Schmiedebergs Arch Pharmacol 255:10

Dross K, Kewitz H (1972) Concentration and origin of choline in the rat brain. Naunyn-Schmiedebergs Arch Pharmacol 274:91-106

Hattori H, Bansal VS, Orihel D, Kanfer JN (1984) Presence of phospholipid-N-methyltransferases and base-exchange enzymes in rat central nervous system axolemma-enriched fractions. J Neurochem 43:1018-1024

Hattori H., Kanfer JN (1984) Synaptosomal phospholipase D: potential role in providing choline for acetylcholine synthesis. Biochem Biophys Res Commun 24:945-949

Illingworth DR, Portman OW (1972) The uptake and metabolism of plasma lysophosphatidylcholine in vivo by the brain of squirrel monkeys. Biochem J 130:557-567

Kanfer JN (1972) Base exchange reactions in the phospholipids in rat brain particles. J Lipid Res 13:468-476

Kanfer JN (1982) The base exchange enzymes and phospholipase D of rat brain microsomes. In: Horrocks, L.A., Ansell, G.B., and Porcellati, G., (eds) Phospholipids in the nervous system, Vol I. Metabolism. Raven Press, New York

Kewitz H, Dross K and Pleul O (1973) Choline and its metabolic successors in brain. In: Genazzani E, Herken H (ed) Central Nervous system. Studies on metabolic regulation and function. Springer Verlag, Berlin Heidelberg New York

Kewitz H, Pleul O (1975) Synthesis of choline from ethanolamine in rat brain. Proc Natl Acad Sci U.S.A. 73:2181-2185

Kewitz H, Pleul O, Dross K, Schwarzkopf T (1975) The supply of choline in rat brain. In: Waser P (ed) Cholinergic Mechanisms, Raven Press New York

Massarelli R, Dainou F, Louis JC, Dreyfus H, Mozzi R, Porcellati G, Freysz L (1982) Choline synthesis in neurons and glial cells. In: Horrocks, L.A., Ansell, G.B., and Porcellati, G. (ed) Phospholipids in the Nervous System, Vol.I Metabolism. Raven Press New York, p 348

Mozzi R, Goracci G, Siepi D, Francescangeli E, Andreoli V, Horrocks LA, Porcellati G (1982) Phospholipid synthesis by interconversion reactions in brain tissue. In: Horrocks, L.A., Ansell, G.B., and Porcellati, G. (ed) Phospholipids in the Nervous System, Vol. I Metabolism. Raven Press New York, p 1-12

Mozzi R, Porcellati G (1979) N-methylation of phosphatidyl ethanolamine in rat brain. FEBS Lett 100:363-366

Pardridge WM, Cornford EM, Braun LD, Oldendorf WH (1979) Transport of choline and choline analogues through the blood-brain barrier. In: Barbeau A, Growdon JH, Wurtman RJ (ed) Nutrition and the brain, Vol 5, Raven Press, New York

Saito M, Kanfer JN (1975) Phosphatidohydrolase activity in a solubilized preparation from rat brain particulate fraction. Arch Biochem Biophys 169:318-323

Zeisel SH (1985) Formation of unesterified choline by rat brain. Biochim Biophys Acta 835:331-343

THE HOMEOSTASIS OF BRAIN CHOLINE

K. Löffelholz, J. Klein, A. Köppen and J. Schmitthenner
Department of Pharmacology
University of Mainz
Obere Zahlbacher Str. 67
W-6500 Mainz
Germany

Abstract

The interest in the homeostasis of brain choline is reinforced by the role of choline as immediate precursor of acetylcholine, phosphatidylcholine and other phospholipids in the brain. In order to obtain a comprehensive view of the mechanisms of homeostasis it appeared necessary to elucidate the negative arterio-venous difference of choline across the brain (net release), a phenomenon that has been known for 20 years and is present in mammals and in man. This finding prompted an intense search for a de novo synthesis of choline in the brain. We detected in anaesthetized rats a reversal of the net release into a net uptake (positive arterio-venous difference), when the plasma level of choline was elevated spontaneously (presumably due to dietary intake) or after i.p. injection of choline. In these experiments, the newly taken up choline was rapidly removed from the extracellular space of the brain by cellular uptake and subsequent phosphorylation. The newly generated phosphocholine was trapped in brain cells and was only slowly incorporated into phosphatidylcholine, which in turn appears to serve as a reservoir for free choline. Surplus choline in a free or bound form may be cleared from the brain by net release.

NATO ASI Series, Vol. H 70
Phospholipids and Signal Transmission
Edited by R. Massarelli, L. A. Horrocks,
J. N. Kanfer, and K. Löffelholz
© Springer-Verlag Berlin Heidelberg 1993

Introduction: The sources of brain choline

Although the choline moiety is biosynthesized de novo via sequential methylation of phosphatidylethanolamine, choline is considered as an essential nutrient in mammals and man (Zeisel et al., 1991). The needs of the tissues for choline are met from both metabolic and dietary sources. Hence, the plasma choline level which is basically about 10 μM increases up to 20 μM or more upon absorption of choline or lecithin from the gut (Zeisel, 1981). Nevertheless, these fluctuations are small when we realize the large amounts of choline bound in phospholipids of blood plasma and of cell membranes. The plasma concentration of phosphatidylcholine is about 100-fold higher than that of free choline.

Until 1971, it was erroneously believed that choline does not penetrate the blood-brain barrier (BBB). In that year, Ivan Diamond provided evidence for a rapid and specific transport of labelled choline into the mouse brain. He reported that this "carrier-mediated process" is saturated at choline concentrations several times the basic plasma level. After entry into brain, choline appeared to be metabolized quickly into phosphocholine and acetylcholine (ACh) and was gradually incorporated into membrane phospholipids (Diamond, 1971).

In the following decade, Diamond's conclusions were fully confirmed by more direct evidence of a BBB choline transporter characterized by a K_m of 442 μM and a V_{max} of 10.0 nmol/g min (Cornford et al., 1978). As predicted, dietary fluctuations of the plasma choline level should lead to parallel fluctuations in the uptake of choline into the brain. The newly taken up choline was, indeed, found to be phosphorylated, as the enzyme catalyzing phosphorylation (choline kinase, EC 2.7.1.32; ATP:choline transferase) also is unsaturated under physiological conditions (Millington and Wurtman, 1982).

One year after Diamond's important findings Dross and Kewitz (1972) observed that there is a net release of choline from the brain into the blood, i.e. more choline leaves the brain than enters it. This phenomenon maintains a negative arterio-venous difference (AVD) of brain choline. In the following, the negative AVD of anaesthetized rats was confirmed (Freeman et al., 1975; Brehm et al., 1985, 1987; Klein et al., 1990) and was also shown in conscious rats (Choi et al., 1975) and in conscious man (Aquilonius et al., 1975).

The negative AVD which indicated a net release rate of several nmol/g min in rats (Dross and Kewitz, 1972; Klein et al., 1990) raised the question of the source of this choline. The answer to this question has been considered crucial for the understanding of the homeostasis of brain choline and thereby for the biosynthesis of phospholipids and of ACh. Three hypotheses have been, and still are, discussed. (1) De novo synthesis of the choline moiety. (2) Uptake of bound choline into the brain, e.g. as lysophosphatidylcholine or phosphatidylcholine, and subsequent hydrolysis yielding choline. (3) Net release of brain choline is a transient phenomenon and is balanced by net uptake. This hypothesis is the result of our own experimental work during the last two years (Klein et al., 1990, 1991). In the following, these hypotheses are briefly discussed.

1. De novo-synthesis of choline in the rat brain by sequential methylation of ethanolamine has been suggested by Kewitz and Pleul (1976). These authors injected intravenously [^{14}C]ethanolamine and detected some radioactivity in brain choline. As the specific radioactivity of brain choline was higher than that of phosphocholine at all time-points after injection of ethanolamine, the authors concluded that choline was synthesized directly from ethanolamine via methylation. Recently this pathway was confirmed in vivo by intraventricular injection of [^{3}H]ethanolamine and in vitro using fetal rat neurons in culture (Andriamampandry et al., 1989, 1990). According to these studies, methylation of phospho-ethanolamine to phosphomonomethylethanolamine, phosphodimethylethanolamine

and finally to phosphocholine was catalyzed by a cytosolic N-methyltransferase that is present in rat brain. This pathway may contribute to the large phosphocholine pool of the brain (308 nmol/g; Klein et al., in press). The physiological importance of this pathway is still unknown. It will be necessary to estimate its quantitative significance for de novo synthesis and the homeostasis of brain choline.

De novo synthesis of the choline moiety of phosphatidylcholine by sequential N-methylation of phosphatidylethanolamine has been known for more than 10 years (for references see Tucek, 1988). However, the rate of choline formation (within phosphatidylcholine) was found to be as low as 6 pmol/g min (Blusztajn and Wurtman, 1979). Thus this pathway does not appear to contribute essentially to the net release of brain choline (several nmol/g min; see above), as determined from the negative AVD.

2. Uptake of bound choline in the form of phospholipids into the brain has been suggested very early (Hoelzl and Franck, 1969; Ansell and Spanner, 1971; Illingworth and Portman, 1972). In fact, these research groups obtained evidence that labelled phosphatidylcholine (Hoelzl and Franck, 1969) or lysophosphatidyl-choline (Illingworth and Portman, 1972), when injected intravenously, appeared in the brain homogenate to a small extent. The relevance of these data in the sense of a possible role of circulating phospholipids as precursors of brain choline is rather doubtful, as, in the blood, phospholipids are very tightly bound to albumin and/or are incorporated in lipoproteins and the permeability of the BBB for these molecules is similar to known nondiffusible compounds such as dextran (Pardridge et al., 1979).

3. Net release of choline is balanced by net uptake. Under normal experimental conditions, the anaesthetized rat exhibits a net release of choline from the brain, as stated above. However, we found that about 10% of the animals show positive AVD values indicating net uptake of brain choline; in these rats, the

plasma level was found to be higher than 15 µM (Klein et al., 1990). Intraperitoneal injection of choline not only elevated the plasma level of choline, but
reversed the negative AVD of brain choline to positive values. Both observations
indicated that choline is taken up into the brain at elevated plasma levels
("reversal point" about 20 µM) and is released, when the plasma level returns to
the basic (fasting) level of about 10 µM (see above). We hypothesized that uptake
and release of brain choline may be balanced depending on the dietary intake of
choline-containing food and, consequently, on the fluctuations of the plasma
choline level.

The presumptive regulation of brain choline movements may indicate that
circulating choline is the major source of brain choline. However, most newly
taken up choline is not immediately available in the free form, because
essentially all of the intraperitoneally injected, unlabelled choline had been
shown to be rapidly metabolized after its uptake into the brain of anaesthetized
rats (Klein et al., 1990, in press). It is possible that phospholipids serve as a
reservoir for free choline (see Discussion). In the present study, we analyzed
the metabolism of unlabelled and labelled choline taken up into the rat brain.

Methods

Male Wistar rats (150-250g) were anaesthetized with pentobarbital (40-80 mg/kg,
i.p.) and injected intraperitoneally with unlabelled and tritiated choline
chloride (60 mg/kg containing 50 µCi [^{3}H]choline; Dupont NEN, Dreieich, Germany).
Anaesthesia was carried out 10 minutes before the animals were sacrificed and at
various times after administration of labelled choline (10 and 30 min; 2 and 24
hours). Total brain tissue (1.54 g) was rapidly removed and homogenized in
methanol:chloroform 2:1 after addition of [^{14}H]choline and [^{14}H]phosphocholine as
internal standards. Following the addition of water and chloroform according to
Bligh and Dyer (1959), the homogenate was separated into a hydrophilic and a
lipophilic phase by centrifugation.

The hydrophilic extract of the brain homogenate was lyophilized, taken up in 1 ml water and separated by TLC essentially according to Yavin (1976). Whatman LK6 plates were spotted with extract and with external standards of choline, phosphocholine and glycerophosphocholine and were developed using 1.5% NaCl/MeOH/NH$_3$ 10:10:1. Spots were detected with iodine vapor and scraped off. Choline (R$_f$=0.18) was separated from silica in 50% methanol/water, taken to dryness and analyzed as described below. Phosphocholine (R$_f$=0.50) was digested for one hour with 1.25 units bovine alkaline phosphatase (Sigma, Deisenhofen, Germany) in 50 mM glycine buffer, pH 9.5. After addition of two volumes of ice-cold ethanol, the incubation mixtures were centrifuged, and aliquots of the supernatant were dried and analyzed as choline. The dried extracts were dissolved in eluent and analyzed by HPLC. Choline was separated on a cation-exchange column (Baker Sulfopropyl 5 μm, 60 x 4.6 mm) using 0.1 M Na phosphate buffer, pH 7.5, with 5 mM tetramethylammonium chloride as eluent. Choline was oxidized to hydrogen peroxide by choline oxidase which was immobilized in a reactor column (Biometra, Göttingen, Germany), and hydrogen peroxide was detected with an electrochemical detector equipped with a platinum electrode operating at +0.5 V. The detection limit of the analytical system was less than 1 pmol. Corrections for recoveries were made using external standards. Total recoveries were 78% for choline and 36% for phosphocholine under these conditions. Glycerophosphocholine (R$_f$=0.61) which was sometimes found to overlap with phosphocholine after TLC separation was not hydrolyzed by alkaline phosphatase in significant amounts (<5%).

The lipophilic extract of the brain homogenate was separated by TLC (Merck plates N°5721) using MeOH/CHCl$_3$/NH$_3$ 18:22:1. Phosphatidylcholine, lysophosphatidylcholine and sphingomyelin (Sigma, Deisenhofen, Germany) were used as external standards. After detection by iodine vapour, spots were scraped off, digested with 6M methanolic HCl for one h with the exception of sphingomyelin (24 h), taken to dryness and analyzed as choline as described above.

Results are presented as means $\pm$ SEM from at least 4 independent experiments. Significance was evaluated with Student's t-test.

Results: Choline metabolism in the brain.

A. Administration of unlabelled choline.

Under control conditions, the rat brain contained 82 $\pm$ 11 nmol/g choline, 308 $\pm$ 58 nmol/g phosphocholine, 19.07 $\pm$ 0.90 μmol/g phosphatidylcholine, 4.53 $\pm$ 0.51 μmol/g sphingomyelin and 0.157 $\pm$ 0.019 μmol/g lysophosphatidylcholine (N=9).

Intraperitoneal injection of 60 mg/kg choline chloride caused an increase of choline and phosphocholine, whereas the choline-containing phospholipids were unchanged. The increase of phosphocholine (+477 or +727 nmol/g after 10 or 30 min, respectively) was up to 10 times greater than the increase of free choline (+71 or +73 nmol/g after 10 or 30 min, respectively). The average net uptake of choline, as determined from the positive AVD (Klein et al., 1990), was 409 and 567 nmol/g of brain tissue during 10 and 30 min, respectively. It is concluded that essentially all of the newly taken up choline was rapidly phosphorylated.

After 2 hours, more than half of the newly formed phosphocholine was still present in the brain.

B. Administration of labelled choline.

After intraperitoneal injection of 60 mg/kg [³H]choline, radioactivity rapidly appeared in the brain homogenate. The maximum was reached after 30 min and this level was maintained for at least 24 hours. Similar to the above findings on unlabelled choline and its metabolites, [³H]choline accounted for less than 10% of

the rapidly taken up labelled material, whereas the appearance of [³H]phospho-
choline (>90%) was not significantly different from the total uptake of tritium.

Two hours after the administration of [³H]choline, the label in the brain was
equally distributed among phosphocholine and phosphatidylcholine. The label was
shifted gradually to phosphatidylcholine within 24 hours and finally reached 80%
which is almost exactly the share of phosphatidylcholine of the total brain
choline pool. The incorporation of choline into sphingomyelin and
lysophosphatidylcholine was significant, but little above the limit of detection.

Discussion: The homeostasis of brain choline.

The neurochemical interest in choline is due to its role as precursor of
acetylcholine and of phospholipids. Therefore it is important to study choline
availability in the brain and the mechanisms that regulate the concentrations of
free choline. A simplified representation of the choline compartments is shown in
Figure 1.

The choline content of the whole rat brain averages about 25 μmol/g (Dross and
Kewitz, 1972; Tucek, 1988). However, choline is unequally distributed between the
extra- and intracellular compartments. The extracellular choline concentration of
the brain, as reflected by the choline concentration of the cerebrospinal fluid of
the cisterna magna or the ventricles, appears to be only about 6 μM (Schuberth and
Sundwall, 1971; Schuberth and Jenden, 1975; Brehm et al., 1987; Klein et al.,
1990), whereas the intracellular choline concentration has been estimated to be
50 μM (Tucek, 1988), the gradient being 50:6. The electrochemical gradient across
the brain cell membrane may represent the driving force for the transport of
choline through the membrane.

There is also a concentration gradient (in μM) across the BBB, i.e. between blood
plasma (10 μM) and extracellular space of the brain (6 μM). The low extracellular

choline concentration is probably caused by a relatively efficient transport of choline out of the cerebrospinal fluid back into the blood. The active outward transport mechanisms may be located in the choroid plexus (Lanman and Schanker, 1980) and in the arachnoid (Ehrlich and Wright, 1982).

Obviously, there is a constant exchange of choline between brain and blood plasma as shown by unidirectional movements of labelled choline (see Introduction). However, only recently has the quantitative balance of brain choline movements been studied (Klein et al., 1990, 1991). We found that small increases of the plasma choline led to a net uptake of choline into the brain of anaesthetized rats. Intraperitoneal injection of 60 mg/kg choline chloride drastically increased the plasma level of choline, which in turn caused a reversal of the negative AVD of brain choline into the positive range indicating a very prominent net uptake of choline (see Introduction). From this observation we calculated the total amount of choline uptake under this experimental condition (Klein et al., 1990). This uptake of choline was not significantly different from the increase of phosphocholine content of the brain. The increase of free choline in the brain was negligible in comparison with the increase in phosphocholine. It is concluded that administration of choline leads not only to an increase of the plasma choline level (see Introduction), but also to a net uptake of choline into the brain of the anaesthetized rat. The newly taken up choline was found to be phosphorylated rapidly and quantitatively.

The fate of the phosphocholine formed after uptake of choline into the brain is only partially known. The present experiments demonstrate that a large fraction of the newly formed [^{14}H]phosphocholine is slowly incorporated into phospholipids.

What happens to the surplus choline that is phosphorylated or incorporated into phospholipids? We found that the positive AVD of brain choline, which was observed shortly after injection of choline (indicating net uptake of choline), was reversed back to negative values (i.e., back to net release) as early as 30 min after the injection. At the same time, the plasma level was approaching the fasting level of about 10 μM, while the phosphocholine content (labelled and unlabelled) in the brain had reached its maximum. After this 30 min-period, phosphocholine declined very slowly mostly because of incorporation into phosphatidylcholine. It is likely that the surplus of bound choline was finally cleared from the brain in the free form. It has been observed, indeed, that the strongest net release of brain choline occurred 30 min after administration of choline ("afternegativity" of AVD; Klein et al., 1990). The release of phosphocholine itself, if it occurs, does not seem to play a quantitatively important role (Klein and Gonzales, unpublished).

Newly formed phosphocholine appears to be trapped in the cytosol of brain cells for many hours. Therefore the question was raised whether a certain type of brain cell is involved in the removal of extracellular surplus choline and the subsequent trapping in the form of phosphorylcholine. A preferential uptake of labelled choline has been observed for glial cells in the rabbit (Francescangeli et al., 1977) and recently in the leech nervous system (Wuttke and Pentreath, 1990).

The average turnover of phosphatidylcholine in the brain is in the range of days. It seems, however, that choline can be released from phospholipids through the activity of various phospholipases and base exchange enzymes. The activity of these enzymes may be enhanced by pathophysiological factors and also by receptor stimulation. For example, choline is mobilized by stimulation of muscarinic receptors. This has been demonstrated in vitro using striatal slices (Dolezal and Tucek, 1984) and cortical synaptosomes and membranes (Qian and Drewes, 1989, 1990). Furthermore we have detected a muscarinic mobilization of brain choline in vivo in the rat cortex and whole brain (Corradetti et al., 1983; Brehm et al.,

1987). The fact that acetylcholine mobilizes its own precursor prompted the hypothesis of a positive feedback regulation of acetylcholine synthesis in brain (Löffelholz, 1989, 1990).

In conclusion, <u>homeostasis of free choline</u> in the brain is maintained, firstly, by a balance between uptake of circulating choline into the brain and release of choline back into the circulation. Secondly, cellular (glial?) uptake with subsequent trapping in the form of phosphocholine is certainly the major intracerebral mechanism of homeostasis. Finally, it must be noted that the mechanisms delineated above for the whole brain may be complemented by regional factors which would not have been detected in our studies. Thus, the extracellular choline concentration in the vicinity of cholinergic neurons may be essentially influenced by neuronal activity, i.e., by release of ACh and acceleration of neuronal uptake of choline.

Acknowledgements. This work was supported by the Deutsche Forschungsgemeinschaft.

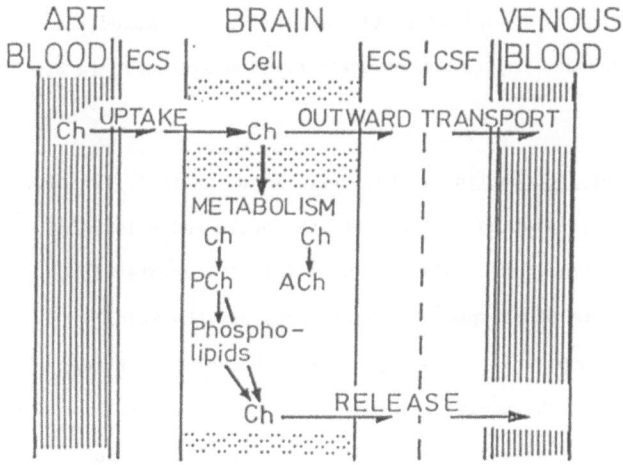

Fig.1. Choline kinetics in the brain, a schematic representation. Choline (Ch) is transported from the arterial (art) blood plasma by a carrier-mediated mechanism through the BBB into the extracellular space (ECS). Ch is rapidly cleared from the ECS by two mechanisms: cellular <u>uptake</u> and metabolism <u>or</u> diffusion into the cerebrospinal fluid (CSF) compartment with subsequent active <u>outward transport</u> back into the circulation. Rapid phosphorylation to phosphorylcholine (PCh) followed by slow incorporation into phospholipids seems to be the major metabolic mechanism of Ch homeostasis in brain. Acetylation of Ch to form acetylcholine (ACh) may be important for the extracellular Ch concentration only in regions of cholinergic innervation. Net release of brain choline (negative AVD) was observed at low plasma levels (fasting conditions); elevation of plasma choline level lead to a transient net uptake of choline (positive AVD).

References

Andriamampandry C, Freysz L, Kanfer JN, Dreyfus H, Massarelli R (1989) Conversions of ethanolamine, monomethylethanolamine and dimethylethanolmine to choline-containing compounds by neurons in culture and by the rat brain. Biochem J 264:555-562

Andriamampandry C, Massarelli R, Freysz L, Kanfer JN (1990) A rat brain cytosolic N-methyltransferase(s) activity converting phosphorylethanolamine into phosphorylcholine. Biochem Biophys Res Commun 171:758-763

Ansell GB, Spanner S (1971) Studies on the origin of choline in the brain of the rat. Biochem J 122:741-750

Aquilonius SM, Ceder G, Lying-Tunell U, Malmlund HO, Schuberth J (1975) The arteriovenous difference of choline across the brain of man. Brain Res 99: 430-433

Bligh EG, Dyer WJ (1959) A rapid method of total lipid extraction and purfication. Can J Biochem Physiol 37:911-917

Blusztajn JK, Zeisel SH, Wurtman RJ (1979) Synthesis of lecithin (phosphatidylcholine) from phosphatidylethanolamine in bovine brain. Brain Res 179:319-327

Brehm R, Corradetti R, Krahn V, Löffelholz K, Pepeu G (1985) Muscarinic mobilization of choline in rat cerebral cortex does not involve alterations of blood-brain barrier. Brain Res 345:306-314

Brehm R, Lindmar R, Löffelholz K (1987) Muscarinic mobilization of choline in rat brain in vivo as shown by the cerebral arterio-venous difference of choline. J Neurochem 48:1480-1485

Choi RL, Freeman JJ, Jenden DJ (1975) Kinetics of plasma choline in relation to turnover of brain choline and formation of acetylcholine. J Neurochem 24: 735-741

Cornford EM, Braun LD, Oldendorf WH (1978) Carrier mediated blood-brain barrier transport of choline and certain choline analogs. J Neurochem 30:299-308

Corradetti R, Lindmar R, Löffelholz K (1983) Mobilization of cellular choline by stimulation of muscarine receptors in isolated chicken heart and rat cortex in vivo. J Pharmacol Exp Ther 226:826-832

Diamond I (1971) Choline metabolism in brain. Arch Neurol 24: 333-339

Dolezal V, Tucek S (1984) Activation of muscarinic receptors stimulates the release of choline from brain slices. Biochem Biophys Res Commun 120:1002-1007

Dross K, Kewitz H (1972) Concentration and origin of choline in the rat brain. Naunyn-Schmiedeberg's Arch Pharmacol 274:91-106

Ehrlich BE, Wright EM (1982) Choline and PAH transport across blood-CSF barriers: the effect of lithium. Brain Res 250: 245-249

Francescangeli E, Goracci G, Piccinin GL, Mozzi R, Woelk H, Porcellati G (1977) The metabolism of labelled choline in neuronal glial cells of the rabbit in vivo. J Neurochem 28:171-176

Freeman JJ, Choi RL, Jenden DJ (1975) Plasma choline: its turnover and exchange with brain choline. J Neurochem 24:729-734

Hoelzl J, Franck HP (1969) In: Proc 2nd Int Meet Internat Soc Neurochem, (Paoletti R, Fumagalli R, Galli C, eds), p 219, Tamburini Editore Milan

Illingworth DR, Portman OW (1972) The uptake and metabolism of plasma lysophosphatidylcholine in vivo by the brain of squirrel monkeys. Biochem J 130:557-567

Kewitz H, Pleul O (1976) Synthesis of choline from ethanolamine in rat brain. Proc Natl Acad Sci USA 73:2181-2185

Klein J, Köppen A, Löffelholz K (1990) Small rises in plasma choline reverse the negative arteriovenous difference of brain choline. J Neurochem 55:1231-1236

Klein J, Köppen A, Löffelholz K (1991) Uptake and storage of choline by rat brain: influence of dietary choline supplementation. J Neurochem 57:370-375

Klein J, Köppen A, Löffelholz K, Schmitthenner J (1992) Uptake and metabolism of choline by rat brain after acute choline administration. J Neurochem, in press

Lanman RC, Schanker LS (1980) Transport of choline out of the cranial cerebrospinal fluid spaces of the rabbit. J Pharmacol Exp Ther 215:563-568

Löffelholz K (1989) Receptor regulation of choline phospholipid hydrolysis. Biochem Pharmacol 38:1543-1549

Löffelholz K (1990) Receptors linked to hydrolysis of choline phospholipids: the role of phospholipase D in a putative mechanism of signal transduction. In: Current Aspects of the Neurosciences, vol 2 (Osborne NN, ed), pp 49-76, MacMillan Press London

Millington WR, Wurtman RJ (1982) Choline administration elevates brain phosphoryl-choline concentrations. J Neurochem 38:1748-1752

Pardridge WM, Cornford EM, Braun LD, Oldendorf WH (1979) Transport of choline and choline analogues through the blood-brain barrier. In: Nutrition and the Brain, vol 5 (Barbeau A, Growdon JH, Wurtman RJ, eds), pp 25-34, Raven Press New York

Qian Z, Drewes LR (1989) Muscarinic acetylcholine receptor regulates phosphatidyl-choline phospholipase D in canine brain. J Biol Chem 264:21720-21724

Qian Z, Drewes LR (1990) A novel mechanism for acetylcholine to generate diacyl-glycerol in brain. J Biol Chem 265:3607-3610

Schuberth J, Jenden DJ (1975) Transport of choline from plasma to cerebrospinal fluid in the rabbit with reference to the origin of choline and to acetylcholine metabolism in brain. Brain Res 84:245-256

Schuberth J, Sundwall A (1971) A method for the determination of choline in biological materials. Acta Pharmacol 29(S4):51

Tucek S (1988) Choline acetyltransferase and the synthesis of acetylcholine. In: The Cholinergic Synapse, Handbook of Experimental Pharmacology, vol 86 (Whittaker VP, ed), pp 125-165, Springer Verlag Berlin, Heidelberg, New York

Wuttke WA, Pentreath VW (1990) Evidence for the uptake of neuronally derived choline by glial cells in the leech central nervous system. J Physiol 420: 387-408

Yavin E (1976) Regulation of phospholipid metabolism in differentiating cells from rat brain cerebral hemisphere in culture. J Biol Chem 251:1392-1397

Zeisel SH (1981) Dietary choline: biochemistry, physiology, and pharmacology. Ann Rev Nutr 1:95-121

Zeisel SH, Da Costa KA, Franklin PD, Alexander EA, Lamont JT, Sheard NF, Beiser A (1991) Choline, an essential nutrient for humans. FASEB J 5:2093-2098

TACRINE (TETRAHYDROAMINOACRIDINE) AND THE METABOLISM OF ACETYLCHOLINE AND CHOLINE

Stanislav Tuček and Vladimír Doležal
Institute of Physiology
Czechoslovak Academy of Sciences
142 20 Prague
Czechoslovakia

Tacrine (9-amino-1,2,3,4-tetrahydroacridine) is a reversible noncompetitive inhibitor of cholinesterases (Shaw and Bentley, 1953; Heilbronn, 1961; Patocka et al., 1976; Bajgar et al., 1979; Marquis, 1990) which has been described to have a marked positive effect in the treatment of patients with Alzheimer's disease (Summers et al., 1986). Since the attempts to treat such patients with other potent inhibitors of cholinesterases had been less successful or not successful at all, the possibility appears likely that tacrine acts not only by inhibiting cholinesterases, but also by producing some additional pharmacological effects.

Several mechanisms have indeed been characterized in recent years by which tacrine may influence the activity of the nerve cells, in addition to the inhibition of cholinesterases. The most important among them seem to be the binding of tacrine to muscarinic receptors (Perry et al., 1988; Pearce and Potter, 1988; Flynn and Mash, 1989; Potter et al., 1989; Hunter et al., 1989; Musilkova and Tucek, 1991) associated with their inhibition, the blockade by tacrine of certain types of ion channels (Osterrieder, 1987; Rogawski, 1987; Drukarch et al., 1987; Stevens and Cotman, 1987; Elinder et al., 1989; Marsh et al., 1990; Griffith and Sim, 1990), and alterations in the release or uptake of several neurotransmitters (Drukarch et al., 1988; de Belleroche and Gardiner, 1988; Robinson et al., 1989; Gardiner and Belleroche, 1990).

Investigations performed in our laboratory were mainly devoted to the question whether tacrine has direct effects on the synthesis, storage and release of acetylcholine (ACh) in brain cholinergic neurons and on the metabolism of choline in the brain (Dolezal and Tucek, 1991, and unpublished observations; Tucek and Dolezal, 1991).

NATO ASI Series, Vol. H 70
Phospholipids and Signal Transmission
Edited by R. Massarelli, L. A. Horrocks,
J. N. Kanfer, and K. Löffelholz
© Springer-Verlag Berlin Heidelberg 1993

METHODS

Experiments were performed on cerebrocortical prisms prepared from Wistar-type white rats of both sexes (160-220 g body weight) using McIlwain's tissue slicer set at 0.35 mm slice thickness, and chopping the tissue in two perpendicular directions. Since we wanted to see the effects of tacrine other than (additional to) those caused by the inhibition of cholinesterases, the prisms were preincubated for 1 h in the presence of 58 μmol/l paraoxon, which produced an irreversible and complete inhibition of cholinesterases. In the usual arrangement of experiments, the prisms were first washed three times, using mild centrifugation, then preincubated for 60 min in standard Krebs-Ringer buffer (KRB) containing paraoxon, washed again, resuspended in appropriate incubation buffer (variable as to the content of K^+ and of drugs), and incubated for 30 min at 38°C. Standard KRB contained (mmol/l) NaCl 123, KCl 3, $CaCl_2$ 1.3, $MgCl_2$ 1, Na_2HPO_4 1.2, $NaHCO_3$ 25, and glucose 5. The concentrations of ACh and choline in the tissue and incubation media were determined radioenzymatically (Goldberg and McCaman, 1973; Dolezal and Tucek, 1982, 1983, 1988). In some experiments, the preincubation or incubation proceeded in the presence of [14C]choline and the fate of label was followed. To distinguish between the radioactivity present in [14C]ACh and in [14C]choline, tissue extracts or incubation media were incubated either with choline oxidase (removing choline but leaving ACh) or with choline oxidase and acetylcholinesterase (removing both choline and ACh), and quaternary amines were then extracted with tetraphenylboron (Fonnum, 1969; Boksa and Collier, 1980).

EFFECTS OF TACRINE ON THE SYNTHESIS AND RELEASE OF ACh DURING INCUBATION UNDER RESTING CONDITIONS

In cortical prisms incubated for 30 min in the presence of 100 μmol/l tacrine, the content of ACh in the prisms and the amount of ACh released into the medium were increased; the amount of ACh synthesized during the incubation was augmented 2.6 times (Table 1). A significant positive effect of tacrine on the content of ACh in the tissue and on the synthesis of ACh was already apparent at 10 μmol/l tacrine.

In cortical prisms incubated for 30 min in the presence of 100 μmol/l tacrine and 1.85 μmol/l [14C]choline, the content of [14C]ACh in the tissue and the amount of [14C]ACh released into the medium were strongly diminished, and so was the total amount of [14C]ACh synthesized during incubation (Table 2).

If, however, the prisms were preincubated for 60 min with [^{14}C]choline and only afterwards incubated for 30 min with 100 μmol/l tacrine, the content of [^{14}C]ACh recovered in the tissue and in the tissue plus medium at the end of incubation was augmented (Table 3); consequently, more [^{14}C]ACh was synthesized from endogenous [^{14}C]choline during the 30 min incubation.

These data indicate that tacrine increases the amounts of ACh synthesized in the prisms, stored in them and released from them during incubation under 'resting' conditions. At the same time, tacrine inhibits the uptake of choline from the medium (Table 2) and thereby the synthesis of ACh from the choline which had been freshly supplied to the nerve terminals via high-affinity choline carriers. This observation accords with data by Buyukuysal and Wurtman (1989) indicating that tacrine inhibits high-affinity choline carriers. It seems apparent that tacrine promotes the synthesis of ACh from an intraterminal source of choline, circumventing the need for high-affinity carrier-mediated supply of this substrate.

TABLE 1. Effect of tacrine (100 μmol/l) on the content of ACh in cerebrocortical prisms and on its release and synthesis during incubation at 3 mmol/l K$^+$.

	Control	Tacrine	% Change
Content (pmol/mg prot.)	473.9 ± 21.3	653.0 ± 23.4	+ 38
Release (pmol/mg prot./30 min)	128.9 ± 6.7	174.6 ± 11.1	+ 35
Synthesis (pmol/mg prot./30 min)	137.2 ± 19.8	358.6 ± 22.7	+161

TABLE 2. Effect of tacrine (100 μmol/l) on the content of [^{14}C]ACh in cerebrocortical prisms and on its release and synthesis during incubation at 3 mmol/l K$^+$, in the presence of 1.85 μmol/l [^{14}C]choline

	Control	Tacrine	% Change
Content (pmol/mg prot.)	9.59 ± 0.38	5.21 ± 0.73	- 46
Release (pmol/mg prot./30 min)	1.59 ± 0.04	0.51 ± 0.15	- 68
Synthesis (pmol/mg prot./30 min)	11.18 ± 0.40	5.73 ± 0.72	- 49

EFFECT OF TACRINE ON THE SYNTHESIS OF ACh IN THE PRESENCE OF HEMICHOLINIUM-3

Hemicholinium-3 (HC-3) is a strong and specific inhibitor of choline carriers in neuronal membranes. We wanted to see if tacrine retains its positive effect on the synthesis of ACh in cortical prisms in the case that the high-affinity choline carriers are blocked by 10 μmol/l HC-3. It is apparent from Table 4 that this was so: while the synthesis of ACh was diminished by 81% during incubation with HC-3 alone, it was augmented by 81% if HC-3 and tacrine were present simultaneously.

EFFECT OF TACRINE ON THE SYNTHESIS AND RELEASE OF ACh UNDER CONDITIONS STIMULATING THE RELEASE OF ACh

In prisms incubated in the presence of a depolarizing concentration (50 mmol/l) of K^+ ions, 100 μmol/l tacrine had no effect on the content of ACh in the tissue at the end of 30 min incubation, but the amount of ACh released into the medium and the amount of ACh synthesized during the incubation were diminished (Table 5).

In prisms incubated with 50 mmol/l K^+ and [^{14}C]choline, the content of [^{14}C]ACh in the tissue and in the medium and the synthesis of [^{14}C]ACh occurring during the incubation period were diminished both at 100 and at 10 μmol/l tacrine (data not shown). In experiments in which the prisms were exposed to 0.1 mM 4-aminopyridine (stimulating the release of ACh) during incubation, 100 μmol/l tacrine diminished the release of ACh while the content of ACh in the tissue and its synthesis were not substantially changed.

It is a question whether the negative effect of tacrine on the synthesis and release of ACh was completely due to the impairment of ACh synthesis, caused by the inhibition of the transport of choline into the nerve terminals, or whether tacrine also exerted a direct inhibitory action on the mechanism of ACh release. To distinguish between these possibilities, we performed experiments with short depolarizations. Brain prisms were preincubated at 3 mmol/l K^+ with [^{14}C]choline (enabling the formation of [^{14}C]ACh in the prisms), and then incubated in the presence of 3 mmol/l K^+, tacrine and HC-3; after 10 min, the concentration of K^+ was raised to 50 mmol/l and the incubation was stopped 5 min later (Table 6). Under these conditions, depolarization strongly diminished the store of [^{14}C]ACh in the tissue, while it augmented the amount of [^{14}C]ACh in the medium. In the presence of 100 μmol/l tacrine, however, the depletion of [^{14}C]ACh from the tissue was prevented, while the amount of [^{14}C]ACh released was diminished. It is apparent that the inhibition of K$^+$-stimulated ACh release by tacrine was not due to a lack of ACh in the terminals (its insufficient synthesis) but, rather, to a more direct interference of tacrine with the mechanism responsible for ACh release.

TABLE 3. Effect of tacrine (100 μmol/1) on the content of [^{14}C]ACh in cerebrocortical prisms and on its release and total content in the tissue and medium in experiments in which [^{14}C]ACh had been preformed during 60 min preincubation with [^{14}C]choline.

	Control	Tacrine	% Change
Content (pmol/mg prot.)	20.50 ± 0.88	26.16 ± 0.83	+ 28
Release (pmol/mg prot./30 min)	2.29 ± 0.21	2.66 ± 0.24	+ 16
Total content, tissue + medium (pmol/mg prot.)	22.79 ± 0.86	28.83 ± 0.67	+ 26

TABLE 4. Effect of tacrine (100 μmol/1) on the content of ACh in cerebrocortical prisms and on its release and synthesis during incubation without and with 10 μmol/1 HC-3.

	pmol ACh/mg prot.	% Change
ACh content		
Control	544.6 ± 8.2	
HC-3	451.8 ± 8.2	- 17
Tacrine	702.6 ± 43.7	+ 29
Tacrine + HC-3	633.9 ± 18.9	+ 16
ACh released during 30 min		
Control	137.9 ± 6.0	
HC-3	127.7 ± 8.6	- 7
Tacrine	158.0 ± 8.2	+ 15
Tacrine + HC-3	150.9 ± 9.6	+ 9
ACh synthesized during 30 min		
Control	126.3 ± 15.8	
HC-3	23.5 ± 11.6	- 81
Tacrine	304.5 ± 37.8	+ 141
Tacrine + HC-3	228.8 ± 15.1	+ 81

TABLE 5. Effect of tacrine (100 μmol/l) on the content of ACh in cerebrocortical prisms and on its release and synthesis during incubation at 50 mmol/l K$^+$.

	Control	Tacrine	% Change
Content (pmol/mg prot.)	252.5 ± 5.9	228.3 ± 11.8	- 10
Release (pmol/mg prot./30min)	709.3 ± 19.6	506.2 ± 21.0	- 29
Synthesis (pmol/mg prot./30 min)	420.4 ± 27.4	198.1 ± 41.6	- 53

EFFECT OF TACRINE ON THE METABOLISM OF CHOLINE AND ITS RELEASE INTO THE MEDIUM

Tacrine had no effect on the formation of [^{32}P]phosphorylcholine during incubation of homogenates with choline and [gamma-^{32}P]ATP (i.e., on the activity of choline kinase) and of [^{14}C]ACh during incubation of homogenates with choline and [^{14}C]acetyl-CoA (i.e., on the activity of choline acetyltransferase) (data not shown). The content of choline in cortical prisms and in incubation media was also not changed by tacrine under the usual incubation conditions.

We have found, however, that tacrine diminished the release of choline from the prisms into the medium under specific conditions. In experiments in which cortical prisms were preincubated with 0.5 mmol/l choline, tacrine strongly diminished the release of choline from the prisms during subsequent incubation in a medium with no added choline; the decrease in the release was accompanied by an increase in the content of choline in the prisms (Table 7). These effects of tacrine were also apparent in the presence of 10 μmol/l HC-3, i.e. under conditions when the high-affinity carriers for choline were blocked. It appears likely that tacrine influenced the release of choline by inhibiting the low-affinity choline carriers since its effect disappeared in the presence of 100 μmol/l HC-3 (known to block also the low-affinity carriers).

The release of choline was also diminished in prisms that had been preincubated for 60 min under standard conditions and then incubated for 10 or 30 min in the presence of 100 μmol tacrine and 10 μmol/l HC-3 (Table 8). It is not easy to offer a straightforward explanation of why the effect of tacrine was insignificant during incubation without HC-3; perhaps there are two parallel pathways for the efflux of choline, one sensitive to tacrine and the other to HC-3, and the effect of closing the tacrine-sensitive pathway is not visible unless the HC-3-sensitive pathway is also closed. The observed effect of tacrine accords with the idea that it inhibits the transport of choline from cells to the extracellular fluid. Perhaps this is the way in which tacrine helps to preserve intracellular stores of choline which can subsequently be utilized for the synthesis of ACh.

TABLE 6. Effect of short depolarization with 50 mmol/l K^+ on the amounts of $[^{14}C]ACh$ released and remaining in the tissue in the absence and presence of tacrine

	$[^{14}C]ACh$ released (pmol/mg prot.)	$[^{14}C]ACh$ in tissue (pmol/mg prot.)	% remaining in tissue
3 mmol/l K^+	7.9 ± 0.3	36.2 ± 1.4	82.0
50 mmol/l K^+	30.2 ± 1.9	19.1 ± 0.2	38.9
50 mmol/l K^+ +10 μmol/l HC-3	29.9 ± 1.8	17.3 ± 0.8	36.8
50 mmol/l K^+ +10 μmol/l HC-3 + 100 μmol/l tacrine	13.5 ± 0.2	33.1 ± 2.0	71.3

TABLE 7. Effect of tacrine and HC-3 on the release of choline from cerebrocortical prisms that had been preincubated for 60 min in the presence of 0.5 mM choline; both the preincubation and the incubation were at 5 mmol/l K^+

	Choline released per 30 min (pmol/mg prot.)	% of control without HC-3	% of control with HC-3
No addition (control)	2877 ± 110	100	
Tacrine 100 μmol/l	2054 ± 118	74*	
HC-3 10 μmol/l	2667 ± 201	96	100
HC-3 10 μmol/l +Tacrine 100 μmol/l	1982 ± 202	71*	71*
HC-3 100 μmol/l	2165 ± 138	78*	100
HC-3 100 μmol/l +Tacrine 100 μmol/l	1731 ± 140	62*	80

*$p < 0.05$

TABLE 8. Effect of tacrine on the release of choline from cerebrocortical prisms that had been preincubated for 60 min at 5 mmol/l K$^+$ without any addition of choline, and then incubated for 30 min with or without tacrine and HC-3

Medium	Choline released per 30 min (pmol/mg protein)	% of control without HC-3	% of control with HC-3
No addition	1385.4 ± 144.4	100	
Tacrine 100 μmol/l	1254.8 ± 131.3	91	
HC-3 10 μmol/l	1391.4 ± 167.7	100	100
HC-3 10 μmol/l + Tacrine 100 μmol/l	1190.3 ± 109.0	86*	86*

*p < 0.05

CONCLUSIONS

Several effects of tacrine on the metabolism of ACh and choline have been observed in our experiments, independent of its cholinesterase-inhibiting activity:

(1) Tacrine increased the content and the synthesis of ACh in cortical prisms incubated at 3 mmol/l K$^+$. The enhanced synthesis was associated with an enhanced utilization of choline from an intracellular source since the uptake of choline from the medium was inhibited, in accord with data by Buyukuysal and Wurtman (1989), and tacrine had a positive effect on the rate of ACh synthesis even in the presence of 10 μmol/l HC-3.

(2) Tacrine increased the release of ACh from cortical prisms incubated at 3 mmol/l K$^+$. The mechanism of this increase has not been investigated; the possibilities are that tacrine acted by blocking presynaptic muscarinic autoreceptors, or by increasing the frequency of the spontaneous release of single and giant quanta of ACh, as described by Thesleff et al. (1990) for neuromuscular junctions, or otherwise. A similar increase in the release of ACh from the heart under the influence of tacrine, additional to that obtained after the inhibition of cholinesterases, has been described by Lindmar and Löffelholz (1990).

(3) Tacrine strongly diminished the release of ACh from the prisms evoked by depolarization with 50 mmol/l K$^+$. It could be shown that the

inhibition of the evoked ACh release was not a consequence of the inhibition of ACh synthesis. An inhibition of stimulated neurotransmitter release by tacrine has also been observed by de Belleroche and Gardiner (1988), Hallak and Giacobini (1989) and Gardiner and de Belleroche (1990). It seems possible that tacrine acted by blocking the voltage-sensitive Ca^{2+}-channels since an inhibition of such channels by tacrine has been observed in electrophysiological experiments by Marsh et al. (1990).

(4) Tacrine inhibited the output of choline from cortical prisms into incubation media in experiments in which the prisms had been preincubated with a high concentration of choline, or in experiments in which the high-affinity uptake of choline had been blocked by HC-3. By restricting the efflux of choline from the cells, tacrine possibly increases the availability of intracellular choline for the synthesis of ACh, as observed in experiments with tissue incubation under 'resting' conditions.

REFERENCES

Bajgar J, Fusek J, Patocka J, Hrdina V (1979) In vivo kinetics of blood cholinesterase inhibition by 9-amino-1,2,3,4-tetrahydroacridine, its 7-methoxy derivative and physostigmine in rats. Physiol Bohemoslov 28:31-34

Boksa P, Collier B (1980) Spontaneous and evoked release of acetylcholine and a cholinergic false transmitter from brain slices: comparison to true and false transmitter in subcellular stores. Neuroscience 5:1517-1532

Buyukuysal RL, Wurtman RJ (1989) Tetrahydroaminoacridine but not 4-aminopyridine inhibits high-affinity choline uptake in striatal and hippocampal synaptosomes. Brain Res 482:371-375

de Belleroche J, Gardiner IM (1988) Inhibitory effect of 1,2,3,4-tetrahydro-9-aminoacridine on the depolarization induced release of GABA from cerebral cortex. Brit J Pharmacol 94:1017-1019

Dolezal V, Tucek S (1982) Effects of choline and glucose on atropine-induced alterations of acetylcholine synthesis and content in the caudate nuclei of rats. Brain Res 240:285-293

Dolezal V, Tucek S (1983) The synthesis and release of acetylcholine in normal and denervated rat diaphragms during incubation in vitro. J Physiol 334:461-474

Dolezal V, Tucek S (1988) Acetylcholine and choline in rat adrenals and brain cortex prisms incubated at elevated concentrations of choline in the medium. Brain Res 449:244-252

Dolezal V, Tucek S (1991) Positive and negative effects of tacrine (tetrahydroaminoacridine) and methoxytacrine on the metabolism of acetylcholine in brain cortical prisms incubated under "resting" conditions. J Neurochem 56:1207-1215

Drukarch B, Kits KS, Van der Meer EG, Lodder JC, Stoof JC (1987) 9-Amino-1,2,3,4-tetrahydroacridine (THA), an alleged drug for the treatment of Alzheimer's disease, inhibits acetylcholinesterase activity and slow outward K$^+$ current. Eur J Pharmacol 141:153-157

Drukarch B, Leysen JE, Stoof JC (1988) Further analysis of the neuropharmacological profile of 9-amino-1,2,3,4-tetra-hydroacridine (THA), an alleged drug for the treatment of Alzheimer's disease. Life Sci 42:1011-1017

Elinder F, Mohammed AK, Winblad B, Arhem P (1989) Effects of THA on ionic currents in myelinated axons of *Xenopus laevis*. Eur J Pharmacol 164:599-602

Flynn DD, Mash DC (1989) Multiple in vitro interactions with and differential in vivo regulation of muscarinic receptor subtypes by tetrahydroaminoacridine. J Pharmacol Exp Ther 250:573-581

Fonnum F (1969) Isolation of choline esters from aqueous solutions by extraction with sodium tetraphenylboron in organic solvents. Biochem J 113:291-298

Gardiner IM, de Belleroche J (1990) Modulation of gamma-aminobutyric acid release in cerebral cortex by fluoride, phorbol ester, and phosphodiesterase inhibitors: differential sensitivity of acetylcholine release to fluoride and K$^+$ channel blockers. J Neurochem 54:1130-1135

Goldberg AM, MacCaman RE (1973) The determination of picomole amounts of acetylcholine in mammalian brain. J Neurochem 20:1-8

Griffith WH, Sim JA (1990) Comparison of 4-aminopyridine and tetrahydroaminoacridine on basal forebrain neurons. J Pharmacol Exp Ther 255:986-993

Hallak M, Giacobini E (1989) Physostigmine, tacrine and metrifonate: the effect of multiple doses on acetylcholine metabolism in rat brain. Neuropharmacology 28:199-206

Heilbronn E (1961) Inhibition of cholinesterases by tetrahydroaminoacridine. Acta Chem Scand 15:1386-1390

Hunter AJ, Murray TK, Jones JA, Cross AJ, Green AR (1989) The cholinergic pharmacology of tetrahydroaminoacridine in vivo and in vitro. Br J Pharmacol 98:79-86

Lindmar R, Löffelholz K (1990) The effect of tacrine on acetylcholine overflow in the heart. Eur J Pharmacol 190: 251-254

Marquis JK (1990) Pharmacological significance of acetylcholinesterase inhibition by tetrahydroaminoacridine. Biochem Pharmacol 40:1071-1076

Marsh SJ, Hubbard A, Brown DA (1990) Some actions of 9-amino-1,2,3,4-tetrahydroacridine (THA) on cholinergic transmission and membrane currents in rat sympathetic ganglia. Eur J Neurosci 2:1127-1134

Musilková J, Tucek S (1991) The binding of cholinesterase inhibitors tacrine (tetrahydroaminoacridine) and 7-methoxytacrine to muscarinic acetylcholine receptors in rat brain in the presence of eserine. Neurosci Lett 125:113-116

Osterrieder W (1987) 9-Amino-1,2,3,4-tetrahydroacridine (THA) is a potent blocker of cardiac potassium channels. Brit J Pharmacol 92:521-527

Patocka J, Bajgar J, Fusek J, Bielavsky J (1976) Kinetics of inhibition of cholinesterase by 1,2,3,4-tetrahydro-9-amino-acridine in vitro. Coll Czechoslov Chem Commun 41:816-824

Pearce BD, Potter LT (1988) Effects of tetrahydroaminoacridine on M_1 and M_2 muscarine receptors. Neurosci Lett 88:281-285

Perry EK, Smith CJ, Court JA, Bonham JR, Rodway M, Atack JR (1988) Interaction of 9-amino-1,2,3,4-tetrahydroaminoacridine (THA) with human cortical nicotinic and muscarinic receptor binding in vitro. Neurosci Lett 91:211-216

Potter LT, Ferrendelli CA, Hanchett HE, Hollfield MA, Lorenzi MV (1989) Tetrahydroaminoacridine and other allosteric antagonists of hippocampal M1 muscarine receptors. Mol Pharmacol 35:652-660

Robinson TN, De Souza RJ, Cross AJ, Green AR (1989) The mechanism of tetrahydroaminoacridine-evoked release of endogenous 5-hydroxytryptamine and dopamine from rat brain tissue prisms. Br J Pharmacol. 98:1127-1136

Rogawski MA (1987) Tetrahydroaminoacridine blocks voltage-dependent ion channels in hippocampal neurons. Eur J Pharmacol 142:169-172

Shaw FH, Bentley GA (1953) The pharmacology of some new anti-cholinesterases. Austral J Exp Biol 31:573-576

Stevens DR, Cotman CW (1987) Excitatory actions of tetrahydro-9-aminoacridine (THA) on hippocampal pyramidal neurons. Neurosci Lett 79:301-305

Summers WK, Majovski LV, Marsh GM, Tachiki K, Kling A (1986) Oral tetrahydroaminoacridine in long-term treatment of senile dementia, Alzheimer type. New Engl J Med 315:1241-1245

Thesleff S, Sellin LC, Tagerud S (1990) Tetrahydroaminoacridine (tacrine) stimulates neurosecretion at mammalian motor endplates. Br J Pharmacol 100:487-490

Tucek S, Dolezal V (1991) Negative effects of tacrine (tetrahydroaminoacridine) and methoxytacrine on the metabolism of acetylcholine in brain slices incubated under conditions stimulating neurotransmitter release. J Neurochem 56:1216-1221

CDPCHOLINE, CDPETHANOLAMINE, LIPID METABOLISM AND DISORDERS OF THE CENTRAL NERVOUS SYSTEM

Eric J. Murphy and Lloyd A. Horrocks
Dept. of Medical Biochemistry
and Neuroscience Program
The Ohio State University
Columbus, Ohio 43210

I. Introduction

CDPamines are essential precursors for the biosynthesis de novo of phosphatidylcholine and phosphatidylethanolamine. The phosphoamine moiety is enzymically transferred to the diacylglycerol or other diradylglycerol by the choline or ethanolamine specific phosphotransferase. As described in the following sections, CDPamines, especially CDPcholine, have been used to reverse increases in fatty acid levels following ischemia. CDPcholine has been used successfully to treat several types of central nervous system diseases including ischemia, Parkinson disease, cerebrovascular disease, senile dementia, and tardive dyskinesia. CDPcholine effects in these diseases may be related in part to its ability to increase dopamine levels.

This review will describe the phosphotransferase reactions and the biochemical processes involved in utilizing CDPcholine. The pharmacological metabolism of CDPcholine and its toxicological properties will also be discussed. Finally, the effects of CDPcholine on monoamine neurotransmitter levels and ischemia will be examined as well as the use of CDPcholine treatment in other disease states.

II. Enzyme Reactions

A. CTP: phosphocholine cytidylyltransferase

Phosphatidylcholine is synthesized using two possible pathways. First is the methylation of phosphatidylethanolamine by phosphatidylethanolamine N-methyltransferase (EC 2.1.1.17) to form phosphatidylcholine. The second pathway, often referred to as the Kennedy pathway, utilizes CDPcholine as a phosphocholine donor to diglyceride. The initial rate-limiting step in this pathway is the biosynthesis of the CDPcholine from phosphocholine and

NATO ASI Series, Vol. H 70
Phospholipids and Signal Transmission
Edited by R. Massarelli, L. A. Horrocks,
J. N. Kanfer, and K. Löffelholz
© Springer-Verlag Berlin Heidelberg 1993

CTP catalyzed by CTP: phosphocholine cytidylyltransferase (EC 2.7.7.15). Rat brain CTP: phosphocholine cytidylyltransferase has an absolute requirement for Mg^{2+} and is inhibited by CDPcholine (Ki 0.090 mM) (Mages et al., 1988). The K_m values for CTP and phosphocholine are 2-10 mM and 0.3 mM respectively. The enzyme exists in both cytosolic and microsomal forms, with the microsomal enzyme being more active. Increases in fatty acids cause the cytosolic form to translocate to endoplasmic reticulum increasing the microsomal form of the enzyme. This activation is stimulated more by unsaturated than saturated fatty acids, although bond position and degree of unsaturation appear not to be important (Mages et al., 1988).

Ca^{2+} also stimulates increased phosphatidylcholine synthesis in hepatocytes (Sanghera, Vance, 1989). High levels of Ca^{2+} stimulated translocation and increased the activity of cytidylyltransferase, possibly by increased phospholipase activity, because Ca^{2+} stimulation increased not only formation of but also degradation of PtdCho. Another possibility is protein kinase C activation by the elevated Ca^{2+} resulting in the phosphorylation of the cytidylyltransferase. Previous work has shown that phorbol esters increase the formation of phosphatidylcholine through the postulated stimulation of protein kinase C (Paddon, Vance, 1980), although semi-purified protein kinase C did not affect cytidylyltransferase activity *in vitro* (Cook, Vance, 1985). Stimulation of protein kinase A has decreased cytidylyltransferase activity both *in vitro* (Alemany et al., 1982) and *in vivo* (Pelech et al., 1981). Thus, the role of Ca^{2+} in cytidylyltransferase activity is complicated and may involve one or several possible mechanisms.

B. CDPamine Phosphotransferase

Both ethanolamine phosphotransferase (EC 2.7.8.1) and choline phosphotransferase (EC 2.7.8.2) have an absolute requirement for either Mn^{2+} or Mg^{2+} as a cofactor (Goracci et al., 1986; Taniguchi et al., 1986; Woodard et al., 1987).

The reaction requires diacylglycerol, CDPcholine or CDPethanolamine, and ATP. ATP is required to rephosphorylate CMP to limit the reversal of the reaction due to increased levels of CMP (Goracci et al., 1985). For brain choline phosphotransferase using Mg^{2+} as a cofactor, the K_m for CDPcholine is 15.8 μM with a Vmax of 76 pmol/min/mg protein (Goldberg et al., 1985). For brain ethanolamine phosphotransferase using Mg^{2+} as a cofactor, the K_m for CDPethanolamine is 9.8 μM with a Vmax of 105 pmol/min/mg protein (Goldberg et al., 1985). The activities of both the ethanolamine and choline phosphotransferases are inhibited by increased Ca^{2+} concentrations (Taniguchi et al., 1986; Woodard et al., 1987). The inhibition by Ca^{2+} is competitive and reversible by chelation of the Ca^{2+} from the transferase by EGTA (Taniguchi et al., 1986). Maximum rates of both transferase activities are achieved using 5 mM Mg^{2+} in an EGTA buffered system. Without EGTA, maximal rates occur with 50 mM Mg^{2+} indicating both phosphotransferase activities are susceptible to Ca^{2+} inhibition (Taniguchi et al., 1986). This indicates that even low concentrations of Ca^{2+} inhibit phosphotransferase activity and suggest a role for Ca^{2+} in regulating PtdEtn and PtdCho synthesis.

The phosphotransferase reactions (Scheme 1) are reversible (Kanoh, Ohno, 1973; Goracci et al., 1985; Goracci et al., 1986). Increased levels of CMP can stimulate the backward reaction (Goracci et al., 1985). CMP increases PtdCho breakdown and to a lesser extent PtdEtn breakdown. The K_m values of CMP for the reverse reaction are between 125 and 190 μM for choline phosphotransferase from rat brain (Kanoh, Ohno, 1973; Goracci et al., 1986) with a Vmax of 21.6 nmol/h/mg protein (Goracci et al., 1986). The K_m value of CMP is 40 μM for ethanolamine phosphotransferase from rat brain with a Vmax of 2.3 nmol/h/mg protein (Goracci et al., 1986). The apparent Ki values for CDPcholine and CDPethanolamine are 1.0 and 0.05 mM respectively (Kanoh, Ohno, 1973). The inhibition constant for CDPcholine is 10-fold higher than its endogenous pool size indicating that endogenous levels of CDPcholine do not inhibit the reverse reaction. However, the CDPethanolamine Ki is equivalent to its endogenous pool size indicating that only the forward reaction is favored *in vivo* under normal conditions (Kanoh, Ohno, 1973). These conclusions are supported by the Vmax values of the two reactions.

CDPCho + DAG

forward ↓ ↑ reverse

PtdCho + CMP

Scheme 1. The reactions of the choline phosphotransferase in the forward and reverse directions.

During brain ischemia, both pathways are reversed, liberating diacylglycerol (DAG) from both choline and ethanolamine glycerophospholipids (Goldberg et al., 1983). The diacylglycerol formed by the reaction can be further degraded by DAG lipase, releasing glycerol and free fatty acids (Goldberg et al., 1983; Goracci et al., 1985). The use of CDPcholine and CDPethanolamine should be able to stimulate the forward reaction by mass action. During cerebral ischemia, CMP levels are increased and ATP levels drop, producing conditions favorable for reversing the phosphotransferase reactions (Goldberg et al., 1983; Ondera et al., 1986). Large increases in CDPamines would force the forward reaction thereby decreasing free fatty acid levels following cerebral ischemia. This indeed has been shown in several models of ischemia both *in vivo* and *in vitro* (Trovarelli et al., 1982; Dorman et al., 1982; Horrocks, Dorman, 1985; Goracci et al., 1986; Damron, Dorman, 1988).

$$DAG + 2H_2O \rightarrow 2 \text{ fatty acid} + Gro$$

$$CDPCho + H_2O \rightarrow CMP + PCho$$

Scheme 2. The reactions of products of the reverse reaction of choline phosphotransferase. Gro is glycerol and *P*Cho is phosphocholine.

The phosphotransferases are membrane-bound on the cytoplasmic side of the microsomal membrane (Binaglia et al., 1985; Morikawa et al., 1987; Woodard et al., 1987). Activities of the enzymes are dependent on the lipid environment.

Both phosphotransferases exhibit substrate specificity. The choline phosphotransferase apparently uses a specific pool of DAG which has a rapid turnover versus the slow pool formed by the bulk DAG pool (Binaglia et al., 1985). Ethanolamine phosphotransferase shows a marked selectivity for endogenous hexaenoic DAG (Kanoh, Ohno, 1975). Choline phosphotransferase shows less selectivity, but favors a DAG with a greater degree of unsaturation (Kanoh, Ohno, 1975). Maximum choline phosphotransferase activity occurs using a DAG with an unsaturated fatty acid in the *sn*-2 position with no requirement for the C-1 position (Possmayer et al., 1977; Arthur et al., 1985). The degree of unsaturation affects substrate specificity and preference is for a monounsaturated fatty acid (Possmayer et al., 1977). Long chain saturated fatty acids in both the *sn*-1 and *sn*-2 positions did not stimulate the formation of ChoGpl (Possmayer et al., 1977).

There are apparently two choline phosphotransferases, one which utilizes diacylglycerols and the other which uses 1-alkyl-2-acetyl-*sn*-glycerol (Galletti et al., 1985). The choline phosphotransferase using the 1-alkyl-2-acetyl-*sn*-glycerol requires Mn^{2+} or Mg^{2+}, is inhibited by Ca^{2+}, and is located on the cytoplasmic side of the microsomal membrane. Unlike the diacylglycerol enzyme, it is not sensitive to dithiothreitol inactivation. The two enzymes differ in pH optima, substrate specificities and their sensitivities to temperature and deoxycholate. The DTT-insensitive enzyme prefers a substrate with a 16:0 or 18:1 alkyl chain at the *sn*-1 position and a *sn*-2 short chain ester. CDPcholine is preferred over CDPethanolamine as the phosphoamine donor. Thus, it appears that there is a separate choline phosphotransferase which uses endogenous 1-alkyl-2-acetylglycerol to form platelet activating factor *de novo*.

Both the ethanolamine phosphotransferase and choline phosphotransferase have substrate specificities relative to the type of *sn*-1 linkage (Morikawa et al., 1987). For the ethanolamine phosphotransferase, the K_m values are 2.2 μM and 8.1 μM for the 1-alkenyl-2-acylglycerol and diacylglycerol substrates, respectively. For the choline phosphotransferase, the K_m values are 24 μM and 4.3 μM for the 1-alkenyl-2-acylglycerol and diacylglycerol substrates, respectively. An increase in CDPcholine stimulates increased use of the 1-alkenyl-2-acylglycerol substrate by the ethanolamine phosphotransferase resulting in increased ethanolamine plasmalogen (PlsEtn) synthesis. Under these conditions, there is an increase in diacylglycerol use by the choline phosphotransferase. Thus, under conditions utilizing endogenous substrates, the formation of PtdCho, PlsEtn, and PtdEtn are favored.

These results are supported by the evidence indicating that 1-alkyl-2-acylglycerol stimulates both the choline and ethanolamine phosphotransferases (Radominska-Pyrek et al., 1976). In the presence of free fatty acids, the choline phosphotransferase activity using 1-alkyl-2-acylglycerol is inhibited but the activity using 1,2-diacylglycerol is stimulated. These data suggests the presence of two choline phosphotransferases. The same results are seen with the ethanolamine phosphotransferase suggesting the possibility of two ethanolamine phosphotransferases. These results, combined with the previously mentioned studies, support the idea of two choline phosphotransferases and the possible existence of two ethanolamine phosphotransferases. Both of these types of phosphotransferases show specificity for the type of *sn*-1 linkage, namely whether it is an ether or ester linkage.

C. CDPcholine and Phosphatidic Acid

CDPcholine stimulates the synthesis *de novo* of phosphatidic acid by an unknown mechanism (Possmayer et al., 1977; Goracci et al., 1985). The increased levels of phosphatidic acid increase not only the amount of substrate available for phosphatidylinositol synthesis, but also diacylglycerol formation by phosphatidate phosphatase. Thus, the use of CDPcholine may not only increase endogenous formation of PtdCho, but also an increase in phosphatidic acid increasing the substrate pool for phosphatidylinositol synthesis.

III. CDPcholine

A. Metabolism

CDPcholine is rapidly absorbed by the liver and kidneys (Galletti et al.,

1985; Augt et al., 1983c) with over 60% of the administered dose absorbed in 30 minutes. Absorption of orally ingested CDPcholine is slow but complete and is readily bio-available (Augt et al., 1983c). Orally ingested CDPcholine has two peaks of radioactivity in blood levels (Dinsdale et al., 1983b). The first peak is seen 1 hr after ingestion, with the second peak occurring 24 h post-ingestion. The second peak may be due to slow absorption by the gut followed by hepatic metabolism.

Very little CDPcholine is lost by excretion (Galletti et al., 1985; Dinsdale et al., 1983b). The major route of elimination is through expired CO_2, accounting for 15% of the total dose administered (Dinsdale et al., 1983b). Only 1% of orally administered CDPcholine is excreted by fecal elimination. Very little is excreted by urinary elimination. For both orally and intravenously administered CDPcholine, there is minimal loss of the drug, and both are equally bio-available.

Once in the liver, CDPcholine is hydrolyzed to choline and cytidine (Galletti et al., 1985). The choline is metabolized to betaine which can be methylated further to methionine. Methionine, in the form of S-adenosylmethionine, can act as a methyl group donor to methylate PtdEtn to form PtdCho. Also, the choline phosphotransferase in the liver is very active leading to a large amount of CDPcholine used for PtdCho. The most probable form for transport of ingested CDPcholine is PtdCho formed in the liver (Augt et al., 1983c; Galletti et al., 1985).

Labeled choline in CDPcholine is incorporated into brain lipids. Of the total dose, only 0.2 - 0.25% is incorporated into the brain (Augt et al., 1983a; Burlina, Galzigna, 1989). Of this 0.25%, 62.8% is incorporated into brain PtdCho and sphingomyelin (Augt et al., 1983a). If CDPcholine is first incorporated into egg lecithin liposomes, 21% of the dose is absorbed by the brain (Burlina, Galzigna, 1989). This 100-fold increase in incorporation could be useful for targeting CDPcholine for central nervous system absorption. Unfortunately, the metabolism of CDPcholine incorporated into egg lecithin liposomes is unknown.

Once in the brain, high resolution autoradiography indicates the radioactivity is distributed throughout the cerebral cortex (Romero et al., 1983c). There appears to be no selectivity for cell type, with neurons, oligodendroglia and astroglia all being equally labeled. The labeled choline from CDPcholine is in all membranes including microsomal, and other subcellular organelles, but is not incorporated into the nuclear membrane (Romero et al., 1983c). Microsomal and cytoplasmic membrane fractions both had high levels of radioactive choline presumably incorporated into PtdCho and sphingomyelin (Aguilar et al., 1983). In synaptosomes there is elevated radioactivity, but it is not known whether this is in acetylcholine or PtdCho.

B. Toxicology

The toxicity of CDPcholine has been studied in several animal models. Unlike choline, high doses of CDPcholine have no cholinergic intoxicating effects.   Even though the administered doses contained the equivalent amounts of choline, CDPcholine had no toxic effects (Augt et al., 1983b). This indicates CDPcholine does not act as a free choline donor and further supports the idea that CDPcholine is transported in the blood as PtdCho, not as free choline.   The LD_{50} for CDPcholine in this study was 4600 mg/kg compared with 53 mg/kg for choline when administered by intravenous injection.   Orally administered CDPcholine had an undeterminable LD_{50}, but the LD_{50} for choline is 3900 mg/kg. Another study found the LD_{50} for intravenously injected CDPcholine to be 4600 mg/kg and 4150 mg/kg in mice and rats respectively (Grau et al., 1983).   The LD_{50} for oral ingestion is 8 g/kg in either rats or mice.

All indications are that CDPcholine has a high therapeutic ratio and a low clearance (Ciaceri, 1985).   Maximum uptake for orally administered CDPcholine is kidney > liver > lungs > spleen > heart >> brain. The levels in the heart and brain continue to increase 24 h after receiving CDPcholine.

Short term (Romero et al., 1983a) and long term high dose (Romero et al., 1983b) studies have shown no toxic effects of CDPcholine.   Beagles received 1.5 g/kg (p.o.) of CDPcholine over 6 months with no apparent hematological or histopathological changes (Romero et al., 1983b).   Low dose (100 or 150 mg/kg p.o.) studies with CDPcholine had no adverse effects on rats (Romero et al., 1983a). There are no biochemical changes nor are there any ulcerative changes in the stomach indicating minimal to no irritation of the stomach wall.

Human trials have shown no adverse effects of orally administered CDPcholine (Dinsdale et al., 1983a; Fernandez, 1983). Two doses were used, 600 mg/day and 1 g/day for 5 consecutive days. Kidney and cardiac function were normal with no changes in neurological tests (Dinsdale et al., 1983a).   A large clinical trial containing 2,817 patients, 1,492 males and 1,306 females, showed few side effects from chronic CDPcholine treatment.   Only 141 (5.0%) of the patients reported side effects. The major reported side effect was digestive trouble reported by 101 (3.6%) of the total patients (Fernandez, 1983).

In summary, CDPcholine is an easily tolerable drug, with a large therapeutic ratio and low clearance. The drug is readily bio-available when administered orally, which is suitable for long-term chronic usage.

IV. Effects of CDPcholine

A. Phospholipases

CDPcholine reduces the apparent increase in phospholipase A_2 (Plase A2) activity following cryogenic brain injury in rabbit brain (Arrigoni et al., 1987). In the rabbit cryogenic brain injury model, the CDPcholine was administered orally, bringing into question the nature of the actual agent changing Plase A_2 activity. Regardless, the Plase A_2 activity dropped to near control levels indicating the drug only affected the mechanism for elevating Plase A_2 activity because basal levels remained constant. CDPcholine has been reported to decrease phospholipase A_1 and phospholipase A_2 activities in neuronal cell cultures (Freysz et al., 1985). The ability of CDPcholine to alter phospholipase activity may be an important function in pathological injury *in vivo*.

B. Neurotransmitter Effects

CDPamines have known effects on learning and memory which may be due to altered neurotransmitter levels. In rats treated with 50 mg/kg CDPcholine (i.p.), biogenic amines are increased in several brain regions (Petkov et al., 1990). Noradrenaline is increased in the cerebral cortex and hippocampus while serotonin levels are increased in the cerebral cortex, striatum and hippocampus. Dopamine levels are increased in the striatum only.

Following hypoxic injury in rats, CDPcholine treatment markedly increases levels of noradrenaline and reduces dopamine to normal levels. In the hypothalamus, CDPcholine increases noradrenaline and decreases dopamine levels, but increases dopamine metabolites (Saligaut et al., 1987). In the striatum, CDPcholine affects the dopaminergic system only, causing a reduction in dopamine. Hypoxia decreases monoamine oxidase activity, resulting in elevated dopamine levels due to the inability to break it down. CDPcholine increases the amounts of metabolites increasing the turnover of dopamine. CDPcholine had no effect on basal neurotransmitter levels.

CDPcholine increases plasma levels of homovanillic acid, a dopamine metabolite, and dopamine in Parkinson disease patients (Cubells, Hernando, 1988). These results suggest that CDPcholine stimulates tyrosine hydroxylase increasing dopamine synthesis. Because no 3-O-methyldopamine is found, indications are that dopamine catabolism follows the monoamine oxidase pathway instead of the methylation pathway.

Hence, CDPcholine administration affects neurotransmitter levels. These neurochemical changes may have a large role in some of CDPcholine's affects in pathological diseases.

IV. CDPamines and Pathological Diseases

A. Ischemia

Choline and ethanolamine phosphotransferase activities are changed during cerebral ischemia (Goldberg et al., 1983). Choline phosphotransferase activity after 5 min of global ischemia in rats results in a 31% increase in Vmax and a 21% decrease in the K_m for CDPcholine. For ethanolamine phosphotransferase only the K_m for CDPethanolamine is decreased by 35%, however, this change is only seen when using Mg^{2+} as the metal cofactor. Incubation of the enzyme preparations with didecanoylglycerol produces similar results for both enzymes, suggesting that the diacylglycerol released during ischemia may be responsible for these changes in phosphotransferase activity. With elevated diacylglycerol levels, the choline phosphotransferase is activated and the forward reaction can occur at lower CDPcholine concentrations. Hence, addition of CDPcholine would cause a shift of the pathway in the forward direction, limiting increases in free fatty acids.

CDPamines decrease the release of fatty acids seen in cerebral ischemia (Dorman et al., 1982; Dorman et al., 1983; Horrocks, Dorman, 1985; Trovarelli et al., 1982). CDPcholine treatment decreases free fatty acid levels including arachidonic acid following ischemia in gerbils (Horrocks, Dorman, 1985). Exogenously added CDPcholine reacts favorably with the DAG pool containing arachidonic acid, thereby, preserving levels of arachidonoyl groups in phospholipids.

CDPcholine limited the breakdown of PtdCho during ischemia (Horrocks, Dorman, 1985). CDPethanolamine combined with CDPcholine administered by intraventricular injection following global ischemia also decreased fatty acid release from the phospholipid pool, specifically choline and ethanolamine glycerophospholipids (Dorman et al., 1983). CDPamines decrease the loss of arachidonic acid from PtdCho and PlsEtn following ischemia (Dorman et al., 1982). In rat brain minces, using an *in vitro* model of ischemia, CDPcholine decreases arachidonic acid increases in the free fatty acid pool and decreases PtdCho catabolism (Damron, Dorman, 1988). Preservation of the phospholipids in the membrane appears to be one action of CDPamine treatment following brain ischemia.

More importantly, injection of CDPamines during the reperfusion phase of reversible cerebral ischemia causes a drop in fatty acid levels.

CDPcholine causes a reduction in levels of 16:0, 18:1 and 18:2 (Horrocks, Dorman, 1985). CDPethanolamine causes a reduction in levels of 18:0 and 20:4 n-6. Both equally decrease 22:6 n-3 levels following ischemia (Horrocks, Dorman, 1985). These reductions in different pools of fatty acids indicate substrate specificity by each phosphotransferase. Post-ischemic administration of CDPamines is a clinically relevant treatment. In either pre- or post-ischemic administration, free fatty acid levels are decreased and membrane phospholipids are preserved. Furthermore, use of both compounds gives a greater effect than when used separately, indicating that CDPamines act synergistically and augment the action of each other.

CDPamine treatment also increases labeling of phospholipids following intracerebral injections of [^{3}H]acetate (Dorman et al., 1982). CDPamines increase labeling of phosphatidylserine and phosphatidylinositol as well as the choline and ethanolamine glycerophospholipids. This increased labeling comes from the CDPamines stimulation of phosphatidic acid formation by the acylation of sn-glycerol-3-phosphate (Possmayer et al., 1977). Phosphatidic acid can be dephosphorylated to form diacylglycerol for use in ChoGpl or EtnGpl synthesis. This is seen in the increased labeling of DAG following CDPamine treatment (Dorman et al., 1983). CDPamines not only increase the forward phosphotransferase reaction following ischemia, but also increase the formation of phosphatidic acid and DAG to be used in phospholipid synthesis.

Pretreatment of rats with CDPamines did not alter Na^+,K^+-ATPase activities following global ischemia (Goldberg et al., 1985). Striatal Na^+,K^+-ATPase activity falls by 43% and cerebral Na^+,K^+-ATPase activity decreases 30% following 5 min of global ischemia. Treatment with CDPamines did not alter the decreased activities in either region. Incubation of striatal membrane preparations with dihexadecanoylglycerol caused a 26% decrease in a Na^+,K^+-ATPase activity, a 20% decrease in Mg^{2+}-ATPase activity and an overall 21% decrease in total ATPase activity. Didecanoylglycerol has similar results although total ATPase activity is decreased only 13% and Mg^{2+}-ATPase activity remains at control levels. These results suggest that increases in DAG, not in free fatty acids, alter the rate of ATPase activity.

CDPcholine alters functional recovery following ischemia in rats. Following an ischemic/anoxic insult, rats were treated with CDPcholine (100-300 mg/kg i.p.) and neurological function was studied (Yamamoto et al., 1990). Treated animals had less neurological deficits than control animals. The effects of CDPcholine on neurological recovery were studied in rats following permanent vertebral artery occlusion and transient common artery occlusion (Kakihana et al., 1988). During the reperfusion phase, two doses of CDPcholine (50 and 250 mg/kg i.p.) were given. Both doses of CDPcholine attenuated neurological deficits, with the larger dose being more effective. CDPcholine also increased acetylcholine levels, acetylcholine synthesis, choline levels, and glucose levels. These results indicate CPDcholine acts as a choline donor for acetylcholine synthesis as indicated by elevated choline and acetylcholine levels. CDPcholine (8 mg/kg i.p.) also restores evoked cortical potentials following cerebral ischemia in cats (Boismare et al., 1978). CDPcholine administered in control animals had no effect. The only effect on evoked potential was seen in ischemic animals. These results support the idea that the effect of CDPcholine may be due to an effect other than effects on neurotransmitters. We postulate that following ischemia the main effect is in reducing the levels of fatty acids released and preserving the cell membrane.

Treatment of human patients with CDPcholine following mild to moderate ischemic insult has reduced neurological problems normally seen with these types of strokes (Tazaki et al., 1988). CDPcholine aids in the recovery of tissue which has undergone reversible damage and has no effects on strokes producing large infarctions. This is especially true if edema is present. Using a permanent focal ischemia model in rats, CDPcholine treatment (250 mg/kg i.p.) had no effect on edema or infarct volume (Murphy et al., 1991). These results correlate well with the results from the human study, because the infarcted region is quite large and edema was present. Thus, further studies need to be conducted to fully understand the effects of CDPcholine on increased neurological function following stroke and the relationship of recovery to infarct size, location of the infarct, and the extent of cerebral edema.

B. Hypoxia

Following hypoxic injury in guinea pigs, CDPcholine increased the incorporation of $[2-^3H]$glycerol and $[1-^{14}C]$palmitate into PtdCho and PtdEtn in mitochondrial membranes (Alberghina et al., 1981). Total phospholipid radioactivity also increased, showing that phospholipid synthesis was increased following hypoxia by CDPcholine. For hypoxic injury, these increases occur primarily in mitochondrial membranes, which is the fraction most affected by hypoxic damage. In rat heart, hypoxia causes a decrease in the formation of CDPcholine due to lower CTP levels (Hatch, Choy, 1990). ATP levels drop to 39% of control. This does not affect the synthesis of phosphocholine. However, CTP levels drop to 28% of control which decreases the rate of CDPcholine synthesis. During hypoxia, the cytidylyltranferase translocates to the microsomal membrane. This translocation increases the activity of the enzyme, to act as a compensatory mechanism for lower CTP levels. This translocation appears to be facilitated by increased fatty acid levels.

Hypoxia causes rats to change behavioral reactions indicative of decreased vigilance (Hamdorf, Cervos-Navarre, 1990). CDPcholine (100 mg/kg) administration shows protective effects by attenuating this deterioration following mild hypoxia. These results are indicative of increases in the dopaminergic system which may very well change the cholinergic balance, resulting in maintenance of normal behavior under these mild conditions.

Both biochemically and behaviorally, CDPcholine has been shown to restore function following hypoxic injury. Biochemically, this may in part be due to increased phospholipid synthesis. The effects of CDPcholine on increasing the dopaminergic system may alter cholinergic balance to restore near normal behavioral patterns.

C. Spinal Cord Injury

CDPcholine increase Mg^{2+}-ATPase levels to near normal following spinal cord injury (Clendenon et al., 1985). Na^+,K^+-ATPase activities are not changed. This may be due to the elevated DAG levels seen in spinal cord injury (Demediuk et al., 1985). However, Mg^{2+}-ATPase is a mitochondrial ATPase, and restoration of its activity strengthens the idea that CDPcholine alters mitochondrial phospholipid synthesis. Spinal cord injury may be very much like hypoxia in regard to preservation of the mitochondrial membrane.

D. Vascular Effects

CDPcholine decreases lesions seen in turbulent and non-turbulent areas of the vasculature in both hypercholesterolaemic and normal rabbits (Weber et al., 1989). By some unknown means, CDPcholine appears to reduce intimal involvement, limiting lesion size. Chronic CDPcholine treatment also decreases aggregatory activity of the vessel wall (Masi et al., 1986). This reduced aggregatory activity may be linked to reduced lesion size. In animals treated with acute doses of CDPcholine, platelet aggregation itself is reduced.

CDPcholine has been used clinically to treat cerebrovascular disease. Large doses of CDPcholine are effective in resolving the mental state and focal condition of the patients (Centrone et al., 1986). In this study, CDPcholine increased consciousness in 90% of the patients in the group. In another study, CDPcholine (1000 mg/day i.p.) was administered to 58 patients suffering from cerebrovascular disease (Sinforiani et al., 1986). CDPcholine increased awareness and perceptive motor skills in these patients. Mechanisms involved in these actions of CDPcholine are unknown, but may be related to the ability to evoke changes in the vasculature.

E. Parkinson Disease

CDPcholine has been used effectively in the treatment of Parkinson disease. CDPcholine treatment increases homovanillic acid and dopamine levels in patient's plasma, and stimulates an increase in dopamine receptors on lymphocytes (Cubells, Hernando, 1988). The lack of 3-0-methyldopamine indicates dopamine is not being metabolized by methylation, but through monoamine oxidase due to elevated plasma homovanillic acid levels. These authors suggest that CDPcholine stimulates an increase in tyrosine hydroxylase activity as well as an increase in sensitivity of the dopaminergic system induced by elevated receptor numbers. Also, CDPcholine permits a reduction in L-dopamine use in Parkinson patients without a reduction in therapeutic benefit (Eberhardt et al., 1990). These authors also suggest a stimulation of tyrosine hydroxylase activity as a means to increase dopamine levels. In the striatum, CDPcholine increases the activity of tyrosine hydroxylase increasing dopamine synthesis (Martinet et al., 1978). However, a decrease in the uptake of dopamine by synaptosomes following CDPcholine treatment has also been seen (Martinet et al., 1978). These results indicate dopamine remains in the synaptic cleft longer. This, in a sense, would increase the effective levels of dopamine in the central nervous system, and may account for the effect of CDPcholine in Parkinson patients.

F. Senile Dementia

CDPcholine (100 mg/day) was administered to patients suffering mild to moderate senile dementia (Serra et al., 1990). These patients had improved symptomatology even during the two weeks between the three week treatment periods. Cognitive and behavioral parameters also changed, improving the overall well-being of the patient.

G. Tardive Dyskinesia

In a limited study, CDPcholine reduced behavioral test scores significantly indicating improved behavior in patients with tardive dyskinesia (Arranz, Ganoza, 1983). CDPcholine (500-1200 mg/day) was given over four weeks with little or no side-effects reported.

H. Glaucoma

CDPcholine (1 g/day) was given to patients suffering perimetric problems resulting from glaucoma (Giraldi et al., 1989). Treatment with α-blockers often is useful in reducing pressure within the eye due to glaucoma, but does not reduce glaucomatous perimetric problems. CDPcholine was given for ten days resulting in reduction of these problems in 75% of the

patients.   This marked improvement lasts for three months. At this time,
a   second   treatment   causes   an   even   greater   decrease   in   perimetric
problems.   In   glaucoma,   the   effects of CDPcholine are long lasting and
therapeutically effective.

V. Conclusion

In summary, CDPamines may be useful compounds in treating several types of
central nervous system disorders and diseases. All of the mechanisms of
action are not yet known, however, CDPcholine does reduce fatty acid
release in ischemia by forcing the forward phosphotransferase reaction.
CDPcholine also increases tyrosine hydroxylase activity through an unknown
mechanism thereby increasing dopamine synthesis. These effects, increased
phospholipid and dopamine synthesis, may be the mechanism involved in
CDPcholine's effects on other central nervous system disorders.

References

Aguilar   J,   Giménez   R,   Bachs   O,   Enrich   C,   Augt   J   (1983)   Cerebral
subcellular   distribution   of   CDPcholine and/or its metabolites after oral
administration of methyl-14C CDPcholine. Drug Res 33:1051-1053

Alberghina M, Viola M, Serra I, Mistretta A, Giuffrida AM (1981) Effect of
CDP-choline on the biosynthesis of phospholipids in brain regions during
hypoxic treatment. J Neurosci Res 6:421-433

Alemany S, Varela I, Harper JF, Mato JM (1982) Calmodulin regulation of
phospholipid and fatty acid methylation by rat liver microsomes. J Biol
Chem 257:9249-9251

Arranz J, Ganoza C (1983) Treatment of chronic dyskinesia with CDPcholine.
Drug Res 33:1071-1073

Arrigoni   E,   Averet   N,   Cohadon   F   (1987)   Effects   of   CDP-choline   on
phospholipase A2 and cholinephosphotransferase activities following a
cryogenic brain injury in the rabbit. Biochem Pharmacol 36:3697-3700

Arthur G, Covic L, Wientzek M, Choy PC (1985) Plasmalogenase in hamster
heart. Biochim Biophys Acta 833:189-195

Augt   J,   Font   E,   Sacristán A, Ortiz JA (1983a) Radioactivity incorporated
into   different   cerebral   phospholipids   after   oral   administration   of
14C-methyl-CDPcholine. Drug Res 33:1048-1050

Augt J, Font E, Sacristán A, Ortiz JA (1983b) Dissimilar effects on acute toxicity studies of CDPcholine and choline. Drug Res 33:1016-1018

Augt J, Font E, Sacristán A, Ortiz JA (1983c) Bioavailability of methyl-14C-CDPcholine by oral route. Drug Res 33:1045-1047

Binaglia L, Roberti R, Corazzi L, Freysz L, Arienti G, Porcellati G (1985) Evidence for a compartmentation of the enzymes involved in CDPcholine metabolism at membrane level. In: Zappia V, Kennedy EP, Nilsson BI, Galletti P (eds) Novel Biochemical, Pharmacological and Clinical Aspects of Cytidinediphosphocholine. Elsevier, Amsterdam, p 131-136

Boismare F, Le Poncin M, Lefrancois J, Lecordier JC (1978) Action of cytidine diphosphocholine on functional and hemodynamic effects of cerebral ischemia in cats. Pharmacology 17:15-20

Burlina AP, Galzigna L (1989) Preparazione, proprietà e potenzialità terapeutiche della citidin-difosfocolina associata a liposomi. Rivista di Neurologia 59:26-31

Centrone G, Ragno G, Calicchio G (1986) Uso della citicoline ad alti dosaggi nelle affezioni acute cerebro-vascolari. Min Med 77:371-373

Ciaceri G (1985) Toxicological studies on CDPcholine. In: Zappia V, Kennedy EP, Nilsson BI, Galletti P (eds) Novel Biochemical, Pharmacological and Clinical Aspects of Cytidinediphosphocholine. Elsevier, Amsterdam, p 159-167

Clendenon NR, Palayoor ST, Gordon WA (1985) Influence of CDPcholine on ATPase activity in acute experimental spinal cord trauma. In: Zappia V, Kennedy EP, Nilsson BI, Galletii P (eds) Novel Biochemical, Pharmacological and Clinical Aspects of Cytidinediphosphocholine. Elsevier, Amsterdam, p 275-284

Cook HW, Vance DE (1985) Evaluation of possible mechanisms of phorbol ester stimulation of phosphatidylcholine synthesis in HeLa cells. Can J Biochem Cell Biol 63:145-151

Cubells JM, Hernando C (1988) Clinical trial on the use of cytidine diphosphate choline in Parkinson's disease. Clin Thera 10:664-671

Damron DS, Dorman RV (1988) [³H]Arachidonic acid metabolism in rat brain minces: Effects of nucleotide triphosphates, CDPcholine and CMP. Neurochem Res 13:777-783

Demediuk P, Saunders RD, Anderson DK, Means ED, Horrocks LA (1985) Membrane lipid changes in laminectomized and traumatized cat spinal cord. Proc Natl Acad Sci USA 82:7071-7075

Dinsdale JRM, Griffiths GK, Castello J, Maddock J, Ortiz JA, Aylward M (1983a) CDPcholine: repeated oral dose tolerance studies in adult healthy volunteers. Drug Res 33:1061-1065

Dinsdale JRM, Griffiths GK, Rowlands C, Castello J, Ortiz JA, Maddock J, Aylward M (1983b) Pharmacokinetics of 14C CDPcholine. Drug Res 33:1066-1070

Dorman RV, Dabrowiecki Z, DeMedio GE, Porcellati G, Horrocks LA (1982) Effects of cytidine nucleotides on CNS membranes during ischemia. In: Grossman RG, Gildenberg PL (eds) Head Injury: Basic and Clinical Aspects. Raven Press, New York, p 93-101

Dorman RV, Dabrowiecki Z, Horrocks LA (1983) Effects of CDPcholine and CDPethanolamine on the alterations in rat brain lipid metabolism induced by global ischemia. J Neurochem 40:276-279

Eberhardt R, Birbamer G, Gerstenbrand F, Rainer E, Traegner H (1990) Citicoline in the treatment of Parkinson's disease. Clin Thera 12:489-495

Fernandez RL (1983) Efficacy and safety of oral CDPcholine: drug surveillance study in 2817 cases. Drug Res 33:1073-1080

Freysz L, Golly F, Mykita S, Avola R, Dreyfus H, Massarelli R (1985) Metabolism of neuronal cell cultures: Modification induced by CDPcholine. In: Zappia V, Kennedy EP, Nilsson BI, Galletti P (eds) Novel Biochemical, Pharmacological and Clinical Aspects of Cytidinediphosphocholine. Elsevier, Amsterdam, p 117-129

Galletti P, De Rosa M, Nappi MA, Pontoni G, del Piano L, Salluzzo A, Zappia V (1985) Transport and metabolism of double-labelled CDPcholine in mammalian tissues. Biochem Pharmacol 34:4121-4130

Giraldi JP, Virno M, Covelli G, Grechi G, De Gregorio F (1989) Therapeutic value of citicoline in the treatment of glaucoma (computerized and automated perimetric investigation). Intl Ophthalm 13:109-112

Goldberg WJ, Dorman RV, Horrocks LA (1983) Effects of ischemia and diglycerides on ethanolamine and choline phosphotransferase activities from rat brain. Neurochem Pathol 1:225-234

Goldberg WJ, Dorman RV, Dabrowiecki Z, Horrocks LA (1985) The effects of ischemia and CDPamines on Na^+,K^+-ATPase and acetylcholinesterase activities in rat brain. Neurochem Pathol 3:237-248

Goracci G, Francescangeli E, Mozzi R, Porcellati S, Porcellati G (1985) Regulation of phospholipid metabolism by nucleotides in brain transport of CDPcholine into brain. In: Zappia V, Kennedy EP, Nilsson BI, Galletti P (eds) Novel Biochemical, Pharmacological and Clinical Aspects of Cytidinediphosphocholine. Elsevier, Amsterdam, p 105-116

Goracci G, Francescangeli E, Horrocks LA, Porcellati G (1986) A comparison of the reversibility of phosphoethanolamine transferase and phosphocholine transferase in rat brain microsomes. Biochim Biophys Acta 876:387-391

Grau T, Romero A, Sacristán A, Ortiz JA (1983) CDPcholine: acute toxicity study. Drug Res 33:1033-1034

Hamdorf G, Cervos-Navarre J (1990) Study of the effects of oral administration of CDPcholine on open-field behaviour under conditions of chronic hypoxia. Drug Res 40:519-522

Hatch GM, Choy PC (1990) Effect of hypoxia on phosphatidylcholine biosynthesis in the isolated hamster heart. Biochem J 268:47-54

Horrocks LA, Dorman RV (1985) Prevention by CDPcholine and CDPethanolamine of lipid changes during brain ischemia. In: Zappia V, Kennedy EP, Nilsson BI, Galletti P (eds) Novel Biochemical, Pharmacological and Clinical Aspects of Cytidinediphosphocholine. Elsevier, Amsterdam, p 205-215

Kakihana M, Fukuda N, Suno M, Nagaoka A (1988) Effects of CDP-choline on neurologic deficits and cerebral glucose metabolism in a rat model of cerebral ischemia. Stroke 19:217-222

Kanoh H, Ohno K (1973) Utilization of endogenous phospholipids by the backreaction of CDP-choline (-ethanolamine): 1,2-diglyceride choline (ethanolamine)-phosphotransferase in rat liver microsomes. Biochim Biophys Acta 306:203-217

Kanoh H, Ohno K (1975) Substrate-selectivity of rat liver microsomal 1,2-diacylglycerol: CDPcholine (ethanolamine) choline (ethanolamine) phosphotransferase in utilizing endogenous substrates. Biochim Biophys Acta 380:199-207

Mages F, Rey C, Fonlupt P, Pacheco H (1988) Kinetic and biochemical properties of CTP: choline-phosphate cytidylyltransferase from the rat brain. Eur J Biochem 178:367-372

Martinet M, Fonlupt P, Pacheco M (1978) Interaction of CDPcholine with synaptosomal transport of biogenic amines and their precursors in vitro and in vivo in the rat corpus striatum. Experientia 34:1197-1199

Masi I, Giani E, Galli C (1986) Effects of CDPcholine on platelet aggregation and the antiaggregatory activity of arterial wall in the rat. Pharm Res Commun 18:273-281

Morikawa S, Taniguchi S, Fujii K, Mori H, Kumada K, Fujiwara M (1987) Preferential synthesis of diacyl and alkenylacyl ethanolamine and choline glycerophospholipids in rabbit platelet membranes. J Biol Chem 262:1213-1217

Murphy EJ, Slivka AP, Horrocks LA (1991) Effect of methylprednisolone and CDPcholine on ischemic infarct volume. Am Soc Neurochem 22:150(Abstract)

Ondera H, Iijima K, Kogure K (1986) Mononucleotide metabolism in the rat brain after transient ischemia. J Neurochem 46:1704-1710

Paddon HB, Vance DE (1980) Tetradecanoyl-phorbol acetate stimulates phosphatidylcholine biosynthesis in HeLa cells by an increase in the rate of the reaction catalyzed by CTP: phosphocholine cytidylyltransferase. Biochim Biophys Acta 620:636-640

Pelech SL, Pritchard PH, Vance DE (1981) cAMP analogues inhibit phosphatidylcholine biosynthesis in cultured rat hepatocytes. J Biol Chem 256:8283-8286

Petkov VD, Stancheva SL, Tocuschieva L, Petkov VV (1990) Changes in brain biogenic monoamines induced by the nootropic drugs adafenoxate and meclofenoxate and by citicholine (experiments on rats). Gen Pharmac 21:71-75

Possmayer F, Duwe G, Hahn M, Buchnea D (1977) Acyl specificity of CDPcholine: 1,2-diacylglycerol cholinephosphotransferase in rat lung. Can J Biochem 55:609-617

Radominska-Pyrek A, Strosznajder J, Dabrowiecki Z, Chojnacki T, Horrocks LA (1976) Effects of free fatty acids on the enzymic synthesis of diacyl and ether types of choline and ethanolamine phosphoglycerides. J Lipid Res 17:657-662

Romero A, Grau T, Sacristán A, Ortiz JA (1983a) Study of subacute toxicity of CDPcholine after 30 days of oral administration to rats. Drug Res 33:1035-1038

Romero A, Grau T, Sacristán A, Ortiz JA (1983b) CDPcholine: 6 month study on toxicity in dogs. Drug Res 33:1038-1042

Romero A, Serratosa J, Sacristán A, Ortiz JA (1983c) High resolution outerradiography in mouse brain 24 h after radiolabelled CDPcholine administration. Drug Res 33:1056-1058

Saligaut C, Daoust M, Moore N, Boismare F (1987) Effects of hypoxia and cytidine (5') diphosphocholine on the concenrations of dopamine, norepinephrine and metabolites in rat hypothalamus and striatum. Arch Int Pharmacodyn 285:25-33

Sanghera JS, Vance DE (1989) Stimulation of CTP: phosphocholine cytidylyltransferase and phosphatidylcholine synthesis by calcium in rat hepatocytes. Biochim Biophys Acta 1003:284-292

Serra F, Diaspri GP, Gasbarrini A, Giancane S, Rimondi A, Tame MR, Sakellaridis E, Bernardi M, Gasbarrini G (1990) Effetto della CDPcolina sul decadimento mentale senile. Min Med 81:465-470

Sinforiani E, Trucco M, Pacchetti C, Gualtieri S (1986) Valutazione degli effetti della citicolina nella malattia cerebro-vascolare cronica. Min Med 77:51-57

Taniguchi S, Morikawa S, Hayashi H, Fujii K, Mori H, Fujiwara M (1986) Effects of Ca^{2+} on ethanolaminephosphotransferase and cholinephosphotransferase in rabbit platelets. J Biochem 100:485-491

Tazaki Y, Sakai F, Otomo E, Kutsuzawa T, Kameyama M, Omae T, Fujishima M, Sakuma A (1988) Treatment of acute cerebral infarction with a choline precursor in a multicenter double-blind placebo-controlled study. Stroke 19:211-216

Trovarelli G, DeMedio GE, Montanini I (1982) The influence of CDP-choline on brain lipid metabolism during ischemia. Il Farmaco 10:664-668

Weber G, Auteri A, Bianciardi G, Fabrini P, Resi L, Salvi M, Toti P, Tanganelli P (1989) Influence of CDPcholine administration on the aortic wall lesions in dietically hypercholesterolaemic rabbits: A morphometric evaluation. Drugs Exptl Clin Res 15:321-323

Woodard DS, Lee T-C, Snyder F (1987) The final step in the de novo biosynthesis of platelet-activating factor. J Biol Chem 262:2520-2527

Yamamoto M, Shimizu M, Okamiya H (1990) Pharmacological actions of the new TRH analogue, YM-14673, in rats subjected to cerebral ischemia and anoxia. Euro J Pharm 181:207-214

BIOSYNTHESIS OF 1-ALKYL-2-ACETYL-sn-GLYCERO-3-PHOSPHOCHOLINE (PLATELET ACTIVATING FACTOR) IN CULTURED NEURONAL AND GLIAL CELLS

E. Francescangeli, L. Freysz*, H. Dreyfus*, A. Boila and G. Goracci
Istituto di Biochimica e Chimica Medica,
Università di Perugia,
Via del Giochetto,
06100 Perugia, Italy

INTRODUCTION

Platelet Activating Factor (PAF; 1-alkyl-2-acetyl-sn-glycero-3-phospho-choline) has been detected in the nervous tissue (Tokumura et al., 1987). In this tissue, much evidence supports the hypothesis that PAF is involved in physiological functions and in pathological situations (for review see Goracci, 1990). In fact, PAF induces neuronal differentiation (Kornecki and Ehrlich, 1988) and the stimulation of chick retina with neurotransmitters led to the production of PAF (Bussolino et al., 1986; Bussolino et al., 1988a). In addition, PAF levels in brain increased during ischemia and convulsions (Feuerstein and Yue, 1989; Kumar et al., 1988). These observations also support the concept that this lipid mediator may have a role in brain dysfunction.

The contribution of different cell types of the nervous tissue and that of different brain areas to PAF synthesis remains still unknown. However, it is very likely that PAF can originate from neural cells because its production has been demonstrated in rat cultured cerebellar cells (Yue et al., 1990), in chick retina (Bussolino et al., 1986) and in human fetal brain cells in culture (Sogos et al., 1990).

Two metabolic routes have been described for the synthesis of PAF in various tissues and cell types: the de novo pathway and the remodeling pathway (Snyder, 1987). Previous studies have demonstrated that rat brain possesses the substrates and the enzymes required for PAF synthesis by both pathways (Fig.1) but their distribution among different cell types is still unknown.

*Centre de Neurochimie,
Antenne de Cronenbourg,
Strasbourg,
France

NATO ASI Series, Vol. H 70
Phospholipids and Signal Transmission
Edited by R. Massarelli, L. A. Horrocks,
J. N. Kanfer, and K. Löffelholz
© Springer-Verlag Berlin Heidelberg 1993

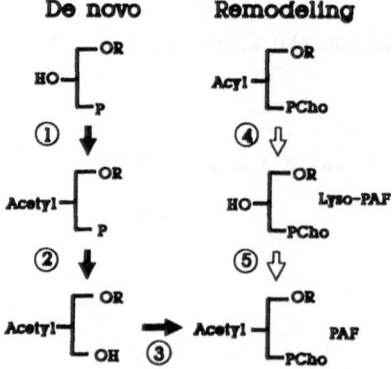

Fig. 1. Metabolic pathways for the synthesis of PAF in the brain. 1 is acetylCoA : 1-alkyl-2-lyso-*sn*-glycero-3-phosphate acetyltransferase (Lee et al., 1986); 2 is 1-alkyl-2-acetyl-*sn*-glycero-3-phosphate phospho-hydrolase (Lee et al., 1988); 3 is CDPcholine : 1-alkyl-2-acetyl-*sn*-glycerol phosphocholinetransferase (Francescangeli and Goracci, 1989); 4 is phospholipase A_2 (Woelk et al., 1974); 5 is acetylCoA: 1-alkyl-2-lyso-*sn*-glycero-3-phosphocholine acetyltransferase (lyso-PAF acetyltransferase) (Goracci and Francescangeli, 1991).

The contribution of the different brain cell populations to the synthesis of PAF and the mechanism involved in the regulation of its metabolism can be approached by the study of the enzyme activities in the different cell types during their development. This experimental model allows to relate the changes of the enzymic activities to the morphological and functional cell maturation.

In the present report, we investigated the activities of PAF-synthesizing phosphocholinetransferase and lyso-PAF acetyltransferase, the enzymes catalyzing the last steps of the *de novo* and of the *remodeling* pathways, in chick brain hemispheres and in primary cultures of chick neurons and glial cells during development.

MATERIALS AND METHODS

Materials
1-O-hexadecyl-2-acetyl-*sn*-glycerol,   1-O-hexadecyl-*sn*-glycero-3-phospho-choline (lyso-PAF) and 1-O-hexadecyl-2-acetyl-*sn*-glycero-3-phosphocholine (PAF) were from Novabiochem (Switzerland). Cytidine-5'diphospho-[methyl-^{14}C]choline and [1-^{14}C]acetyl-CoA were from Amersham (U.K.).

Chick brain hemispheres
Cerebral hemispheres of 8 to 20 day-old chick embryos were freed of their meninges, immediately frozen and lyophylized. The lyophylized material was stored at -20°C until use.

Neuronal primary cultures
Neuronal cultures from 8-day-old chick embryo hemispheres were obtained following the procedure of Pettmann et al. (1979). Briefly, the cell suspension was seeded on plastic Petri dishes precoated with poly-L-lysine and the cultures were grown in Dulbecco's Modified Eagle Medium (DMEM) containing 10% fetal calf serum (FCS) in a CO_2 (5%)- air (95%) atmosphere at 37°C. The medium was changed at 1 and 3 days in vitro. At the end of the incubation the culture medium was discarded and the cultures were washed 3 times with 0.9% NaCl solution prewarmed at 37°C. The cells were scraped into 3 ml of ice-cold bidistilled water, homogenized, lyophylized and stored at -20°C until enzyme assay.

Glial primary cultures
Glial cell cultures were prepared from 14-day-old chick embryo cerebral hemispheres as previously described (Booher et al. 1972). Cells were seeded on plastic Petri dishes and grown during 16 days in DMEM containing 10% FCS. The medium was changed every 4 days. The cells were collected and stored as described for neurons.

Enzyme assays
The lyophylized hemispheres, neurons or glial cells were resuspended in ice-cold water and sonicated in an ice bath (three pulses 30 sec each, 150 watts) using a MSE ultrasonic disintegrator Mk2. Phosphocholinetransferase and lyso-PAF acetyltransferase activities were assayed as previously reported (Francescangeli and Goracci, 1989; Goracci and Francescangeli 1991). Lipids were extracted by the method of Bligh and Dyer (1959) and labeled phospholipids were isolated by TLC on silica gel G plates with chloroform/methanol/acetic acid/water (50:25:8:4 by vol) as developing solvent and identified by co-chromatography with known standards (Francescangeli et al., 1989). The radioactivity of the products was measured by liquid scintillation spectrometry. Protein concentration was determined according to Lowry et al. (1951).

RESULTS AND DISCUSSION

Chick embryo brain homogenates
The ontogenesis of the chick brain is well known (Romanoff, 1960) and its development can be easily followed. Three major steps can be

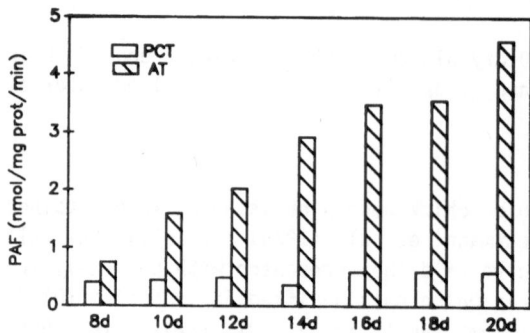

Fig. 2. Specific activities of PAF-synthesizing phosphocholinetransferase and lyso-PAF acetyltransferase in chick embryo brain hemispheres during development. PAF-synthesizing phosphocholinetransferase (PCT) assay: 60 mM Tris-HCl pH 8, 20 mM $MgCl_2$, 0.5 mM 1-hexadecyl-2-acetyl-sn-glycerol, 0.005% Tween 20, 20 mM DTT, 0.1mM CDP-[methyl-^{14}C]choline (spec. rad. 1 and 1.47 nCi/nmol), 0.1-0.16 mg protein, at 37°C for 15 min in a final volume of 0.3 ml. Lyso-PAF acetyltransferase (AT) assay: 100 mM Tris-HCl pH 7.4, 0.03 mM lyso-PAF, 0.2 mM [1-^{14}C]acetyl-CoA (spec. rad. 1.5 and 1.93 nCi/nmol), 0.2-0.4 mg protein, at 37°C for 10 min in a final volume of 0.5ml. Results are expressed as nmol/mg prot/min and are the average of duplicate samples of two different experiments.

distinguished. The first period before the 12th day of embryonic life corresponds essentially to the proliferation and differentiation of neurons (Altman, 1969). Near the 12th day the number of neurons reaches its maximum and then remains constant until adulthood (Nurenberger, 1958, Judes et al., 1968). The second period begins around the 12th day of embryonic life up to the 30th day of postnatal life.

This step is characterized by the proliferation and differentiation of astrocytes and oligodendrocytes. Furthermore, myelination begins around the 16th day and continues intensively up to the 30th day (Reddich, 1951, Bensted et al., 1957, Kurihara and Tsukada, 1968). The third period after the 30th day of postnatal life corresponds to the end of the maturation of the brain (Freysz et al., 1980).

Freysz et al. (1972, 1980) have shown that the embryonic chick brain contains phosphocholinetransferase activity. This enzyme is able to catalyze the synthesis of long chain diacyl and alkylacyl glycerophospho-choline using as substrate the corresponding diradylglycerols. These observations indicate that the embryonic chick brain possesses the enzymatic machinery for the synthesis of alkylacyl glycerophosphocholine, which is the precursor of the PAF synthesis through the *remodeling* pathway. On the other hand, Bussolino et al. (1986) have reported that the retina of the chick possesses PAF-synthesizing phosphocholine-transferase and lyso-PAF acetyltransferase activities. These observations suggest that chick brain may also synthesize PAF by the two pathways and that this lipid mediator may play a role during its development. This hypothesis has been tested by investigating the activities of lyso-PAF acetyltransferase and PAF-synthesizing phosphocholinetransferase in brain hemispheres of chick embryos during development.

Before the 12th day of the embryonic life the specific activity of lyso-PAF acetyltransferase undergoes a rapid increase whereas that of PAF-synthesizing phosphocholinetransferase remains almost unchanged (Fig. 2). Taking into account that during this period the number of neuronal cells increases and reaches its maximum, one might speculate that the activity of lyso-PAF acetyltransferase is associated with the proliferation and/or the differentiation of neuronal cells. However, between the 12th and the 20th day of the embryonal development of the chick, the specific activity of the brain lyso-PAF acetyltransferase further increases whereas that of the PAF-synthesizing phosphocholinetrans-ferase increased only slightly. These results suggest that the lyso-PAF acetyltransferase may be also associated with glial cells (astrocytes and oligodendrocytes) because during this period they continue to proliferate and to differentiate.

Human endothelial cells produce PAF (Camussi et al., 1983) essentially by the *remodeling* pathway and possesses a relatively high lyso-PAF acetyl-transferase activity (Ghigo et al., 1988). Thus a contribution of these cells to this enzymic activity cannot be excluded because the development of blood vessels also takes place in the hemispheres of the embryos. However, between 8 and 21 days of chick embryo brain development, the ratio of the brain tissue's volume over the length of contained blood vessels remains practically constant (Romanoff,1960). Therefore, the increase of the lyso-PAF acetyltransferase specific activity seems to be mostly due to the proliferation of the neural cells. This hypothesis prompted us to investigate the activities of PAF-synthesizing enzymes in neuronal and glial cell primary cultures of chick hemispheres in order to evaluate their distribution in these cell types and the changes of their specific activities during the development of the cultures.

Neuronal primary cultures
On the basis of morphological and biochemical observations, the purity of the primary cultures of neurons from cerebral hemispheres of 8 day-old chick embryos was higher than 95% (Dreyfus et al., 1981). Two periods can be distinguished in the development of these cultures; a first period of

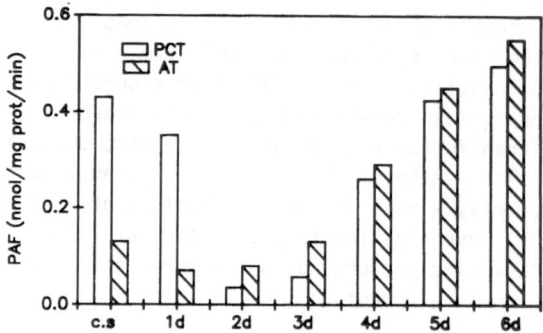

Fig. 3. Specific activities of PAF- synthesizing phosphocholinetransferase and lyso-PAF acetyltransferase in chick neuronal primary cultures. PAF-synthesizing phosphocholinetransferase (PCT) assay: 60 mM Tris-HCl pH 8, 0.5 mM 1-hexadecyl-2-acetyl-*sn*-glycerol, 0.005% Tween 20, 10 mM MgCl$_2$, 20 mM DTT, 0.1 mM CDP[methyl-^{14}C]-choline (spec. rad. 1.22 nCi/nmol), 0.045-0.15 mg protein, at 37°C for 15 min in a final volume of 0.3 ml. Lyso-PAF acetyltransferase (AT) assay: 100 mM Tris-HCl pH 6.9, 0.06 mM lyso-PAF, 0.2 mM [1-^{14}C]acetyl-CoA (spec. rad. 1.55 nCi/nmol), 0.1-0.3 mg protein, at 37°C for 10 min in a final volume of 0.5 ml. Results are expressed as nmol/mg prot/min and are the average of duplicate determinations of one experiment representative of five others. c.s is cell suspension

intense cell division from day 1 to 3 and thereafter a period of cell maturation with a constant cell number and DNA content but with an increase of protein content. During this period the formation of cellular contacts and synapse-like junctions are observed (Louis et al., 1981).

As shown in Fig. 3, in these cultures the specific activity of PAF-synthesizing phosphocholinetransferase decreased between seeding and the 2nd day and remained constant thereafter during the period of cell proliferation. During the same period the specific activity of lyso-PAF acetyltransferase did not change significantly. Between the 3rd and 6th day *in vitro* the specific activity of the PAF-synthesizing phosphocholine-transferase and lyso-PAF acetyltransferase increased near 8-fold and 4-fold, respectively. These increases were much higher when the results were expressed on a DNA basis since, during this period of development, the DNA content remained constant whereas the amount of protein continued to increase (Louis et al., 1981). The increase of both enzymatic activities, during the period of neuronal maturation and synaptic like-junctions formation, is consistent with the hypothesis that PAF may be involved in some mechanism of neuronal differentiation and in cellular communication.

The observation that, in these cells, both enzymes had similar specific activities which increased in parallel during the development indicates that the synthesis of PAF by the PAF-synthesizing phosphocholinetransferase (*de novo*) and the lyso-PAF acetyltransferase (*remodeling*) may depend on the availability of the substrates (i.e. alkylacetylglycerol and lyso-PAF, respectively) and on the concentration of intracellular Ca^{2+}. In fact, calcium ions inhibit the activity of cholinephosphotransferase (Francescangeli and Goracci, 1989) and are required for acetyltransferase (Ninio et al., 1983; Gomez-Cambronero et al., 1984). Furthermore, it has been shown that the stimulation of chick retina with Ca^{2+} ionophore (A23187) stimulated only the activity of lyso-PAF acetyltransferase and permitted evaluation of its increase from the day 8 of embryonic life to hatching (Bussolino et al., 1988a).

On the other hand it has been also observed that the increase of PAF synthesis by phosphocholinetransferase, which takes place upon stimulation of chick retina with acetylcholine or dopamine, is essentially due to the increased availability of alkylacetylglycerol (Bussolino et al., 1988a).

Glial cell primary cultures
Primary glial cell cultures, obtained from of 14-day-old chick embryo cerebral hemispheres, proliferate during the first two weeks and form a confluent monolayer thereafter. About 86% of these cells are of astroglial type as determined by GFAP and GS immunochemistry (Booher et al., 1972).

During the development of these cultures the specific activity of PAF-synthesizing phosphocholinetransferase remained constant whereas the specific activity of lyso-PAF acetyltransferase increased continuously up to the 11th day and then remained constant. In contrast to primary neuronal cultures, the specific activity of the latter enzyme mainly increased during the period of proliferation and did not change significantly during the period of maturation (Fig. 4).

Moreover, the specific activity of PAF-synthesizing phosphocholine-transferase was similar in 6-day-old neuronal and in 16-day-old glial cell cultures whereas the specific activity of lyso-PAF acetyltransferase was about 4 times higher in the glial cell cultures. These data suggest that, in chick glial cells, PAF may be preferentially synthesized through the *remodeling* pathway. Obviously, the availability of the substrate lyso-PAF, which is produced by the action of phospholipase A_2 on alkylacyl-GPC, may be the limiting factor.

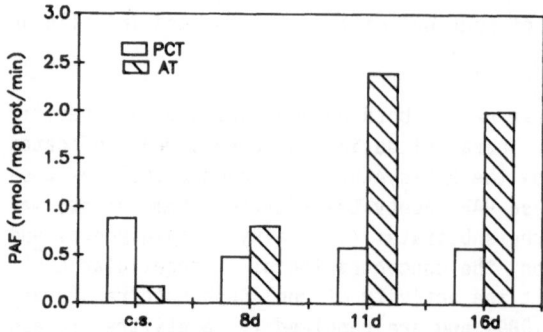

Fig. 4. Specific activities of PAF-synthesizing phosphocholinetransferase and lyso-PAF acetyltransferase in chick glial cell primary cultures. PAF-synthesizing phosphocholinetransferase (PCT) assay: 60 mM Tris-HCl pH 8, 0.5 mM 1-hexadecyl-2-acetyl-*sn*-glycerol, 0.005% Tween 20, 10 mM MgCl$_2$, 20 mM DTT, 0.1 mM CDP-[methyl-^{14}C]choline (spec. rad. 0.87 nCi/nmol), 0.09-0.2 mg protein, at 37°C for 15 min in a final volume of 0.3 ml. Lyso-PAF acetyltransferase (AT) assay: 100 mM Tris-HCl pH 6.9, 0.06 mM lyso-PAF, 0.2 mM [1-^{14}C]acetyl-CoA (spec. rad. 1.55 nCi/nmol), 0.18-0.4 mg protein, at 37°C for 10 min in a final volume of 0.5 ml. Results are expressed as nmol/mg prot/min and are the average of duplicate determinations of two different experiments. c.s. is cell suspension

CONCLUSIONS

The occurrence of PAF synthesis in brain has been reported by Francescangeli and Goracci (1989). They observed that rat brain can synthesize PAF by the *de novo* as well as by the *remodeling* pathways (Goracci and Francescangeli, 1991). The specific activities of lyso-PAF acetyltransferase and PAF-synthesizing phosphocholinetransferase are higher in microsomes than in other cellular fractions as was also reported for other tissues (Woodard et al., 1987; Gomez-Cambronero et al., 1985). Moreover, the ratio of the specific activity of PAF-synthesizing phospho-cholinetransferase versus the acetyltransferase, as measured in rat brain homogenate or in the microsomal fraction, was about 2.5-3.0. This indicates that in this tissue PAF may be better synthesized by the *de novo* pathway when saturating concentrations of the lipid substrates (alkyl-acetylglycerol or lyso-PAF) are available.

Chick brain hemispheres possess the enzymes catalyzing the last step of PAF synthesis. However, in the brain homogenate from 20-day-old embryo the specific activity of lyso-PAF acetyltransferase was about 7-fold higher than that of the phosphocholinetransferase suggesting that the *remodeling* pathway may be more efficient in the synthesis of PAF than the *de novo* one in chick hemispheres. In contrast, Bussolino et al. (1986) reported, for chick retina, a 3-fold higher specific activity for phosphocholinetransferase versus acetyltransferase. These data indicate species and cell differences in the specific activities of the enzymes catalyzing the last step of PAF synthesis in nervous tissue.

The determination of the enzyme activities in the chick hemispheres during development showed a tremendous increase of the lyso-PAF acetyltransferase activity and only a slight one for the PAF-synthesizing phosphocholine-transferase between the 12th and 20th day of embryonic life. Since this developmental period corresponds essentially to the proliferation and differentiation of glial cells it should be assumed that the lyso-PAF acetyltransferase is preferentially located in glial cells. This hypothesis has been confirmed by studies of PAF synthesis in chick neuronal and glial cell primary cultures.

In primary cultures of chick neurons the specific activities of both PAF-synthesizing phosphocholinetransferase and the lyso-PAF acetyl-transferase did not change during the 2nd and the 3rd day (period of proliferation) and increased, at a similar rate, when the cells ceased to divide and begun to mature and to form cellular contacts and synaptic-like junctions. The results suggest the involvement of PAF in cellular communication, a hypothesis which is consistent with the observations that neuronal cells possess PAF receptors (Junier et al., 1988; Kumar et al., 1988; Marcheselli et al., 1990) and that this lipid mediator is able to induce morphological differentiation of NG 108-15 cells (Kornecki et al., 1988), to increase intracellular Ca^{2+} (Kornecki et al., 1988) and to stimulate the release of dopamine in PC12 cells (Bussolino et al., 1988b).

In glial cell cultures the specific activity of PAF-synthesizing phospho-cholinetransferase remained constant during the first two weeks of culture whereas the lyso-PAF acetyltransferase increased during the first 11 days and did not change significantly thereafter indicating that the activity of this enzyme increased mostly during the period of glial cell proliferation.

It has been suggested that the *de novo* synthesis pathway for PAF may be mainly devoted to maintain the physiological levels of PAF for normal cell function, whereas the *remodeling* pathway may be activated under conditions in which PAF production is coupled to a stimulation dependent response (Snyder et al., 1987). This hypothesis is consistent with the observation that the PAF-synthesizing phosphocholinetransferase has similar specific activity in mature chick neuronal and glial cell cultures. On the other hand, the specific activity of lyso-PAF acetyltransferase is much higher in glial cells than in neurons suggesting that astrocytes may be preferentially involved in the synthesis of extracellular PAF. It is obvious that more extensive studies at the cellular level are needed to understand the physiological role of both PAF synthetic pathways in the nervous tissue. Likewise it is necessary to establish (1) the physiological relevance of endogenous levels of PAF in the various nervous cells, (2) whether PAF is released upon cellular stimulation of neurons or glial cells and (3) the cellular responses of these cells when stimulated with PAF.

Nevertheless, regardless of the function of PAF in nervous tissue, the differences in the specific activities of the enzymes involved in its synthesis in neurons and astrocytes suggest that each cell type plays a specific role in the production of PAF and that their interaction with the lipid mediator may produce different responses.

REFERENCES

Altman J (1969) DNA metabolism and cell proliferation. In: Lajtha A (ed) Handbook of Neurochemistry, vol II. Plenum Press, New York, p 157

Bensted JPM, Dobbing J, Morgan RS, Reid RTW, Payling-Wright G (1957) Neurological development and myelination in the spinal cord of chick embryo. J Embryol Exp Morphol 5:428-431

Bligh EG, Dyer WJ (1959) A rapid method of total lipid extraction. Can J Biochem Physiol 37:911-917

Booher J, Sensenbrenner M (1972) Growth and cultivation of dissociated neurons and glial cells from embryonic chick, rat and human brain in flask cultures. Neurobiology 2:97-105

Bussolino F, Gremo F, Tetta C, Pescarmona GP, Camussi G (1986) Production of platelet-activating factor by chick retina. J Biol Chem. 261:16502-16508.

Bussolino F, Pescarmona G, Camussi G, Gremo F (1988a) Acetylcholine and dopamine promote the production of platelet activating factor in immature cells of chick embryonic retina. J Neurochem 51: 1755-1759

Bussolino F, Tessari F, Turrini F, Braquet P, Camussi G, Prosdocimi M, Bosia A (1988b) Platelet-activating factor induces dopamine release in PC-12 cell line. Am J Physiol 255: (Cell Physiol) 24:C559-C565

Camussi G, Aglietta M, Malavasi F, Tetta C, Sanavio F, Piacibello W, Bussolino F (1983) The release of platelet-activating factor from human endothelial cells in culture. J Immunol 131:2397-2401

Dreyfus H, Harth S, Massarelli R, Louis JC (1981) Mechanisms of differentiation in cultured neurons: involvement of gangliosides. In: Rapport MM, Gorio A (eds) Gangliosides in Neurological and Neuromuscular Function, Development and Repair. Raven Press, New York, p 151

Feuerstein G, Yue TL (1989) PAF as a putative mediator in cardiac and cerebrovascular disease. In: Saito K, Hanahan DJ (eds) Platelet-Activating Factor and diseases. International Medical Publ,Tokyo, p 103

Francescangeli E, Goracci G (1989) The *de novo* biosynthesis of Platelet-Activating Factor in rat brain. Biochem Biophys Res Commun 161: 107-112

Freysz L, Lastennet A, Mandel P (1972) Phosphocholine diglyceride transferase activity during development of the chick brain. J Neurochem 19:2599-2605

Freysz L, Horrocks LA, Mandel P (1980) Activities of enzymes synthesizing diacyl, alkylacyl and alkenylacyl glycerophosphocholine during development of chick brain. J Neurochem 34:961-969

Ghigo D, Bussolino F, Gamberino G, Heller R, Turrini F, Pescarmona G, Cragoe EJ, Pegoraro L, Bosia A (1988) Role of Na^+/H^+ exchange in the thrombin-induced platelet-activating factor production by human endothelial cells. J Biol Chem 253:19437-19446

Gomez-Cambronero J, Inarrea P, Alonso F, Sanchez-Crespo M (1984) The role of calcium ions in the process of acetyltransferase activation during the formation of platelet-activating factor (PAF-acether). Biochem J 219:419-424

Gomez-Cambronero J, Velasco S, Maton JM, Sanchez-Crespo M (1985) Modulation of lyso-platelet activating factor:acetylCoA acetyltransferase from rat splenic microsomes. Biochim. Biophys. Acta 845, 515-519.

Goracci G (1990) PAF in the nervous system: biochemistry and pathophysiology. In: Krieglstein J and Oberpichler H (eds) Pharmacology of Cerebral Ischemia, Wissenschaftliche Verlagsgesellshaft, Stuttgart, p 377

Goracci G, Francescangeli E (1991) Properties of PAF-synthesizing phosphocholinetransferase and evidence for Lyso-PAF acetyltransferase activity in rat brain. Lipids, 26:986-991

Judes C, Sensenbrenner M, Mandel P, Jacob M (1968) Differentiating cells of chick embryo hemispheres: Characterization and quantitative evaluation of cell types and bulk preparation of enriched fractions. Z Zellforsch Mikrosk Anat 89:137-150

Kornecki E, Ehrlich YH (1988) Neuroregulatory and neuropathological actions of the ether-phospholipid platelet-activating factor. Science 240:1792-1794

Kumar R, Harvey SAK, Kester M, Hanahan DJ, Olson MS (1988) Production and effects of platelet-activating factor in the rat brain. Biochim Biophys Acta 963:375-383

Kurihara T, Tsukada Y (1968) 2′,3′-Cyclic nucleotide-3′-phosphodiesterase in the developing chick brain and spinal cord. J Neurochem 15:827-832

Junier MP, Tiberghien C, Rougeot C, Fareur V, Dray F (1988) Inhibitory effect of platelet activating factor on LHRH and somatostatin release on rat medial eminence in vitro correlated with the the characterization of specific binding sites in rat hypothalamus. Endocrinology 123: 72-80

Lee T-C, Malone B, Snyder F (1986) A new de novo pathway for the formation of 1-alkyl-2-acetyl-sn-glycerols, precursors of platelet-activating factor. Biochemical characterization of 1-alkyl-2-lyso-sn-glycero-3-P acetyltransferase in rat spleen. J Biol Chem 261:5373-5377

Lee T-C, Malone B, Snyder F (1988) Formation of 1-alkyl-2-acetyl-sn-glycerols via the de novo biosynthetic pathway for platelet-activating factor. Biochemical characterization of 1-alkyl-2-acetyl-snglycero-3-P phosphohydrolase in rat spleen. J Biol Chem 263:9181-9187

Louis JC, Pettmann B, Courageot J, Rumigny JF, Mandel P, Sensenbrenner M (1981) Developmental changes in cultured neurones from chick embryo cerebral hemispheres. Exp Brain Res 42:63-72

Lowry OH, Rosebrough NJ, Farr AL, Randall J (1951) Protein measurement with the Folin phenol reagent. J Biol Chem 193:265-275

Marcheselli VL, Rossowska MJ, Domingo MT, Braquet P, Bazan NG (1990) Distinct platelet-activating factor binding sites in synaptic endings and intracellular membranes of rat cerebral cortex. J Biol Chem 265:9140-9145

Ninio E, Mencia-Huerta JM, Benveniste J (1983) Biosynthesis of platelet-activating factor (PAF-acether). V. Enhancement of acetyltransferase activity in murine peritoneal cells by calcium ionophore A23187. Biochim Biophys Acta 751:298-304

Nurenberger JI (1958) Direct enumeration of cells of the brain. In: Windle WF (ed) Biology Neuroglia, Charles C Thomas, Springfield, Ill, p 193

Pettmann B, Louis JC, Sensenbrenner M (1979) Morphological and biochemical maturation of neurons cultured in the absence of glial cells. Nature 281:378-380

Reddick YL (1951) Histogenesis of the cellular elements in the postodic medulla of the chick embryo. Anat. Rec 109:81-97

Romanoff AL (1960) The nervous system. In: Romanoff AL (ed) Avian Embryo. McMillan, New York, p 209

Snyder F (1987) The significance of dual pathways for the biosynthesis of platelet activating factor: 1-alkyl-2-lyso-sn-glycero-3-phosphate as a branchpoint. In: Winslow CM, Lee ML (eds) New Horizons in Platelet Activating Factor Research. Wiley and Sons Ltd, New York, p 13

Sogos V, Bussolino F, Pilia E, Torelli S, Gremo F (1990) Acetyl-choline-induced production of platelet-activating factor by human fetal brain cells in culture. J Neurosci Res 27:706-711

Tokumura A, Kamiyasu K, Takauchi K, Tsukatami H (1987) Evidence for the existence of various homologoues and analogues of platelet-activating factor in a lipid extract from bovine brain. Biochem Biophys Res Commun 145:415-425

Woelk H, Goracci G, Porcellati G (1974) The action of brain phospholipase A_2 on purified specifically labeled 1,2-diacyl, 2-acyl-1-alk-1'--enyl, and 2-acyl-1-alkyl-*sn*-glycero-3-phosphorylcholine. Hoppe-Seyler's Z Physiol Chem 355:75-81

Woodard DS, Lee T-c, Snyder F (1987) The final step in the de novo biosynthesis of platelet-activating factor. Properties of a unique CDP-choline: 1-alkyl-2-acetyl-*sn*-glycerol cholinephosphotransferase in microsomes from renal inner medulla of rats. J Biol Chem 262:2520-2527

Yue T-L, Lysko PG, Feuerstein J (1990) Production of platelet-activating factor from rat cerebellar granule cells in culture. J Neurochem 54:1809-1811

PAF AND HYPOTHALAMIC SECRETIONS IN THE RAT

Fernand Dray, Catherine Rougeot, Christopher Tiberghien, Marie-Pierre Junier
URIA, Institut Pasteur and U. 207, INSERM
28, rue du Docteur Roux
75724 - Paris cedex 15

Platelet-activating factor (PAF), an ether phospholipid, is recognized as being an effective and potent modulator of secretory processes in various cell types (Snyder, 1982). Physio-pathological studies have shown that PAF participates in inflammatory processes and in immune functions (Braquet et al., Res. Pharmacol. Rev., 1987). In the central nervous system, PAF has been isolated from bovine cerebral tissues (Tokumura and Tsukatani, 1986) and Francescangeli and Goracci (1989) demonstrated that rat brain synthesizes PAF under resting conditions. Moreover, Kumar et al. (1988) reported that the rat cerebrum produces PAF in response to convulsive stimuli.

However, the physiological significance of PAF production in the brain is unclear. It was of interest to investigate whether PAF is implicated in the release of brain neuropeptides. This study, using male adult rats, examined the effects of PAF on the release of hypothalamic releasing hormones, in particular : luteinizing hormone releasing hormone (LHRH), somatostatin (SRIF), growth hormone releasing factor (GRF), corticotropin releasing factor (CRF) and arginine vasopressin (AVP). In addition, the effects of PAF on the release of pituitary hormones secreted under the control of these releasing hormones was examined, i.e. luteinizing hormone (LH), growth hormone (GH), adrenocorticotropin hormone (ACTH) and corticosterone.

The first step of the events triggered by PAF in many tissues is its interaction with a specific receptor. Rat brain was therefore investigated for the presence of specific PAF binding sites. Finally, the effects of characterized potent PAF antagonists on the specific binding of PAF and on the neuroendocrine secretions in which PAF is implicated were analyzed to define the mode of PAF action on neuropeptide release.

NATO ASI Series, Vol. H 70
Phospholipids and Signal Transmission
Edited by R. Massarelli, L. A. Horrocks,
J. N. Kanfer, and K. Löffelholz
© Springer-Verlag Berlin Heidelberg 1993

METHODS

Using adult male rats as a model, the effects of PAF on neuroendocrine secretions were investigated.
- in vitro, in isolated and incubated hypothalamic fragments and pituitary glands and,
- in vivo, in conscious or anesthetized catheterized rats.
In parallel, rat brain was analysed to determine whether it contains PAF specific binding sites using partially purified hypothalamic and hypophyseal membrane preparations.

I - Median eminence (ME) and anterior pituitary (AP) incubation procedures

The ME and the AP were dissected out under a stereomicroscope as previously described (Negro-Vilar et al., 1979 and Fafeur et al., 1985). Tissues were incubated in 0.4 (ME) or 1 ml (AP) of Krebs-Ringer bicarbonate glucose buffer (KRBG) containing 1000 KIU/ml aprotinin and 0.1% bovine serum albumin (BSA), at 37°C under an atmosphere of 95% O_2 and 5% CO_2. At the end of the incubation period in the presence of test substances or their vehicles (30 min for the ME and 3 hours for the AP), the medium was removed and stored at -80°C until radioimmunoassay (RIA) of the various peptides under investigation.

II - Conscious rat experiments

The rats were prepared as described previously (Rougeot et al., 1990). Under anesthesia, an intracardiac silastic catheter was introduced and a left lateral cerebroventricular metal cannula was implanted. Both catheter and cannula were secured to the skull with dental cement. On the day of the experiment, the cerebroventricular cannula and the intracardiac catheter were extended by tubing allowing intracerebroventricular (i.c.v.) infusion of test substances or their vehicles and blood sampling while the rats were freely moving, eating or sleeping. Sequential blood samples (0.3 ml) were withdrawn before and 15, 30, 45, 60, 90 and 120 min after the i.c.v. injection, collected and centrifuged at 4°C. The plasma fractions were removed and stored at -80°C until use for hormone determinations by RIA .

III - Anesthetized rat experiments.

The rats were prepared according to the procedure described by Porter (1985). At the end of the surgical procedure, the rats were heparinized (250 IU heparin in 2 ml saline solution), the pituitary stalk abscissed and the portal blood collected into tubes containing aprotinin (300 KIU) at 0°C.

Blood samples were withdrawn continuously 30 min before and 15, 30, 45 and 60 min after the i.c.v. injection of PAF or its vehicle. The samples were then centrifuged for 15 min at 4000g and 4°C. The plasma fractions were removed and stored at -80°C until the CRF determination by RIA.

IV - PAF binding experiments.

Hypothalamic and hypophyseal membranes were prepared as previously described (Junier et al., 1988). All incubations were carried out in triplicate at 4°C, in polyethylene tubes with aliquots of membrane (about 150 μg), 1 nM [3H]PAF and one of a series of unlabelled PAF concentrations (0.1 nM to 1 μM) in a final volume of 0.2 ml of buffer: 50 mM Tris/HCl, 10 mM MgCl2, 0.25 % BSA, pH 7.4. The samples were incubated for 5 hours at 4°C, and free and bound ligand were separated by a filtration technique. Three milliliters of cold incubation buffer were added to each sample which was immediately filtered under constant vacuum through Whatman glass filters GF/B which had been previously wetted in incubation buffer. Then the filters were washed twice with 3 ml of the same buffer and placed into a plastic vial with 4 ml of scintillation fluid (Atomlight, New England Nuclear/Dupont de Nemours, Boston, MA, USA), and counted in a LKB Wallac 1214 Rackbeta scintillation counter at an efficiency of 56%.

V - Radioimmunoassays.

The peptides rCRF, rGRF, SRIF, LHRH and AVP were radioimunoassayed as previously described (Louis et al, 1983; Rougeot et al, 1986 and 1988).

The concentrations of the hormones LH and GH were measured using the NIDDK kit, while the hormones ACTH and corticosterone were determined by radioimmunoassay as described by Rougeot et al. (1983, 1990).

RESULTS

SPECIFIC BINDING OF [3H]PAF

I - [3H]PAF binding to hypothalamic membranes.

Unlabelled PAF fully displaced specifically bound [3H]PAF (1 nM) as indicated in Figure 1. [3H]PAF-specific binding was saturable and the computerized analysis of the data using a non-linear iterating last square fitted curve program (Munson and Rodbard, 1980) revealed the putative presence of two populations of binding sites. The dissociation constants (Kd) determined were Kd1 = 2.1 ± 0.3 nM and Kd2 = 61.6 ± 16.4 nM with the

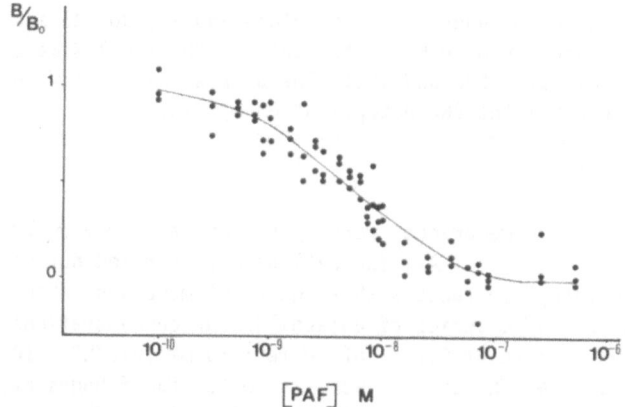

Fig. 1. Competition of specific [³H]PAF (1 nM) binding to rat hypothalamic membranes by unlabelled PAF

binding capacity Bml = 25.4 ± 3.2 and Bm2 = 146.2 ± 47.5 fmol/mg protein, respectively. Bound PAF was displaced by various PAF antagonists: L-652,731, Kadsurenone, BN 52021, BN 50739 and RP 52770, and the displacement correlated with the concentration of the antagonist in each case. The comparative binding inhibition potencies (IC50) of these compounds were PAF = 2.3 ± 0.15 nM < RP 52770 = 12.4 ± 4.1 nM < BN 50739 = 67.7 ± 10.1 nM < L 652,731 = 138 ± 75 nM < Kadsurenone = 283 ± 176 nM < BN 52021 = 780 ± 480 nM.

II - [³H]PAF binding to hypophyseal membranes.

Specific [³H]PAF binding to hypophyseal membranes was assayed over the concentration range of 0.5 to 7 nM. No specific [³H]PAF binding in the pituitary was detected.

LHRH AND LH SECRETIONS

I - Effects of PAF on LHRH and LH release in vitro from the ME and AP, respectively.

The effect of increasing concentrations of PAF on LHRH release from ME nerve terminals in vitro is presented in Figure 2. PAF inhibited LHRH release in a dose-related manner with a maximal effect at 10^{-14} M PAF (79.2 ± 14.2 pg/mg protein, n = 6 vs control 183.3 ± 8.3 pg/mg protein, n = 13).

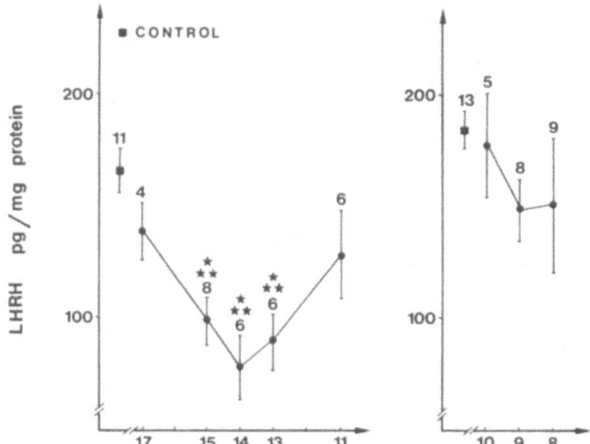

Fig. 2. Effect of increasing concentrations of PAF on the levels of LHRH released from ME during a 30 min incubation period. Values are mean ± SEM, numbers at the top of error bars indicate the number of determinations. Abscissa coordinate is -log [(PAF)M]. *** p < 0.001 vs control (Newman-Keul test)

In contrast, incubation of AP with various PAF concentrations (10^{-14} to 10^{-7}M) in the medium did not significantly affect LH release. However, in the same experiment, LHRH ($5 \cdot 10^{-7}$M) induced a significant rise in LH release (to 5.4 times the control mean) (data not shown).

II - Effect of PAF on the plasma LH level of conscious rats.

The i.c.v. infusion of PAF in conscious rats (5 nmol/5 μl over a 10 min period) did not significantly alter the plasma LH levels during the 2-hour observation period following the injection, compared to plasma LH levels of the same rat before injection. The mean level within 2 h of injection was 4.8 ± 1.5 ng/ml whereas that during the 90 min before injection was 4.6 ± 3.4 ng/ml, n = 3 rats).

SRIF, GRF AND GH SECRETIONS

I - Effects of PAF on SRIF and GRF release from ME and GH release from AP, in vitro.

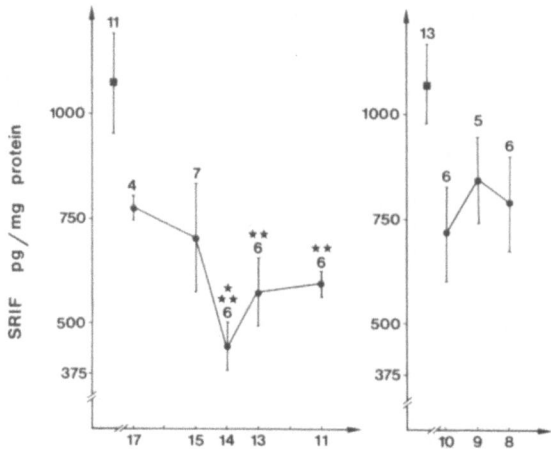

Fig. 3. Effect of increasing concentrations of PAF on the levels of SRIF released from ME during a 30 min incubation period. Values are mean ± SEM, numbers at the top of error bars indicate the number of determinations. Abscissa coordinate is -log [(PAF) M]. ** p < 0.01, *** p < 0.001 vs control (Newman-Keul test).

As illustrated in Figure 3, PAF decreased SRIF release from ME nerve terminals in vitro, with a maximal inhibition at 10^{-14}M (452.2 ± 54.3 pg/mg protein, n = 6 as compared to controls 1073.9 ± 97.8 pg/mg protein, n = 13). In the same experiments, GRF release was not significantly affected by any of the doses of PAF tested (10^{-17}M to 10^{-8}M). Only the highest concentration of PAF used (10^{-8}M) increased the release of GH from AP, that to 1.6 times the mean control value. However, incubation with 10^{-8}M GRF resulted in a GH release 4.3 times higher than the control mean.

II - Effect of PAF on the plasma GH level of conscious rats.

In vivo, the i.c.v. infusion of 5 nmol PAF suppressed systematically the physiological peak of GH secretion in the 7 chronically implanted and treated rats. In the 8 vehicle-treated rats, the physiological peak of GH secretion was not altered. The mean basal value of plasma GH was 9.4 ± 3.2 ng/ml and the mean peak value of plasma GH was 71 ± 6 ng/ml (vehicle) as compared to 7.2 ± 4.7 ng/ml (PAF i.c.v. injection).

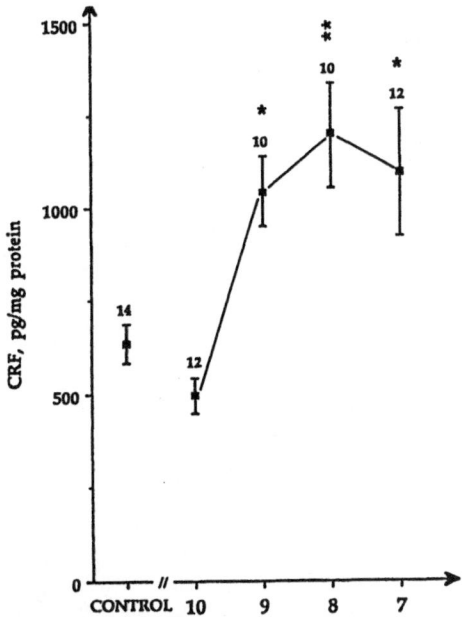

Fig. 4. Effect of increasing concentrations of PAF on the levels of CRF released from ME during a 30 min incubation period. Values are mean ± SEM, numbers at the top of error bars indicate the number of determinations. Abscissa coordinate is -log [(PAF) M]. * p ≤ 0.05, ** p < 0.01, vs control (Newman-Keul test).

CRF, AVP, ACTH AND CORTICOSTERONE SECRETIONS

I - Effects of PAF on CRF and AVP release from ME and ACTH release from AP, in vitro.

PAF from 10^{-9} to 10^{-7}M induced a significant increase in the CRF release from ME, with a maximal effect at 10^{-8}M (1195 ± 143 pg/mg protein, n = 10 as compared to controls 635 ± 52 pg/mg protein, n = 12) (Figure 4). In contrast, none of the doses of PAF tested (10^{-11} to 10^{-7}M) modified the release of AVP from ME (data not shown). Moreover, these PAF concentrations did not alter the amount of ACTH released from AP, whereas 10^{-9}M CRF in the same experiments induced a significant release of ACTH (2.4 times the control mean) (data not shown).

II - Effect of PAF on CRF release, in vivo, in pituitary portal blood of anesthetized rats.

The effect of a 3 nmol PAF i.c.v. injection on the pituitary portal plasma CRF level from anesthetized rats is shown in Figure 5. The plasma CRF level of PAF-injected rats was markedly higher within 45 min of starting

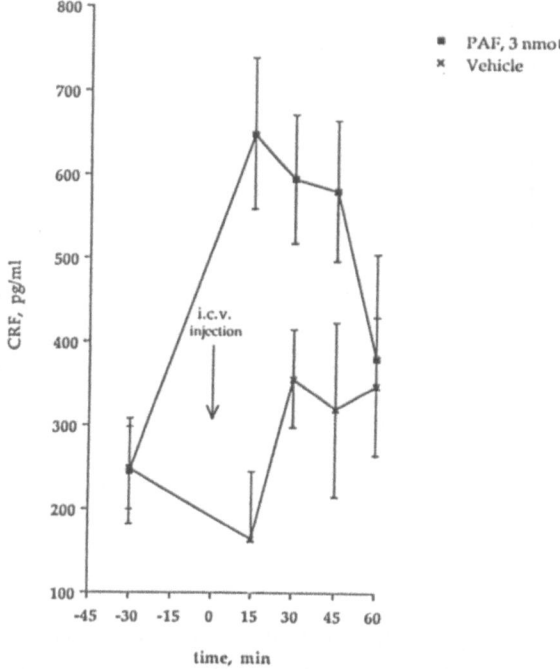

Fig 5. Effect of 3 nmol PAF i.c.v. injection, or its vehicle, on the release of CRF into the pituitary portal blood of anesthetized rats. Each point represents the mean ± SEM of determination from 9 rats (3 nmol PAF) or 7 rats (vehicle). (p < 0.025 vs vehicle by two way ANOVA).

i.c.v. injection (595 ± 77 pg/ml, n = 9 rats) than that of vehicle-injected rats (245 ± 62 pg/ml, n = 7 rats).

III - Effects of PAF on the plasma ACTH and corticosterone levels of conscious rats.

In the 20 treated-conscious rats, the plasma ACTH level increased from the baseline (44 ± 15 pg/ml) to 146 ± 19 pg/ml, within 15 min of the onset of i.c.v. injection of 5 nmol PAF and was maximal at 60 min, 203 ± 29 pg/ml (Figure 6A). This plasma ACTH peak induced by PAF was 3 times higher than that of the vehicle control (66 ± 11 pg/ml, n = 17 rats).

The time course of plasma corticosterone after i.c.v. injection of 5 nmol PAF is shown in Figure 6B. Plasma corticosterone levels increased above basal levels within 15 min of the onset of the injection, however the increase was significantly above that of the vehicle-treated rats only for 45 min, 112 ± 16 ng/ml, n = 20 as compared to vehicle 46 ± 13 ng/ml, n = 17. The corticosterone response was sustained for at least 120 min after the onset of PAF infusion.

395

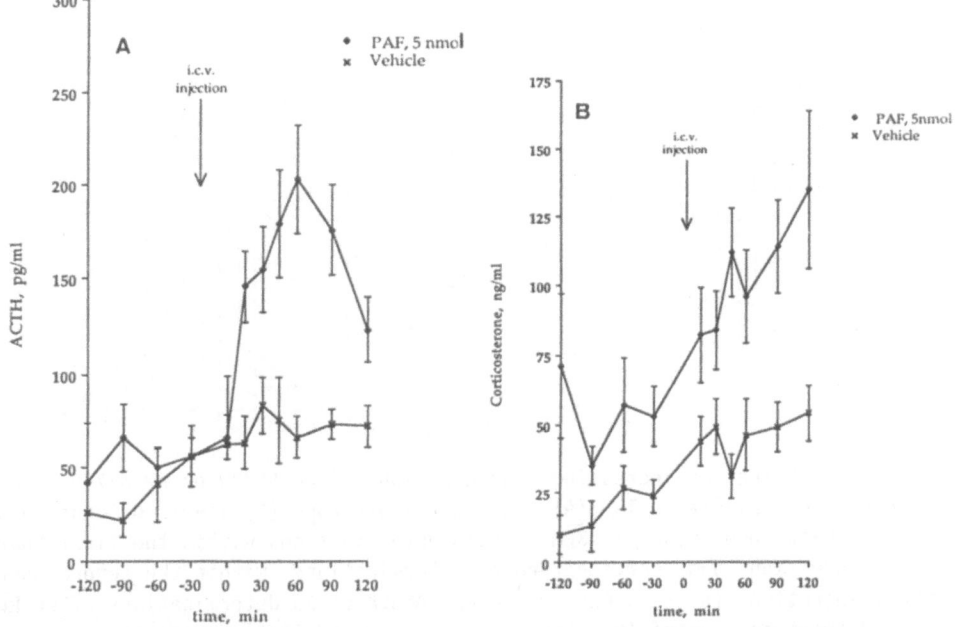

Fig. 6. Effect of 5 nmol i.c.v. injection of PAF or vehicle on plasma ACTH (A) and corticosterone (B) levels in conscious rats. Arrows indicate the start of the injection (5 μl/10 min). Each point represents the mean ± SEM of ACTH and corticosterone determination from 20 rats (PAF) or 17 rats (vehicle) (p < 0.025 vs vehicle by two-way ANOVA).

In addition, the analysis of the dose-response of plasma ACTH (first hour post-injection) and corticosterone (second hour post-injection) levels to PAF, over 60 min observation period, showed that PAF stimulated the in vivo ACTH and corticosterone secretion, in concentration-dependent manner with a minimal effective concentration of 1 nmol PAF for ACTH and 3.6 nmol PAF for corticosterone.

IV - Effects of PAF antagonists on basal plasma ACTH and corticosterone levels of conscious rats.

The effects of i.c.v. injection of BN 50739 and RP 52770 (two potent PAF antagonists) on the ACTH and corticosterone secretion are presented in Figure 7A.B. Doses of either 33 or 50 nmol BN 50739 or 40 or 55 nmol RP 52770, induced significant dose-dependent drops in the basal plasma ACTH and corticosterone levels during the first and the second hour post-injection, respectively.

396

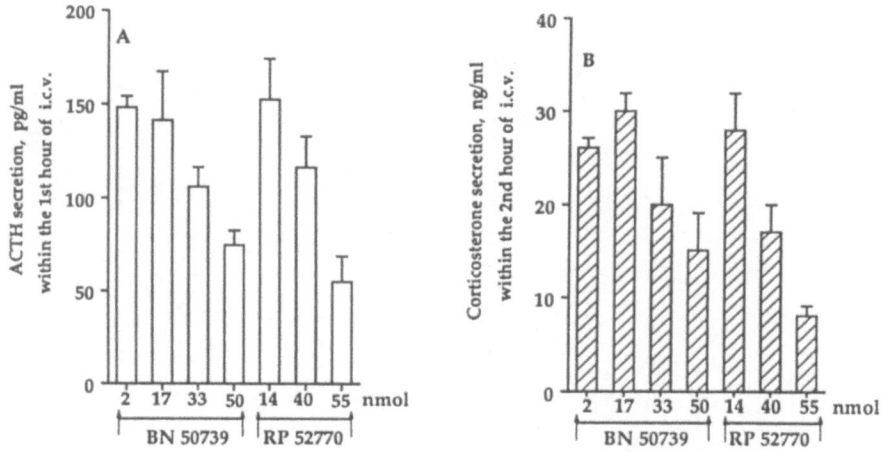

Fig 7. Effect of increasing concentrations of BN 50739 or RP 52770 given
i.c.v., on plasma ACTH (A) and corticosterone (B) levels of conscious
rats. Data are mean ± SEM of ACTH determinations within the first hour
post-injection and corticosterone determinations within the second hour
post-injection (p < 0.025 by one way ANOVA of 29 determinations after RP
52770 treatment and of 40 determinations after BN 50739 treatment).

Additionally, i.c.v. injection of these doses of PAF receptor antagonists
produced a transient drop in the level of circulating corticosterone to
below the level of detectability (< 5 ng/ml) as did a 1 nmol i.c.v.
injection of dexamethasone (data not shown). Moreover, the ability of
these compounds to interact with the secretion of stress-related hormones
in vivo was correlated to their ability to displace [3H]PAF from its
hypothalamic receptor sites. Indeed, the ratio between the effective
concentrations of BN 50739 and PAF was approximately 30 and was identical
for both the biological and the binding activities. This ratio was 40 for
the biological and 6 for the binding activities for RP 52770.

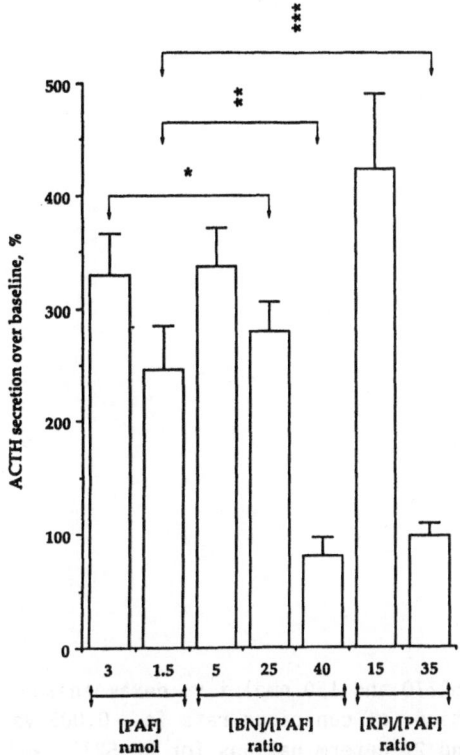

Fig. 8. Effect of i.c.v. pretreatment with BN 50739 or RP 52770, on PAF-induced ACTH secretion of conscious rats. The results are expressed as percentage over baseline of ACTH secretion within 60 min of starting i.c.v. injection (* p ≤ 0.05, ** p ≤ 0.01, *** p ≤ 0.001 vs PAF by Student's t test of 14 to 26 determinations).

V - Effects of PAF antagonists on stimulated ACTH secretion of conscious rats.

1) PAF-induced ACTH secretion

I.c.v. pretreatment with BN 507339 or RP 52770 reversed the plasma ACTH response to PAF (Figure 8) and the inhibition was dependent on the ratio of antagonist to PAF concentration. Indeed, a [BN]/[PAF] ratio of 5, or a [RP]/[PAF] ratio of 14 did not counteract the PAF-stimulated secretion of ACTH in vivo, whilst a [BN]/[PAF] ratio of 25 decreased PAF-stimulated ACTH secretion (15 %), and a [BN]/[PAF] ratio of 40 or a [RP]/[PAF] ratio of 35 abolished the ACTH response to PAF.

2) Ether stress-induced ACTH secretion

Pretreatment with 50 nmol i.c.v. BN 50739 did not prevent the ACTH response to ether stress, while pretreatment with 55 nmol i.c.v. RP 52770

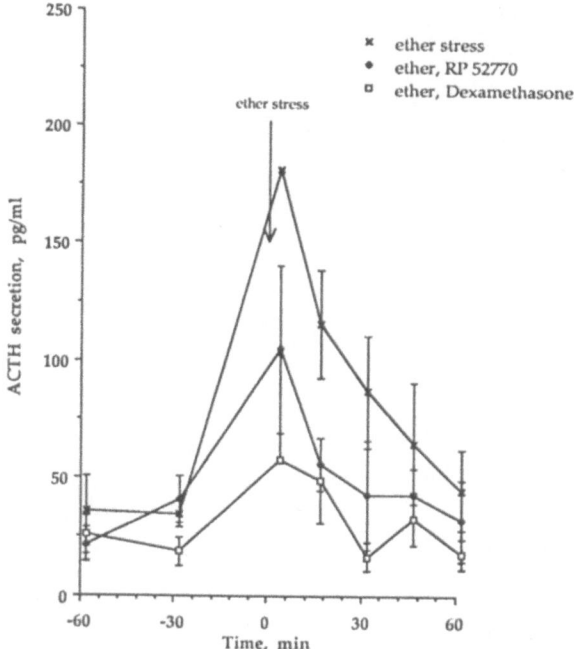

Fig. 9. Effect of 55 nmol i.c.v RP 52770 and 150 nmol i.v. dexamethasone on ether stress-induced ACTH secretion of conscious rats (p ≤ 0.005 vs ether stress alone by ANOVA of 24 and 27 determinations for RP 52770 and dexamethasone respectively).

significantly inhibited the ether stress-induced ACTH activation by 46%. However this inhibition was much less pronounced than that of rats pretreated with dexamethasone (150 nmol i.v.), where the ACTH response to ether was abolished (Figure 9).

3) *Interleukin 1-induced ACTH secretion*

Figure 10 illustrates the effect of 2 nmol IL1β i.c.v. injection on plasma ACTH, expressed as a percentage of the basal value. IL1β induced a 298 ± 24% increase in ACTH secretion over baseline within 120 min. Pre-injection with 50 nmol BN 50739 had no effect on the IL1β-induced ACTH secretion. In contrast, pretreatment with 55 nmol i.c.v. RP 52770 reversed the IL1β-induced ACTH secretion within 60 min of starting cytokine injection. Dexamethasone (1 nmol i.c.v. or 150 nmol i.v.) suppressed the adrenocorticotropin response to cytokine over the whole 2-hour observation period.

DISCUSSION

The characterization of specific [^{3}H]PAF binding to hypothalamic membranes revealed the existence of more than one population of binding sites. The

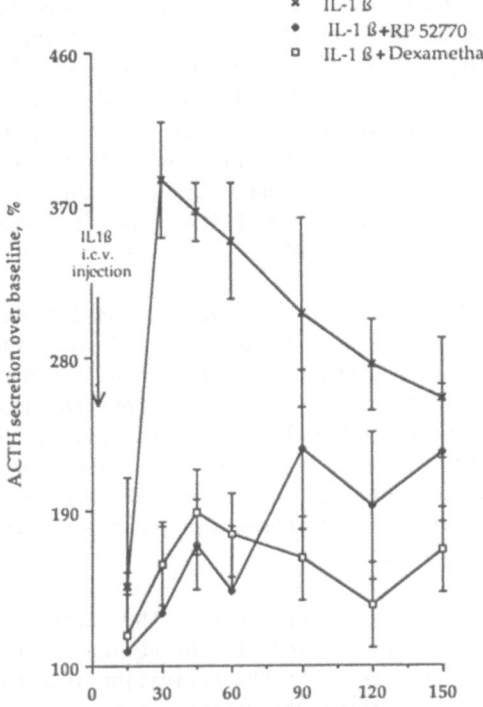

Fig. 10. Effect of 55 nmol i.c.v. RP 52770 and 150 nmol i.v. dexamethasone on ACTH secretion of conscious rats induced by IL1β (2 nmol i.c.v.). The results are expressed as percentage of basal ACTH secretion (p $\leq$ 0.0005 vs IL1β alone by ANOVA of 57 and 48 determinations for dexamathasone and RP 52770 respectively).

affinity (Kd) and the capacity (Bm) of the first class are close to that reported for PAF binding sites in various tissues. The Scatchard plot for PAF binding to the hypothalamus is curvilinear suggesting more than one population. Nevertheless, the binding parameters of the second class of PAF binding sites appear to be variable and the analysis of data cannot determine (or exclude) the existence of several populations.

PAF decreased LHRH release from incubated rat hypothalamic fragments and did not significantly affect pituitary release of LH under the control of LHRH, either in vivo or in vitro. The discrepancy between the ligand affinity (2.10^{-9}M) and the concentration of PAF giving maximal inhibition of LHRH release (10^{-13} to 10^{-15}M), and the absence of inhibitory effect of PAF on LH secretion, in vivo, suggests that the in vitro inhibitory effect of PAF on LHRH is unlikely to have any physiological significance for gonadotroph secretion.

PAF decreased SRIF release from incubated rat hypothalamic fragments, whereas GRF release was not affected. PAF tended to increase GH release from AP in vitro, but suppression, instead of stimulation, of the physiological peak of GH secretion was observed in vivo. This result indicates an indirect effect of PAF, exerted predominantly in vivo on GH secretion, through an endogenous compound independant from SRIF or GRF. On the other hand, there is also a discrepancy between the PAF concentration required for receptor occupancy ($2 \cdot 10^{-9}$M) and for the maximal inhibition of SRIF release (10^{-14} to 10^{-11}M).

In vitro, PAF exerted a stimulatory effect on CRF release, and did not alter the release of AVP from the hypothalamus. In addition, PAF did not affect the release of the pituitary hormone ACTH in vitro, which is under the control of CRF and AVP. The ability of PAF to displace PAF from its binding sites (10^{-10}M to 10^{-7}M) correlated well with its ability to stimulate CRF release (10^{-9} to 10^{-7}M). In addition, the stimulatory effect of PAF on the release of CRF was confirmed in vivo on CRF secretion into the portal circulation, and ACTH and corticosterone secretion into the peripheral circulation of conscious rats.

These results suggest that PAF has a physiological role on the secretion of hypothalamic-pituitary-adrenal (HPA) axis products by acting on the hypothalamus through mediation of CRF release. The HPA activation induced by exogenous PAF is antagonized by two potent PAF receptor antagonists. Moreover, there is a close correlation between the ability of PAF receptor antagonists to displace [³H]PAF from its hypothalamic binding sites and their ability to counteract PAF-stimulated pituitary-adrenocortical secretion in vivo. PAF-induced HPA secretion appears to be pharmacologically specific and exerted through a centrally mechanism mediated by the PAF receptor site(s). The HPA activation induced by acute ether stress or immune mediators, i.e. IL1β is dramatically reduced in rats pretreated with one of both PAF receptor antagonists. These results suggest that endogenous PAF is involved in the modulation of the cellular responses to ether stress and IL1β, leading to activation of the HPA axis.

On the other hand, i.c.v. injection of PAF receptor antagonists results in the concentration of plasma corticosterone falling to an undetectable level. A similar diminution of plasma corticosterone levels has been observed with dexamethasone, a synthetic glucocorticoid. Moreover, in the presence of both dexamethasone and a PAF antagonist, ether stress- and IL1β-induced HPA secretion was inhibited. The PAF antagonist-induced suppression of basal and stimulated plasma corticosterone levels could operate by interfering with the glucocorticoid negative feedback system in the brain. HPA axis secretion could therefore be controlled, in part, by an interaction between PAF and glucocorticoid negative feedback.

Finally, the similarity of the effects of PAF and IL1β, at the same doses, on secretion (1-6 nmol), and the fact that IL1β-induced HPA axis activation is suppressed by pretreatment of rats with a potent PAF antagonist suggests that the effects of IL1β on the central nervous system may be mediated partly through PAF production. These results suggest that PAF is involved in the physiological regulation of HPA secretion and that it has a central role in mediating basal and stimulated HPA secretion.

REFERENCES

Braquet P, Touqui L, Shen TY, Vargaftig BB (1987) Perspectives in platelet-activating factor research. Pharmacol Rev 39:97-145

Fafeur V, Gouin E, Dray F (1985) Growth hormone releasing factor (GRF) stimulates PGE_2 production in rat anterior pituitary. Evidence of a PGE_2 involvement in GRF-induced GH release. Biochem Biophys Res Commun 126:725

Francescangeli E and Goracci G (1989) The de novo biosynthesis of PAF in rat brain. Biochem Biophys Res Commun 161:107-112

Junier MP, Tiberghien C, Rougeot C, Fafeur V, Dray F (1988) Inhibitory effect of PAF on LHRH and SRIF release from rat median eminence in vitro correlated with the characterization of specific PAF receptor sites in rat hypothalamus. Endocrinology 123:72-80

Kumar R, Harvey, SA, Kester M, Hanahan DJ and Olson MS (1988) Production and effects of platelet-activating factor in the rat brain. Biochim Biophys Acta 963:375-383

Louis JC, Rougeot C, Bepoldin O, Vulliez B, Mandel P, Dray F (1983) Presence of somatostatin, enkephalins and substance P-like peptides in cultured neurons from embryonic chick cerebral hemispheres. J Neurochem 41:930

Negro-Vilar A, Ojeda SR, McCann SH (1979) Catecholaminergic modulation of LHRH release by median eminence terminals in vitro. Endocrinology 104:1749

Porter JC (1975) Methods for studying pituitary-hypothalamic axis in situ. Methods in Enzymology. Academic Press New York 39:166-183.

Rougeot C, Guillaume V, Junier MP, Fafeur V, Oliver C and Dray F (1986) Prostaglandin E_2 does not stimulate GRF release in vivo and in vitro studies. Program of the 6th International Conference on Prostaglandins and Related Compounds, Florence, Italy, p 115 (Abstract)

Rougeot C, Junier M-P, Everaere S, Braquet P, Dray F (1988) PAF-acether stimulates CRF and β-endorphin release. In: Braquet P (ed) The Role of Platelet-Activating Factor in Immune Disorders. New Trends Lipid Mediators Res. vol 2. Karger, Basel, pp 79-84

Rougeot C, Junier MP, Minary P, Weidenfeld J, Braquet P, Dray F (1990) Intracerebroventricular injection of PAF induces secretion of ACTH, β-endorphin and corticosterone in conscious rats. A possible link between the immune and nervous systems. Neuroendocrinology 51:267-275

Rougeot C, Trivers GE, Harris CC, Dray F (1983) ACTH immunoassays: Development of ultrasensitive and specific procedures for the measurement in biological materials. In: Avrameas S, Druet P, Masseyeff R, Feldmann G (eds), Immunoenzymatic Techniques, Developments in Immunology, Vol. 18, Elsevier, Amsterdam, pp 179-191

Snyder F (1982) Platelet-activating factor (PAF), a novel type of phospholipid with diverse biological properties. In: Cordes E. (ed), Annual Reports in Medicinal Chemistry, vol 17, p 243

Tokumura A, Tsukatani M (1986) Determination of molecular species of platelet-activating factor in bovine brain lipid extract. Program 2nd Int. Conf. on PAF, Gatlinburg, p 37

ACTIVATION OF PHOSPHOLIPID HYDROLYSIS AND GENERATION OF EICOSANOIDS IN CULTURED RAT ASTROGLIAL CELLS BY PAF, AND MODULATION BY N-3 FATTY ACIDS.

A. Petroni, M. Salami, M. Blasevich and C. Galli
Institute of Pharmacological Sciences
University of Milan
Via Balzaretti, 9
20133, Milan, Italy

INTRODUCTION

Astroglial cells, in addition to providing mechanical and metabolic support to neurons, interact with neuronal function by binding and metabolizing several types of neuromediators. Glial cells are also involved in processes which are activated by brain injury, such as formation of scar tissue and removal of cell debris after cell death. Responses to injury and participation to inflammatory events are based on the production and release of molecules, such as the eicosanoids and other compounds derived from phospholipid hydrolysis, which are typical mediators of inflammation. Eicosanoids are in fact generated in brain, in various pathological conditions, such as trauma, ischemia and convulsions, following the activated release of arachidonic acid (AA) from cell phospholipids.

Platelet-activating-factor (PAF), which is generated through activation of phospholipase A_2, shares with the cyclo and lipoxygenase metabolites derived from AA acid the feature of being potent mediators of inflammation which are synthesized and released from various types of cells. The brain has high concentrations of alkyl-ether phosphoglycerides, which are the precursors of PAF, and PAF appears to be produced by nervous tissue, such as the chick retina after stimulation with neurotransmitters (Bussolino et al., 1986). PAF is produced in brain during pharmacologically induced convulsions (Kumar et al., 1988) and the protective effects of PAF antagonists with respect to the biochemical alterations occurring in brain ischemia (Birkle et al., 1988) suggest that this phospholipid is produced in the CNS under these conditions. PAF is produced also by rat cerebellar granule cells in culture (Yue et al., 1990).

Exogenous PAF activates phospholipases in various types of cells, resulting e.g. in accumulation of inositol phosphates and diacylglycerols (DAG) in macrophages (Uhing et al., 1989), and in phosphoinositide hydrolysis in primary cultures of Kupffer cells (Gandhi et al, 1990) and in neutrophils (Naccache et al., 1985). In rat liver cells, PAF

NATO ASI Series, Vol. H 70
Phospholipids and Signal Transmission
Edited by R. Massarelli, L. A. Horrocks,
J. N. Kanfer, and K. Löffelholz
© Springer-Verlag Berlin Heidelberg 1993

stimulates AA metabolism by a receptor-mediated mechanism (Levine, 1988). PAF acts also on various types of brain cells: in neural cells of the hybrid clones NG108-15 and PC12 it increases intracellular levels of free calcium ions, suggesting that it may play a role in modulating neuronal function (Kornecki and Ehrlich, 1988), whereas in cultured astrocytes it stimulates inositol phosphate production (Murphy and Welk, 1990). PAF may thus be involved in the activation of various cells in brain, especially through phospholipid hydrolysis, which follows trauma or injuries resulting from circulatory events. As astroglial cells are closely associated with the cerebral vasculature and are highly responsive to trauma, these cells should be special targets of PAF action. Astrocytes are eicosanoid producing cells (Keller et al., 1985, Petroni et al., 1990), and thus these cells may activate the AA cascade as part of their responses to PAF.

The aim of our study was to investigate the response of the eicosanoid system to PAF in astroglial cell cultures and to measure the products of phospholipase activation, under these conditions, in comparison with the effects induced by the calcium ionophore A23187, a potent stimulator of phospholipases.

An additional aspect of our study was to investigate the effects of the incorporation of long chain polyunsaturated fatty acids (PUFA) in glial cells on the eicosanoid system. Cells in culture appear, in fact, to be generally depleted of PUFA - which are precursors of the eicosanoids and/or modulate the enzymes involved in eicosanoid formation - in comparison with the situation in tissues. Since our cells were selectively depleted of the n-3 PUFA 22:6 docosahexaenoic acid (DHA) when compared to the brain, we have studied the effects of restoring DHA levels in these cells on the formation of products of the AA cyclo (CO) and lipo (LO) oxygenases after stimulation.

PAF AND EICOSANOID PRODUCTION IN ASTROGLIAL CELLS.

We have studied the effects of PAF (1-O-octadecyl-2-acetyl-sn-glycero-3-phosphocholine) and of the calcium ionophore A23187 on the formation of eicosanoids and of products of phospholipid hydrolysis in primary cultures of astroglial cells. The cells were prepared from newborn rat cerebral hemispheres (Booher and Sensenbrenner, 1972) and used at 12-14 days when they reached confluence. The purity was checked by the presence of glial fibrillary acidic protein (>95% astrocytes) (Gebicke-Haerter et al., 1981). Formation of the CO products, thromboxane and prostacyclin from endogenous precursor pools after cell stimulation was assessed by measuring the stable metabolites TXB_2 and $6\text{-keto-PGF}_{1}\alpha$ by enzymeimmunoassay. AA-labeled cells (6 to 18 h before stimulation) were used for the determination of the LO products hydroxyeicosatetraenoic acids (HETE), by HPLC coupled with on line radiodetection, and for the determination of the products of phospholipase activity, free AA and diacylglycerol (DAG) and separated by one and two dimensional TLC.

Several experiments were carried out with different agonist concentrations and time periods of stimulation. The effects of PAF and A23187 incubated at the concentrations of 1 μM each for 15 min on the production of TXB_2 and 6-keto-$PGF_1\alpha$ are reported in Table 1. PAF stimulated quite actively the accumulation of both TXB_2 and 6-keto-$PGF_1\alpha$, although this effect was somewhat lower than that induced by A23187. The effects of both agonists on the formation of the LO products from endogenous precursor were also tested, but only the stimulation with A23187 at the concentration of 10 μM and for a period of at least one hour resulted in some formation of immunoreactive LTC_4, in the order of about 20% of the amounts of prostaglandins produced (Petroni et al., 1990). The formation of the LO products HETEs after PAF and A23187 stimulation for periods up to 2.5 hr was assessed both in conditions in which endogenous AA acted as precursor, and also with the use of [^{3}H]AA labeled cells. Stimulation with the calcium ionophore resulted in formation of 12-HETE (41 ng/mg P) and some 15-HETE, both measured by HPLC with the use of an internal standard (13-hydroxy octadecadienoic acid). The conversion of AA to these metabolites after ionophore stimulation was confirmed by radiodetection of peaks produced when labeled cells were used.

Table 1.   Levels of TXB_2 and 6-keto-$PGF_1\alpha$ in astroglial cells 15 min after stimulation with PAF and A23187 (1 μM)

	TXB_2	6-keto-$PGF_1\alpha$
	pg / mg Protein	
Non-stimulated	24 ± 8	28 ± 12
PAF	486 ± 78	119 ± 20
A23187	684 ± 72	230 ± 30*

Values are the average ± SEM of determinations carried out on 6 samples. * Significantly greater (p<0.01) than values obtained with PAF stimulation.

In contrast with the finding with A23187, PAF stimulation did not result in detectable formation of HETEs from the endogenous AA pool, but when using AA labeled cells, radioactive peaks corresponding to 12-HETE and

15-HETE, in approximately a 3:1 ratio, were produced. The discrepancy between the effects of A23187 and PAF on the formation of the LO products, in contrast with their comparative activities on the production of CO metabolites, suggests that the activation of the LO pathways may require the mobilization of a different AA pool, which was not released after PAF stimulation. In order to explore this point we have measured the radioactive products formed through phospholipid hydrolysis in AA labeled cells after stimulation with PAF or A23187. It was found that at 2 min after stimulation with PAF the release of AA was minimal, in comparison with the marked elevation induced by A23187, and that some elevation of AA occurred only at 30 min after PAF stimulation. DAG, on the other hand were markedly elevated 2 min after PAF stimulation, whereas no elevation occurred after A23187 stimulation either at 2 or 30 min (Table 2).

Table 2. Percentage increments of labeled AA and DAG at 2 and 30 min after stimulation with Ca ionophore or PAF (1 μM)

	AA		DAG	
	2 min	30 min	2 min	30 min
PAF	7 ± 1	38 ± 5	55 ± 8	6 ± 1
A23187	51 ± 6	86 ± 7	6 ± 1	7 ± 1

Values represent the increments over non-stimulated levels, of radioactivity associated with AA and DAG, and are the average ± SEM of 6 determinations.

These results indicate that the release of AA from labeled astroglial cells after PAF stimulation was much smaller than that induced by A23187, whereas the marked release of DAG suggests predominant activation of phospholipase C, as already described by measurements of inositol phosphate formation (Murphy and Welk, 1990). Only the relatively small AA pool associated with the polyphosphoinosites appears thus to be released, through the DAG pathway, after PAF stimulation, and the size of this substrate pool may not be adequate for the formation of quantities of HETEs, measurable with the UV detector. Only the radioactive products generated from the highly labeled phosphoinositide pools could be detected under these conditions.

EFFECTS OF THE INCORPORATION OF 22:6 N-3 ON EICOSANOID PRODUCTION IN GLIAL CELLS

Concentrations of long chain PUFA in cells grown in culture are generally much lower than those measured in cells and tissues obtained "ex vivo". Although the depletion of long chain PUFA in cultured cells is predictable on the basis of the relatively (and variably) low levels of the essential fatty acids linoleic and linolenic and/or of the long chain products of the n-6 and n-3 series in the media used for growth, it is also a parameter which is frequently neglected in studies with cells in culture. In comparison with brain tissue our cell preparations were indeed depleted of 22:6 n-3 (docosahexaenoic acid, DHA) (Table 3) after 12-14 days of growth, when they were used for the incubation. AA levels, instead, were practically identical to those found in brain. DHA, together with AA, is the major long chain PUFA in brain tissue and in specialized nervous structures, such as the retina.

Table 3. Percentage levels of major fatty acids in rat brain tissue and astroglial cell lipids

Fatty Acids	Total Brain[1]	Glial cells[2]
16:0	36.0 ± 1.8	36.5
18:0	22.7 ± 1.9	20.9
18:1	18.5 ± 1.4	19.5
18.2	0.6 ± 0.1	1.0
20:4 n-6	9.1 ± 0.9	8.5 (7.5 - 9.8)
22:6 n-3	10.8 ± 1.4	2.4 (1.7 - 3.1)

[1]Average ± SEM of determinations made on 6 rat brains; [2]Representative values obtained with several cell preparations. Range of values in parentheses.

Incubation of cultured cells with DHA (5 or 20 μM) for periods up to 72 hr, either as an albumin complex or as free acid dissolved in fetal calf serum, enhanced the percentages of DHA in cell phospholipids from the initial low level of around 2.5 percent, up to 5-6 percent, still lower, however, than in brain tissue. Cells enriched with DHA produced significantly lower amounts of TXB_2 and 6-keto-$PGF_1\alpha$ than control cells, after stimulation with the calcium ionophore A23187 (Table 4).

In addition, when the same cells were labeled with [^{3}H]AA for 6 hr before stimulation with the ionophore and the formation of radioactive LO products was checked, it appeared that the pattern of HETEs was quite different in DHA enriched vs control cells (Table 5).

Table 4.   Levels of TXB_2 and 6-keto-$PGF_1\alpha$ in preparations of control and DHA enriched glial cells after stimulation with the calcium ionophore A23187

Preparations	TXB_2	6-keto-$PGF_1\alpha$
	pg / mg Protein	
Control	918 ± 130	687 ± 80
DHA	554 ± 31[a]	475 ± 16[b]

DHA enriched cells were obtained by growing the cells with 5 μM DHA for 72 hr. Then the medium was replaced and, after 1 hr, preparations were stimulated with 1 μM A23187 for 15 min. Values are the average ± SEM of determinations carried out on 6 samples. Values with a superscript are significantly different from controls at the following levels: a = $p<0.001$, b = $p<0.01$ (Student's t test).

Table 5.   Incorporation of radioactivity in 12- and 15-HETE in [^{3}H]AA-labeled control and DHA-enriched cells after A23187 stimulation.

	Control		+ DHA	
	dpm μg Protein	% of cell radioactivity	dpm/ μg Protein	% of cell radioactivity
12-HETE	105	7.54	29	3.47
15-HETE	35	2.55	134	15.75

Control and DHA enriched cells (20 μM, 72 h) were labeled with [^{3}H]AA for 6 h, and then stimulated with 1 μM A23187 for 2.5 hr. Values represent the amount of radioactivity measured in peaks separated by reverse-phase HPLC coupled with radiodetection (Radiomatic Flow One Beta, Meriden, CT and are expressed on the basis of the protein content and as percentage of the radioactivity incorporated in the cells.

In fact production of labeled 12-HETE was about 3-fold higher than that of 15-HETE in control cells, whereas DHA-enriched cells produced more 15- than 12-HETE.   No peak corresponding to the HETES was observed, instead, in non-stimulated cells.

DISCUSSION AND CONCLUSIONS

Glial cells in culture represent a good model for studies of processes such as phospholipase activation and eicosanoid production. PAF, which may be released in vivo from cells in the vascular bed and/or from neurons and astrocytes, stimulates both PPI turnover (Murphy and Welk, 1990) and eicosanoid production. This last effect seems to be associated with DAG production rather than the direct release of arachidonate. Production of LO metabolites is less efficient under these conditions and only the radioactive products generated from the small pool which was labeled upon incubation with [³H]AA could be detected.

It is important to recognize, however, that from a quantitative and also from a qualitative point of view, eicosanoid production is modulated not only by the levels of the direct fatty acid precursor, AA, in the cells, but also by the long chain n-3 fatty acids, such as DHA, which is present in high concentrations in the nervous system. Since the fatty acid profile of cells in culture may differ from that in cells and tissue obtained "ex vivo", the control of the fatty acid composition of these preparations becomes an important issue in studies of fatty acid metabolism and eicosanoid production in "in vitro" systems.

REFERENCES

Birkle DL, Kurian P, Braquet P, Bazan NG (1988) Platelet-activating-factor antagonist BN52021 decreases accumulation of free polyunsaturated fatty acid in mouse brain during ischemia and electroconvulsive shock. J Neurochem 51:1900-1905

Booher J, Sensenbrenner M (1972) Growth and cultivation of dissociated neurons and glial cells from embryonic rat and human brain in flask cultures. Neurobiology 2:97-105

Bussolino F, Gremo F, Tetta C, Pescarmona GP, Camussi G (1986) Production of platelet-activating factor by chick retina. J Biol Chem 261:16502-16508

Gandhi CR, Hanahan DJ, Olson MS (1990) Two distinct pathways of platelet-activating factor-induced hydrolysis of polyphosphoinositides in primary cultures of rat Kupfer cells. J Biol Chem 265:18234-18241

Gebicke-Haerter PJ, Althaus HH, Schwartz P, Neuhoff V (1981) Oligodendrocytes from postnatal cat brain in cell culture. I. Regeneration and maintenance. Develop Brain Res 1:497-518

Keller M, Jackisch R, Seregi A, Hertting G (1985) Comparison of prostanoid forming capacity of neuronal and astroglial cells in primary cultures. Neurochem Int 7:655-665

Kornecki E, Ehrlich YH (1988) Neuroregulatory and neuropathological actions of the ether-phospholipid platelet-activating factor. Science 240:1792-1794

Kumar R, Harvey SAK, Kester M, Hanahan DJ, Olson MS (1988) Production and effects of platelet-activating factor in rat brain. Biochim Biophys Acta 963:375-383

Levine L (1988) Platelet-activating factor stimulates arachidonic acid metabolism in rat liver cells (C-9 cell line) by a receptor-mediated mechanism. Mol Pharmacol 34:793-799

Murphy S, Welk G (1990) Hydrolysis of polyphosphoinositides in astrocytes by platelet-activating factor. Eur J Pharmacol - Mol Pharmacol Section 188:399-401

Naccache PH, Molschi MM, Volpi M, Becker EL, Shaafi RI (1985) Unique inhibitory profile of platelet activating factor induced calcium mobilization, polyphosphoinositide turnover and granule enzyme-secretion in rabbit neutrophils towards pertussis toxin and phorbol ester. Biochem Biophys Res Commun 130:677-684

Petroni A, Blasevich M, Visioli F, Galli C (1990) Arachidonic acid cyclo and lipoxygenase pathways in astroglial cells In: Samuelsson B and Paoletti R (eds), Advances in Prostaglandin, Thromboxane and Leukotriene Research, Vol 21, Raven Press, Ltd., New York, pp 743-747

Uhing RJ, Prpic V, Hollenbach PW, Adams DO (1989) Involvement of protein inase C in platelet-activating factor stimulated diacylglycerol accumulation in murine peritoneal macrophages. J Biol Chem 264:9224-9230

Yue TL, Lysko PG, Feuerstein G (1990) Production of platelet-activating factor from rat cerebellar granule cells in culture. J Neurochem 54:1809-1811

A NEURAL PRIMARY GENOMIC RESPONSE TO THE LIPID MEDIATOR PLATELET-ACTIVATING FACTOR

John P. Doucet and Nicolas G. Bazan
LSU Neuroscience Center,
LSU Eye Center, and
Department of Biochemistry
  and Molecular Biology
LSU Medical Center
New Orleans, Louisiana 70112, USA

ABSTRACT

Platelet-activating factor (PAF; 1-O-alkyl-2-acetyl-sn-3-phosphocholine) is a cell membrane metabolite generated in response to stimulation and, when released from the cell, becomes a potent cell activator. Tissue of the central nervous system is rich in phospholipid PAF precursors. PAF accumulates in brain following seizures and ischemia. Certain PAF antagonists prevent neural cell damage following ischemia, and other PAF antagonists suppress establishment of long-term potentiation. Several cultured neural cells synthesize PAF and retain it intracellularly. A potent PAF antagonist selective for a high-affinity, intracellular PAF binding site in brain suppresses stimulation of neural immediate-early gene expression in vitro and in vivo. This effect suggests a role for PAF in long-term reparative processes that involve the stimulation of new protein synthesis.

PAF IN NEURAL TISSUE

PAF has been implicated as a modulator of neural activity, particularly during the initiation of various forms of neural trauma. PAF precursors are abundant in neural tissue (Blank et al., 1981). Convulsions, induced either electrically or through the administration of picrotoxin or bicuculline, are accompanied by the synthesis of PAF in brain (Kumar et al., 1988). PAF antagonists reduce cellular damage and improve metabolism in brain following ischemia (Panetta et al., 1987; Spinnewyn et al., 1987; Oberpichler et al., 1990; Lindsberg et al., 1990; Gilboe et al., 1991).

NATO ASI Series, Vol. H 70
Phospholipids and Signal Transmission
Edited by R. Massarelli, L. A. Horrocks,
J. N. Kanfer, and K. Löffelholz
© Springer-Verlag Berlin Heidelberg 1993

PAF apparently does not traverse the blood-brain barrier (Kumar et al., 1988), and hence PAF accumulation in brain following neural trauma must derive from neural metabolism.

PAF induces differentiation in cells of neural origin. PAF induces astrocytic differentiation of immature glial cells and gliosis-type proliferation in astrocytic cells (Kentroti et al., 1991). At low concentrations PAF stimulates neuronal differentiation in neuroglial hybrid cells, and high concentrations of PAF are neurotoxic (Kornecki and Ehrlich, 1988). Neural cells may therefore be affected by low concentrations of PAF, reflecting physiological PAF accumulation following neural stimulation and an ability to induce nascent neural cells to maturity. When PAF accumulates in high concentrations in response to pathological conditions such as neurotrauma, neurotoxic events may arise in affected tissues.

PAF is synthesized by mammalian cells through different routes of biosynthesis (reviewed in Snyder, 1991): (1) a remodelling pathway, initiated by phospholipase A_2 activation that releases the PAF precursor lyso-PAF from membrane 1-alkyl-2-acyl-sn-glycero-3-phosphocholines or (2) a de novo pathway involving choline transfer to 1-alkyl-2-acetyl-sn-glycerols. It has been suggested that brain cells synthesize PAF through the latter de novo pathway (Francescangeli and Goracci, 1989), but, during pathological conditions such as seizure and ischemia, Ca^{2+}-evoked phospholipase A_2 activation (Bazan, 1970; Bazan and Rodriguez de Turco, 1980) is likely a major source of PAF liberated from cell membrane precursors (Bazan et al., 1991).

Certain cells have been shown to accumulate and retain newly-synthesized PAF following stimulation (Henson, 1987). Neural cells, such as cerebellar granule cells (Yue et al., 1990), fetal brain cells (Sogos et al., 1990), and embryonic chick retina, retain PAF after its synthesis (Bussolino et al., 1986). Pharmacologically distinct synaptic and intracellular PAF binding sites exist in rat brain (Marcheselli et al., 1990). The piperidinothieno diazepine BN50730 (Braquet et al., 1990) exhibits high affinity and selectivity for these intracellular sites (Marcheselli and Bazan, 1991). Localization of high-affinity PAF binding sites to intracellular membranes in brain suggests that cellular retention of PAF is functionally linked to metabolic events subsequent to its stimulus-evoked synthesis.

Several lines of evidence suggest that PAF synthesized and retained inside the cell is a component of an intracellular signal transduction mechanism. (1) Synthesis of intracellular PAF follows cell stimulation (Henson, 1987; Bussolino et al., 1988; Sogos et al., 1990). (2) PAF-synthetic enzymes are localized on intracellular membranes (Record et al., 1989; Vallari et al., 1990). (3) Intracellular PAF stimulates phospholipase A$_2$ activity and eicosanoid generation (Stewart and Phillips, 1989; Stewart et al. 1989; Stewart, 1991). Although there is evidence of neural cell release of PAF (Yue et al., 1990), intercellular PAF as a stimulator of neural cells may also derive from glia or other cells (e.g. macrophages) resident in the brain. Generation of PAF through the phospholipase A$_2$ remodelling of neural membranes, in addition to activation of phospholipase A$_2$ by intracellular PAF, suggests an amplification circuit that may account for the protective effects of PAF antagonists against trauma-evoked free fatty acid release and cellular damage (Birkle et al., 1988; Spinnewyn et al., 1987).

PAF AS A TRANSDUCER OF THE PRIMARY GENOMIC RESPONSE

Cell stimulation is transduced to the nucleus through mechanisms involving the generation of second messengers and stimulation of target protein kinases. The primary genomic response is the earliest transcriptional activity following cell stimulation (reviewed in Doucet et al, 1990; Herschman, 1991; and Curran and Morgan, 1991). The stimulus-evoked transcription is transient and is independent of on-going protein synthesis. Nuclear machinery is poised for a burst of new gene expression following cell stimulation. The components of the primary genomic response include members of the fos, jun, fra, myc, and TIS immediate-early gene families (reviewed in Herschman, 1991). The structures of these immediate-early gene products suggest proteins of various inferred functions, but the majority of those discovered and by far the most well-studied are nuclear-localized transcription factors. The transduction of extracellular stimuli to the generation of transcription factors suggests a regulation of long-term protein synthetic mechanism that mediates cellular fate in the event of stimulus.

In the nervous system, the primary genomic response is elicited by physiological and pathological conditions (reviewed in Doucet et al., 1990, and Morgan and Curran, 1991). These eliciting events include ischemia, seizure, injury, and establishment of long-term potentiation. The existence of stimulus-evoked expression of transcription factors in neural tissue suggests that long-term changes in cellular activity and function, including reparative changes following neurotrauma and changes during memory formation, derive from primary, short-term genomic responses.

The best-studied immediate-early genes in neural tissue include fos, jun, and zif/268 (the latter variously known as Egr-1, tis 8, NGFI-A). fos and jun encode DNA-binding transcription factors with leucine-zipper structures that allow formation of mutual heterodimers (Figure 1; reviewed in Ransome and Verma, 1990). Fos-Jun and Jun-Jun dimers are among the proteins that constitute sequence-specific AP-1 transcription regulatory activity (Figure 2). Zif/268 is also a transcription factor, but it binds DNA as a monomer through zinc-finger structures at sequences distinct from AP-1 proteins (Christy et al., 1988; Christy and Nathans, 1989).

A primary genomic response to PAF has been reported in leukocytes (Squinto et al., 1989; Mazer et al., 1991; Shulam et al., 1991; Ho et al., 1987), epidermal carcinoma cells (Tripathi et al., 1991), and neural cells (Squinto et al, 1989; Doucet and Bazan, in preparation). In NG108-15 neurohybrid cells, picomolar PAF elicits expression of zif/268 (Doucet and Bazan, in preparation), and this potency of PAF suggests that it is an important component of signal transduction to the nucleus in neural cells. In neuroblastoma cells (Squinto et al., 1989; Squinto et al., 1990; Bazan et al., 1991), PAF elicits both (1) transcription of both fos and jun and (2) stimulation of transcription from promoters containing multiple AP-1 binding sequences. This indicates that the transcriptional stimulation by PAF is physiologically coupled to corresponding protein activity. In addition, deletion mutagenesis of the fos promoter has revealed the requirement of a calcium-response sequence (CaRE) for PAF-stimulated fos expression (Squinto et al., 1989). The fos CaRE, which is coincident with the cyclic AMP-response sequence, is comprised of a few base pairs upstream of the transcription initiation site of the fos gene and functions as a binding site for a transcription factor during cyclic AMP and Ca^{2+} second messenger cascades that stimulates fos expression (Sheng et al., 1988; Sheng et al., 1990). The finding that PAF utilizes the fos CaRE for transcriptional regulation is supported by studies demonstrating a cytoplasmic influx of Ca^{2+} stimulated by PAF from both intracellular and extracellular stores in neural cells (Yue et al., 1991a and 1991b). The presence of the CaRE in transcriptional promoter regions of other immediate-early genes, in particular zif/268 (Christy et al., 1988), which is co-regulated with fos under many types of cell stimulations, including depolarization of neural cells (Sukhatme et al., 1988; Bartel et al.,1989), suggests that PAF and other stimulators of Ca^{2+} mobilization may regulate multiple components of the primary genomic response through a common mechanism.

PAF is implicated in the regulation of the primary genomic response to seizure. Expression of fos and several other immediate-early genes is stimulated by pharmacologically- or electrically-induced seizures (Morgan et al., 1987; Cole et al., 1989; Saffen et al., 1989; reviewed in Doucet et al., 1990, and Morgan and Curran, 1991). The potent intracellular PAF antagonist BN50730 (Marcheselli and Bazan, 1991) suppresses stimulated expression of immediate-early genes fos and zif/268 induced by

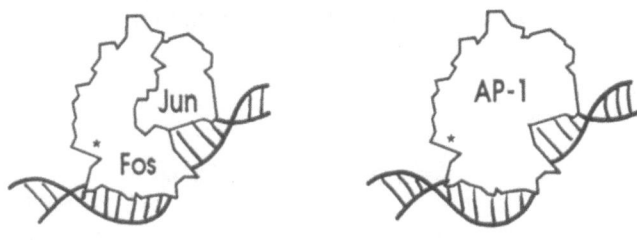

FIGURE 1. Heterodimerization of <u>fos</u> and <u>jun</u> immediate-early gene products as AP-1 transcription-regulatory activity during genomic binding. Protein structures are purely schematic, the geopolitical symbolism notwithstanding. *, Wiesbaden.

electroconvulsive shock (Marcheselli et al., 1991; Marcheselli and Bazan, 1992). This suggests that (1) the endogenous PAF generated through seizure is functionally coupled to the stimulation of a primary genomic response and that (2) the coupling of PAF to immediate-early gene expression may be an intracellular event in neural tissue.

The activation of phospholipase A_2 following ischemic onset (Bazan, 1970; Bazan and Rodriguez de Turco, 1980) and the protective effects of PAF antagonists against cellular damage in brain (Panetta et al., 1987; Spinnewyn et al., 1987; Birkle et al., 1988; Oberpichler et al., 1990; Frerichs et al., 1990; Gilboe et al., 1991) both implicated PAF in the progression of ischemic damage. Primary genomic responses to ischemia have been described (Onodera et al., 1989; Jorgenson et al., 1989; Kindy et al., 1991). In both experimentally evoked seizures and ischemia, increased immediate-early gene activity in dentate gyrus and hippocampal neurons is particularly pronounced, and specific PAF binding sites are concentrated in these brain structures (Domingo et al., 1988). Hippocampal neurons are sensitive to Ca^{2+} fluxes during brain injury (Dienel, 1984; Seisjö, 1989; Seisjö, 1990; Young, 1992), and the accumulation of PAF during pathological conditions may affect extracellular Ca^{2+} mobilization through mechanisms mediated by possibly both specific PAF receptors and by membrane perturbations (Sawyer and Anderson, 1989; Young, 1992) resulting from the remodelling of membrane phospholipids that accompanies the synthesis and accumulation of PAF and its precursors in cell membranes.

TRANSCRIPTIONAL REGULATION OF HUMAN FOS AND JUN GENES

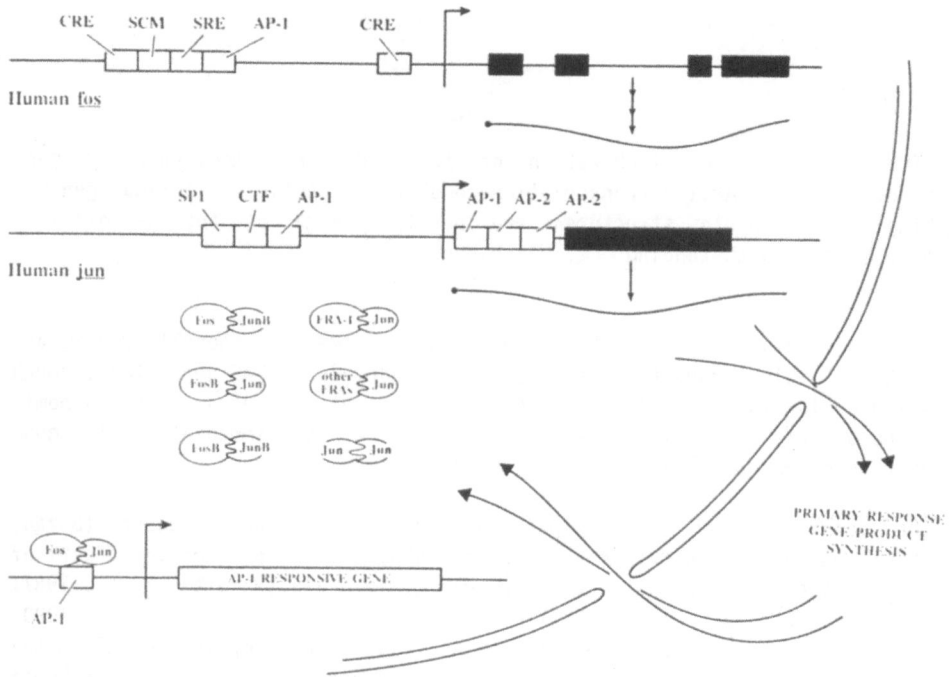

FIGURE 2. Transcriptional regulation of human <u>fos</u> and <u>jun</u> immediate-early genes. Sequences conferring inducibility of human <u>fos</u> and <u>jun</u> are represented schematically. Some regulatory elements are common to other immediate-early genes in addition to <u>fos</u> and <u>jun</u>. Orthologous comparisons of human and rodent clones reveal a general conservation of these regulatory sequences. Transcriptional induction by substrates of second messenger activities is followed by rapid cytoplasmic messenger RNA translation, protein synthesis, and return transport to the nucleus. The diagram includes examples of different <u>Fos</u> and <u>Jun</u> family complexes which have been described in the literature. <u>Fos</u>-<u>Jun</u> complexes regulate transcription from cis AP-1 elements. CRE, cyclic AMP responsive element; SCM, sis-conditioned medium responsive sequence; SRE, serum-responsive element; AP-1, activator protein-1 binding sequence; SP1, binding site for cellular transcription factor Sp1; CTF, CAAT-sequence binding factor; AP-2, activator protein-2 binding site. Reviewed in Morgan and Curran, 1991. Diagram from Doucet et al. (1990).

PAF AND THE PRIMARY GENOMIC RESPONSE IN LONG-TERM POTENTIATION

One of the long-term events that may be established physiologically though the activity of immediate-early transcription factors is memory formation. Long-term potentiation (LTP) is an experimental model of memory that involves activity-dependent changes in synaptic morphology and circuitry. LTP induces a primary genomic response in brain, particularly in cells of the dentate gyrus (reviewed in Abraham et al., 1992). The activity of the zif/268 transcription factor may be most specifically linked to LTP, since its expression is particularly correlative with the method of LTP induction and its duration. Neuronal expression of zif/268 is induced by picomolar concentrations of PAF in vitro (Doucet and Bazan, in preparation), suggesting that small changes in PAF synthesis are sufficiently potent to stimulate a primary genomic response. A role for PAF in the formation of LTP-mediated neuronal changes is suggested by the inhibition of hippocampal synaptic responses during establishment of LTP by both an inhibitor of phospholipase A_2 (Linden et al., 1989), an initiating enzyme in generation of PAF from alkylacylglycerophospho-cholines, and antagonists of PAF receptors (del Cerro et al., 1990; Arai and Lynch, 1992; Koto et al., 1992; Bazan and Cluzel, 1992.). Cell-generated PAF may therefore mediate plastic responses in the neuron, and this plasticity may involve the synthesis of new proteins mediated by products of a PAF-mediated primary genomic response.

MECHANISM AND SIGNIFICANCE OF A NEURAL PRIMARY GENOMIC RESPONSE TO PAF

Exogenous PAF stimulates both Ca^{2+} influx and phosphoinositide turnover in neural cells through a mechanism independent of voltage-gated Ca^{2+} channels (Kornecki and Ehrlich, 1988; Yue et al., 1991a and 1991b). This duality for increasing cytoplasmic Ca2+ suggests a Ca^{2+} signal amplification mechanism following cell stimulation. Although it is probable that a PAF-stimulated Ca^{2+} influx mediates the primary genomic response to PAF, the activation of phospholipase C during cell stimulation with PAF may also participate (Schalasta and Doppler, 1990) through the generation of diacylglycerols and the subsequent activation of protein kinase C. Several lines of evidence suggest a GTPase-mediated linkage of PAF receptor stimulation with activation of phospholipases A_2, C, and D (reviewed in Snyder, 1990, and Shukla, 1992). Intracellular PAF stimulates free fatty acid accumulation through phospholipase A_2 activation and subsequent eicosanoid generation (Bazan et al., 1987a and 1987b; Stewart and Phillips, 1989; Stewart et al. 1989; Stewart, 1991).

This intracellular activity of PAF may transduce cell stimulation as a second messenger (Henson, 1987).

Acetylcholine and dopamine stimulate intracellular PAF accumulation in neural cells (Bussolino et al., 1988; Sogos et al., 1991). The intracellular PAF antagonist BN50730 suppresses the cholinergic stimulation of zif/268 expression in neuronal cells in vitro (Doucet and Bazan, in preparation), suggesting that intracellular PAF modulates the neural primary genomic response to cholinergic stimulation. At the synapse, PAF stimulates phosphoinositide turnover and increased Na^+/Ca^{2+} exchange (Kumar et al., 1988). In fetal hippocampal cell culture, PAF augments excitatory neurotransmission in a PAF antagonist-sensitive manner (Clark and Bazan, 1991), suggesting that PAF is acting presynaptically through a receptor-mediated mechanism.

PAF is a component of a neural mechanism involving the release and binding of neurotransmitters that results in induction of a primary genomic response following neural activity (Figure 3). The effectiveness of PAF antagonists ameliorating cellular damage by neurotrauma suggests that generation of PAF is significant in the onset and progression of pathological consequences of trauma. Neuronal synthesis and retention of PAF following stimulation suggests that PAF participates in the transduction of neural signals. It appears then that PAF is involved in two states of neural activity, a physiological and a pathological state. Indeed in neuronal culture high concentrations of PAF elicit cell death, whereas lower, perhaps more physiological concentrations will stimulate maturation of neural cells (Kornecki and Ehrlich, 1988). The fact that PAF precursors are phospholipid components of cellular membranes suggests that membrane hydrolysis accompanying PAF synthesis, which, during the unregulated cellular activities following neural trauma, may be a significant factor in the loss of cellular integrity. Both receptor-mediated Ca^{2+} flux and phospholipase activations attributed to cellular stimulation by PAF further suggest effects which, during cell over-stimulation in neurotrauma, could lead to cellular damage as a consequence of PAF synthesis. As a phospholipid, it is additionally conceivable that either PAF, or its inactive precursor lyso-PAF, both of which exhibit uncommon chemistries at the sn-2 position, may exert its activities through membrane perturbations and alteration of membrane-bound protein receptors and channels. The accumulation of PAF following neurotrauma may generate high tissue concentrations to establish such a mechanism.

Neuronal transmission, particularly during sustained neurotrauma, depletes neurotransmitter stores, which, without replenishment, may ultimately result in impaired intercellular communication and neural activity. One mechanism of a primary genomic response evoked by neural activity might be

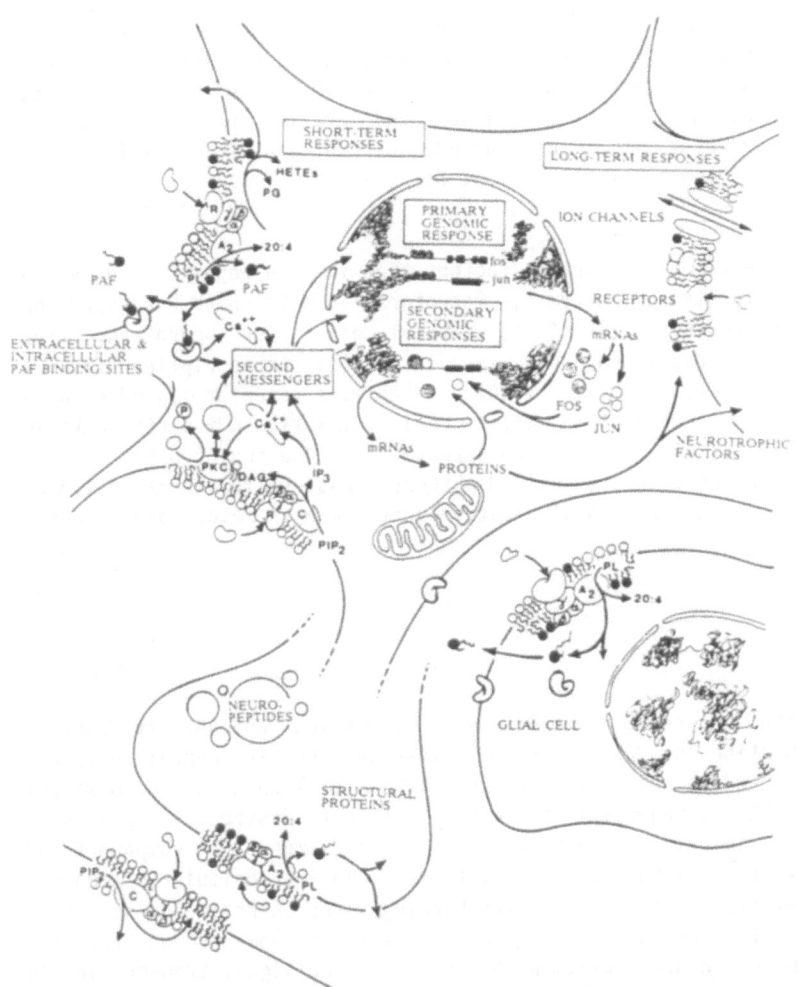

FIGURE 3. Signal transduction of acute, genomic, and long term responses to cell stimulation. Stimulation of a cell by hormones, neurotransmitters, growth factors, or ion fluxes results in the formation of membrane phospholipid-derived second messengers (left), including inositol trisphosphate (IP$_3$), hydroxyeicosatetraenoic acids (HETEs), arachidonic acid (20:4) metabolites (eicosanoids), and PAF, through the activation of selective phospholipases. Of these, PAF, lipoxygenase eicosanoids, and the phospholipase C metabolites have been linked to the stimulation of primary genomic responses (center). Members of the Fos, Jun, FRA (Fos-Related Antigens), and TIS (tetradecanoylphorbol acetate [TPA]-Inducible Sequences, including Zif/268) families are transcription factors. The immediate-early expression of these transcription factors

(FIGURE 3, Continued) couples the generation of second messengers to long-term responses dependent upon synthesis of new proteins. Membrane-derived second-messengers accumulate as a response to injury, ischemia, convulsions, or inflammation and actively participate in the pathological process, which may include aberrant expression of long-term response genes. Diagram from Doucet et al. (1990).

to rapidly generate transcription factors that in turn regulate expression of structural proteins, enzymes, receptors, and trophins necessary for reparative processes. In fact, several genes for neurotransmitter precursors are regulated by immediate-early transcription factors (reviewed in Doucet et al., 1990). PAF as a receptor ligand or as a second messenger accumulating after cell stimulation may participate in parallel or in conjunction with neurotransmitters and neurotrophic factors to stimulate the protein synthetic machinery necessary for normal cellular maintenance or the generation of plastic events such as memory and repair.

SUMMARY

Although ischemia, seizure, and memory formation undoubtedly function through modulation of normal signal transduction in neural cells and effect corresponding physiological and pathological changes, the powerful effects of PAF antagonists to counteract cyto-destructive processes suggests that neural activity of PAF is an important component and modulator of the circuitry of cell-to-cell and intracellular communication. The existence of intracellular binding sites in brain and selective antagonists of intracellular binding reveal a new avenue for the investigation of neuromodulation in particular and signal transduction in general. The ability of biologically relevant concentrations of PAF to stimulate a primary genomic response in neural tissue suggest that the abundance of PAF in neural tissue is a pathway of immediate-early gene stimulation in the nervous system alternative to, or at least in conjunction with, the genomic response to excitatory amino acid and cholinergic transmitters and neurotrophic factors. The two roles for PAF, physiologic and pathologic, suggest an accessible juncture through which neurotoxic effects, as aberrations of normal cellular processes, can be studied.

ACKNOWLEDGEMENT

Supported by NIH grant NS 23002.

REFERENCES

Abraham W, Dragunow M, Tate W (1992) The role of immediate-early genes in the stabilization of long-term potentiation. Mol Neurobiol 5:1-15

Arai A, Lynch G (1992) Antagonists of the platelet-activating factor receptor block long-term potentiation in hippocampal slices. Eur J Neurol 4:411-419

Bazan N (1970) Effects of ischemia and electroconvulsive shock on free fatty acid pool in the brain. Biochim Biophys Acta 218:1-10

Bazan N, Cluzel J (1992, in press) Membrane-derived lipid second messengers as targets for neuroprotection: Platelet-activating factor. In: Marangos P, Lal H (eds), Emerging Strategies in Neuroprotection, Boston, Birkhauser

Bazan N, Rodriguez de Turco E (1980) Membrane lipids in the pathogenesis of brain edema: Phospholipids and arachidonic acid, the earliest membrane components changed at the onset of ischemia. Adv Neurol 28:197-205

Bazan N, Birkle D, Tang W, Reddy T (1986) The accumulation of free arachidonic acid, diacylglycerols, prostaglandins, and lipoxygenase reaction products in the brain during experimental epilepsy. Adv Neurol 44:879-902

Bazan H, Braquet P, Reddy S, and Bazan N (1987a) Inhibition of the alkali burn-induced lipoxygenation of arachidonic acid in the rabbit cornea in vivo by a platelet-activating factor antagonist. J Ocul Pharmacol 3:357-362

Bazan H, Reddy S, Woodland J, Bazan N (1987b) The accumulation of platelet-activating factor in the injured cornea may be interrelated with the synthesis of lipoxygenase products. Biochim Biophys Res Commun 149: 915-919

Bazan N, Squinto S, Braquet P, Panetta T, Marcheselli V (1991) Platelet-activating factor and polyunsaturated fatty acids in cerebral ischemia or convulsions: Intracellular PAF-binding sites and activation of a Fos/Jun/Ap-1 transcriptional signaling system. Lipids 26:1236-1242

Birkle D, Kurian P, Braquet P, Bazan NG (1988) The platelet-activating factor antagonist BN 52021 decreases accumulation of free polyunsaturated fatty acid in mouse brain during ischemia and electroconvulsive shock. J Neurochem 151:88-109

Blank M, Cress E, Wittle T, Snyder F (1981) In vivo metabolism of a new class of biologically active phospholipids: 1-alkyl-2-acetyl-sn-glycero-3-phosphocholine, a platelet-activating, hypotensive phospholipid, Life Sci 29:769-775

Braquet P, Touqui L, Shen T, Vargaftig B (1987) Perspectives in platelet-activating factor research. Pharmacol Rev 39:97-145

Braquet P, Laurent J, Rolland A, Martin C, Pommier J, Hosford D, Esanu A (1990) From ginkgolides to N-substiuted piperidinothieno diazepines, a new series of highly potent PAF antagonists. Adv Prost Thromb Leuk Res 21:929-937

Bussolino F, Gremo F, Tetta C, Pescarmona G, Camussi G (1986) Production of platelet-activating factor by chick retina. J Biol Chem 261:16502-16508

Christy B, Nathans D (1989) DNA binding site of the growth-factor inducible protein Zif/268. Proc Natl Acad Sci USA 86:8737-8741

Christy B, Lau L, Nathans D (1988) A gene activated in mouse 3T3 cells by serum growth factors encodes a protein with "zinc finger" sequences. Proc Natl Acad Sci USA 85:7857-7861

Clark G, Happel L, Zorumski C, Bazan N (1991) Platelet-activating factor augments excitatory synaptic neurotransmission in cultured rat hippocampal neurons. Soc Neurosci Abs 17:951

Cole A, Saffen D, Baraban J, Worley P (1989) Rapid increase of an immediate early gene messenger RNA in hippocampal neurons by synaptic NMDA receptor activation. Nature 340:474-476

del Cerro S, Arai A, Lynch G (1990) Inhibition of long-term potentiation by an antagonist of platelet-activating factor receptors. Behav Neural Biol 54:213-217

Dienel G (1984) Regional accumulation of calcium in post-ischemic brain. J Neurochem 43:913-917

Domingo M, Spinnewyn B, Charbrier P, Braquet P (1988) Presence of specific binding sites for platelet-activating factor (PAF) in brain. Biochem Biophys Res Commun 151:730-736

Doucet J, Squinto S, Bazan NG (1990) Fos-Jun and the primary genomic response in the nervous system: Physiological role and possible pathophysiological significance. Mol Neurobiol 4:27-56

Francescangeli E, Goracci G (1989) The de novo synthesis of platelet-activating factor in rat brain. Biochem Biophys Res Commun 161: 107-112

Frerichs K, Lindsberg P, Hallenbeck J, Feurerstein G (1990) Platelet-activating factor and progressive brain damage following focal brain injury. J Neurosurg 73:223-233

Gilboe D, Kintner D, Fitzpatrick J, Emoto S, Esanu A, Braquet P, Bazan N (1991) Recovery of postischemic brain metabolism and function following treatment with a free radical scavenger and platelet-activating factor antagonist. J Neurochem 56:311-319

Henson P (1987) Extracellular and intracellular activities of PAF. In: Snyder F (ed) Platelet-Activating Factor and Related Lipid Mediators, Plenum, New York, pp. 255-271

Herschman H (1991) Primary response genes induced by growth factors and tumor promoters. Ann Rev Biochem 60:281-319

Ho Y-S, Lee W, Snyderman R (1987) Chemoattractant-induced activation of c-_fos_ gene expression in human monocytes. J Exp Med 165:1524-1538

Jörgenson M, Deckert J, Wright D, Gehlert D (1989) Delayed c-_fos_ proto-oncogene expression in the rat hippocampus induced by transient global cerebral ischemia: An in situ hybridization study. Brain Res 484: 393-398

Junier M, Tiberghien C, Rougeot C, Fauer V, Dray F (1988) Inhibitory effect of PAF on lutenizing hormone-releasing hormone and somatostatin release from rat median eminence _in vitro_ correlated with characterization of specific PAF receptor sites. Endocrinology 123: 72-80

Kentroti S, Baker R, Lee K, Bruce C, Vernadakis A (1991) Platelet-activating factor increases glutamine synthetase activity in early and late passage C-6 glioma cells. J Neurosci Res 28:497-506

Kindy M, Carney J, Dempsy R, Carney J (1991) Ischemic induction of protooncogene expression in gerbil brain. J Molec Neurosci 2:217-228

Kornecki E, Ehrlich Y (1988) Neuroregulatory and neuropathological actions of the ether-phospholipid platelet-activating factor. Science 240:1792-1794

Koto K, Clark GD, Bazan N, Clifford D, Zormuski C Platelet-activating factor as a potential messenger in long-term potentiation, Soc Neurosci Abs 18 (in press)

Kumar R, Harvey S, Kester M, Hanahan D, Olson M (1988) Production and effects of platelet-activating factor in the rat brain. Biochim Biophys Acta 963:375-383

Linden D, Sheu F, Murakami K, Routtenberg A (1987) Enhancement of long-term potentiation by cis-unsaturated fatty acid. Relation to protein kinase C and phospholipase A_2. J Neurosci 7:3783-3792

Lindsberg P, Yue T-L, Frerichs K, Hallenbeck J, Feuerstein G (1990) Evidence for platelet-activating factor as a novel mediator in experimental stroke in rabbits. Stroke 21:1452-1457

Marcheselli VL, Bazan NG (1991) A specific antagonist for intracellular platelet-activating factor (PAF) binding sites lacks activity on synaptic membranes. Trans Amer Soc Neurochem 22:187

Marcheselli VL, Bazan NG (1992) ECS-induced _zif_-268 in hippocampus is inhibited by a PAF antagonist. Trans Amer Soc Neurochem 23

Marcheselli VL, Rossowska M, Domingo, MT, Braquet P, Bazan NG (1990) Distinct platelet-activating factor binding sites in synaptic endings and in intracellular membranes of rat cerebral cortex. J Biol Chem 265: 9140-9145

Marcheselli VL, Doucet JP, Bazan NG (1991) Platelet-activating factor is a mediator of _fos_ expression induced by a single seizure in rat hippocampus. Soc Neurosci Abs 17:349

Mazer B, Domenico J, Sawami H, Gelfand EW (1991) Platelet-activating factor induces an increase in intracellular calcium and expression of regulatory genes in human B lymphoblastoid cells. J Immunol 146: 1914-1918

Morgan J, Curran T (1991) Stimulus-transcription coupling in the nervous system: Involvement of the inducible proto-oncogenes _fos_ and _jun_. Annu Rev Neurosci 14:421-451

Morgan J, Cohen D, Hempstead J, Curran T (1987) Mapping patterns of c-_fos_ in the nervous system. Science 237:192-197

Oberpichler H, Sauer D, Rosberg C, Mennel H, Krieglstein J (1990) PAF antagonist gingkolide-B reduces postischemic neuronal damage in the rat-brain hippocampus. J Cereb Blood Flow Metab 10:133-135

Onodera H, Kogure K, Ono Y, Igarashi K, Kiyota Y, Nagaoka A (1989) Proto-oncogene c-_fos_ is transiently induced in the rat cerebral cortex after forebrain ischemia. Neurosci Lett 98:101-104

Panetta T, Marcheselli VL, Braquet P, Spinnewyn B, Bazan NG (1987) Effects of a platelet-activating factor antagonist (BN 52021) on free fatty acids, diacylglycerols, polyphosphoinositides, and blood flow in the gerbil brain: Inhibition of ischemia-reperfusion-induced cerebral injury. Biochem Biophys Res Commun 149:580-587

Ransone L, Verma I (1990) Nuclear proto-oncogenes _fos_ and _jun_. Annu Rev Cell Biol 6:539-557

Record M, Ribbes G, Terce F, Chap H (1989) Subcellular localization of phospholipids and enzymes involved in PAF-acether metabolism. J Cell Biochem 40:353-358

Rougeot C, Junier M, Minary P, Weidenfeld J, Braquet P, Dray F (1990) Intracerebroventricular injection of platelet-activating factor induces secretion of adrenocorticotropin, beta-endorphin and corticosterone in conscious rats--a possible link between the immune and nervous systems. Neuroendocrinology 51:315-319

Saffen D, Cole A, Worley P, Christy B, Ryder K, Baraban J (1988) Convulsant-induced increase in transcription factor messenger RNAs in rat brain. Proc Natl Acad Sci 85:7795-7799

Sawyer D, Anderson O (1989) Platelet-activating factor is a general membrane perturbant. Biochim Biophys Acta 987:129-132

Schalasta G, Doppler C (1990) Inhibition of c-_fos_ transcription and phosphorylation of the serum response factor by an inhibitor of phospholipase C-type reactions. Mol Cell Biol 10:5558-5561

Schulman PG, Kuruvilla A, Putcha G, Mangus L, F-Johnson J, Shearer WT (1991) Platelet-activating factor induces phospholipid turnover calcium flux, arachidonic acid liberation, eicosanoid generation, and oncogene expression in a human B cell line. J Immunol 146:1642-1647

Sheng M, Dougan S, McFadden G, Greenberg M (1988) Calcium and growth factor pathways of c-_fos_ transcriptional activation require distinct upstream regulatory sequences. Mol Cell Biol 8:2782-2796

Sheng M, McFadden G, Greenberg M (1990) Membrane depolarization and calcium induce c-_fos_ transcription via phosphorylation of transcription factor CREB. Neuron 4:571-582

Shukla S (1992) Platelet-activating factor and signal transduction mechanisms. FASEB J 6:2296-2301

Siesjö B (1990) Calcium in the brain under physiological and pathological conditions. Eur Neurology 30:3-9

Siesjö BK, Bengtsson F (1989) Calcium fluxes, calcium antagonists, and calcium-related pathology in brain ischemia, hypoglycemia, and spreading depression: A unifying hypothesis. J Cereb Blood Flow Metab 9:127-140

Snyder F (1990) Platelet-activating factor and related acetylated lipids as potent biologically active cellular mediators. Amer J Physiol 259:C697-C708

Sogos V, Bussolino E, Pilia E, Torelli S, Gremo F (1990) Acetylcholine-induced production of platelet-activating factor by human fetal brain cells in culture. J Neurosci Res 27:706-711

Spinnewyn B, Blavet N, Clostre F, Bazan N, Braquet P (1987) Involvement of platelet-activating factor (PAF) in cerebral post-ischemic phase in mongolian gerbils. Prostaglandins 34:337-349

Squinto SP, Block AL, Braquet P, Bazan NG (1989) Platelet-activating factor stimulates a _Fos_/_Jun_/AP-1 transcriptional signaling system in human neuroblastoma cells. J Neurosci Res 24:558-566

Squinto SP, Braquet P, Block A, Bazan NG (1990) Platelet-activating factor activated HIV promoter in transfected SH-SY5Y neuroblastoma cells and MOLT-4 T-lymphocytes. J Mol Neurosci 2:79-85

Stewart A, Phillips W (1989) Intracellular platelet-activating factor regulates eicosanoid generation in guinea-pig resident peritoneal macrophages. Br J Pharmacol 98:141-143

Stewart A, Dubbin P, Harris T, Dusting G (1989) Evidence for an intracellular action of platelet-activating factor in bovine cultured aortic endothelial cells. Br J Pharmacol 96:503-505

Stewart A, Dubbin P, Harris T, Dusting G (1990) Platelet-activating factor may act as a second messenger in the release of icosanoids and superoxide anions from leukocytes and endothelial cells. Proc Natl Acad Sci USA 87:3215-3222

Sukhatme V, Cao X, Chang L, Tsai-Morris C, Stamenkovich D, Ferreira P, Cohen D, Edwards S, Shows T, Curran T, Le Beau M, Adamson E (1988) A zinc finger-encoding gene coregulated with c-_fos_ during growth and differentiation, and after cellular depolarization. Cell 53:37-43

Tripathi Y, Kandala J, Guntaka R, Lim R, Shukla S (1991) Platelet-activating factor induces expression of early response genes c-_fos_ and tis-1 in human epidermoid carcinoma cells. Life Sci 49:1761-1767

Vallari D, Record M, Snyder F (1990) Conversion of alkylacetylglycerol to platelet-activating factor in HL-60 cells and subcellular localization of the mediator. Arch Biochem Biophys 276:538-542

Young W (1992) Role of calcium in central nervous system injury. J Neurotrauma 9:S9-S25

Yue T-L, Lysko PG, Feuerstein G (1990) Production of platelet-activating factor from rat cerebellar granule cells in culture. J Neurochem 54: 1809-1811

Yue T-L, Gleason M, Gu J-L, Lysko PG, Hallenbeck J, Feuerstein G (1991a) Platelet-activating factor (PAF) receptor-mediated calcium mobilization and phosphoinositied turnover in neurohybrid NG108-15 cells: Studies with BN50739, a new PAF antagonist. J Pharmacol Exp Therap 257:374-381

Yue T-L, Gleason M, Hallenbeck J, Feuerstein G (1991b) Characterization of platelet-activating factor-induced elevation of cytosolic free-calcium level in neurohybrid NCB-20 cells. Neuroscience 41:177-185

SUBJECT INDEX

438

NATO ASI Series H

NATO ASI Series H

NATO ASI Series H

NATO ASI Series H